PALEOBOTANY, PALEOECOLOGY, AND EVOLUTION

PALEOBOTANY, PALEOECOLOGY, AND EVOLUTION

VOLUME 2

edited by

Karl J. Niklas

PRAEGER

PRAEGER SPECIAL STUDIES • PRAEGER SCIENTIFIC

Library of Congress Cataloging in Publication Data
Main entry under title:

Paleobotany, paleoecology, and evolution.

Includes bibliographical references and index.
1. Paleobotany. 2. Paleoecology. 3. Evolution.
I. Niklas, Karl J.
QE905.P29   560'.45   81-1838
ISBN 0-03-056656-8 (v. 2)   AACR2

Published in 1981 by Praeger Publishers
CBS Educational and Professional Publishing
A Division of CBS, Inc.
521 Fifth Avenue, New York, New York 10175 U.S.A.

© 1981 by Praeger Publishers

123456789   145   987654321

Printed in the United States of America

LIST OF CONTRIBUTORS

Richard K. Bambach, Department of Geological Sciences, Virginia Polytechnic Institute and State University, Blacksburg, Virginia.

Stephen F. Barrett, Department of the Geophysical Sciences, The University of Chicago, Chicago, Illinois.

William L. Crepet, Biological Sciences Group, U-42, University of Connecticut, Storrs, Connecticut.

David L. Dilcher, Departments of Biology and Geology, Indiana University, Bloomington, Indiana.

Elizabeth H. Gierlowski, Department of Earth Sciences, Case Western Reserve University, Cleveland, Ohio.

Karl J. Niklas, Division of Biological Sciences, Section of Plant Biology, Cornell University, Ithaca, New York.

William C. Parker, Department of the Geophysical Sciences, The University of Chicago, Chicago, Illinois.

Judith Totman Parrish, Department of the Geophysical Sciences, The University of Chicago, Chicago, Illinois.

Anne Raymond, Department of the Geophysical Sciences, The University of Chicago, Chicago, Illinois.

Greg Rettalack, Departments of Biology and Geology, Indiana University, Bloomington, Indiana.

Rudolf M. Schuster, Department of Botany, University of Massachusetts, Amherst. Cryptogamic Laboratory, Hadley, Massachusetts.

J. John Sepkoski, Jr., Department of the Geophysical Sciences, The University of Chicago, Chicago, Illinois.

Thomas N. Taylor, Department of Botany. The Ohio State University, Columbus, Ohio.

Bruce H. Tiffney. Department of Biology and Peabody Museum, Yale University, New Haven, Connecticut.

Jack A. Wolfe, U.S. Geological Survey, Menlo Park, California.

Alfred M. Ziegler, Department of the Geophysical Sciences, The University of Chicago, Chicago, Illinois.

Harlan P. Banks

AN APPRECIATION—HARLAN P. BANKS

It is indeed appropriate that this volume consisting of such a broad range of scientific papers should be published in honor and recognition of the continuing contributions of Harlan P. Banks to paleobotany. For if there can be one hallmark that characterizes the broad spectrum of his scientific career, it is the amazing productivity and the wide-ranging ramifications of his energies and interests, which are so well mirrored by the present volume.

Harlan Banks is prolific, not only in the papers he produces, the seminars and lectures he gives, the botanists and paleobotanists he has launched, but also in the large numbers of young minds he has filled with the love of botany. These are only some of the many facets of his impressive career, but they are the activities for which we feel he will be long remembered with respect, admiration, and affection.

Harlan P. Banks is a major worker and synthesizer in the area of earliest vascular plants and their significance to land plant evolution. He early advocated a high standard of rigorous and exhaustive study of plant fossil materials, to wrest the maximum information from recalcitrant materials. Such studies applied by various workers to early vascular plant fossils led in 1968 to what has become a classic paper in paleobotany. In this paper, Dr. Banks reorganized the old Order Psilophytales, which had become a taxonomic repository for many early land plants of diverse characteristics. He excluded a number of taxa that were clearly not referable to the group, and separated the remainder into three categories: the rhyniophytes, the zosterophyllophytes, and the trimerophytes. He established definitive criteria for each category, and for the major subdivisions within these categories. He resolved what was then a bewildering and diverse array of early plants into clearly related groups based on their morphological, anatomical, and reproductive features; and equally clearly, he revealed those areas that were in need of further work and new insights. In 1975 he modified the classification and discussed many of the new studies and contributions that the earlier paper had engendered. Although future studies may bring modifications, these two papers and the concepts they embody are landmarks in Paleozoic paleobotany. Dr. Banks alone, and in collaboration with his colleagues and students, has produced numerous papers that represent many of the major thrusts and significant achievements in the area of early vascular plants. A selected list of such papers is appended.

Harlan Banks has had a strong influence on the field of paleobotany, which extends beyond the papers he has authored or coauthored. He early grasped the importance of stratigraphic control to paleobotanical studies that provide immediate benefits to paleobotanists, paleontologists, and stratigraphers alike, and that eventually accrue to evolutionists, paleoclimatologists, and paleogeographers, as is witnessed by several papers in this symposium. He foresaw the potential that the fledgling field of palynology could provide to independent temporal control of the megafossil record.

Harlan Banks is a plantsman, and, perhaps because of this and his New England background, an inveterate peddler of the love of botany and paleobotany. His Yankee conscience and drive is reflected in the service he has given botany at many levels. He has served the Botanical Society of America in numerous capacities on committees, as Secretary-Treasurer, Vice-President, and President.

Harlan Banks strongly believes in the international cooperation of paleobotanists and, to that end, has promoted cooperation, exchanges, and visits at the international level. His efforts have been recognized by various fellowships abroad such as Fulbright Research Scholar, Liège, Belgium; Corresponding Member, Geological Society of Belgium; Guggenheim Fellow at University of Liège and Cambridge University; and Fellow of Clare Hall, Cambridge University, England. He has served as Vice-President and President of the International Organization of Paleobotany and is a member of various editorial boards.

Besides his direct contributions to the broad range of international paleobotany, Harlan Banks has played a major role in enlisting and training graduate students in both botany and paleobotany. He transmits his own enthusiasm and excitement, and, in an era of sputnik and molecular biology, he has conveyed to students that rigor, accomplishments and intellectual satisfaction in science need not belong only to the current fashion to be significant. He has always encouraged young paleobotanists, while at the same time he exhorts them to excel, to develop their evidences, and to suspend judgement until their evidence is in. His own horizons are not confined and he opens vistas to others. Thirty some prospective students received twenty-one doctorates and thirteen masters degrees under his able guidance. The subjects of these degrees were as wide-ranging as his interests and reflect a broad perspective of botanical approaches encompassing the technical, floristic, taxonomic, and anatomic-morphologic disciplines of living and extinct botany. Many of these thirty students, now active colleagues of Dr. Banks, convey his message and enthusiasm for botany in over twenty different schools or research institutions, and continue active research in paleobotany.

Probably the most intangible legacy accorded the botanical community is the influence Harlan Banks has had, not only on his own graduate students, but on the thousands of undergraduate students who were first introduced to the world of plants in his excellent general botany course. In this course the students were exposed to an extraordinary teacher who could take any topic and make it spring to life, become interesting, exciting, and truly relevant. Many of these young lives were changed and/or redirected because of Harlan Banks's talent as a teacher and storyteller. In some ways this is his most important contribution—one that has touched all of us who are priviliged to know him. For Harlan Banks is a TEACHER in the finest and best sense of the word.

In all aspects of his career, in research, writing papers, and student interactions, Harlan Banks is teaching. He has the talent to hold an audience in the palm of his hand from the moment he begins, whether it is directed to a group of distinguished scholars or to a general botany class. No matter his message—"The Rhyniophytina, Zosterophyllophytina, and Trimerophyllophytina," "The worldwide distribution in the Devonian of toothy cuticles," or "Plants are fun"—he weaves a spell and transmits knowledge, enthusiasm, and motivation. We wish him many more years of continued opportunity to weave such superb spells.

Patricia M. Bonamo

James D. Grierson

SELECTED BIBLIOGRAPHY

Banks, Harlan P. 1944. A new Devonian lycopod genus from southeastern New York. *Amer. J. Bot.* 31:649–659.

Zimmerly, Bessie C. and Harlan P. Banks. 1950. On gametophytes of *Psilotum. J. Bot.* 37(8):668. (abstr.)

Fry, Wayne L. and Harlan P. Banks. 1955. Three new genera of algae from the Upper Devonian of New York. *J. Paleont.* 29:37–44.

Banks, Harlan P. 1959. The stratigraphic occurrence of Devonian plants with applications to phylogeny. *Proc. IX Int. Bot. Congress* II:17–18.

Banks, Harlan P. 1960. Notes on Devonian lycopods. *Senckenbergiana lethaea* 41(1–6):59–88.

Banks, Harlan P. 1961. The stratigraphic occurrence of Devonian plants with applications to phylogeny. In *Recent Advances in Botany.* Univ. Toronto Press, pp. 963–968.

Leclercq, Suzanne and Harlan P. Banks. 1962. *Pseudosporochnus nodosus* sp. nov., a Middle Devonian plant with cladoxylalean affinities. *Palaeontographica B* 110:1–34.

Banks, Harlan P. 1962. *The Invasion of the Land.* Film 7 of Part VII. A.I.B.S. Secondary School Biological Film Series. New York: McGraw-Hill.

Grierson, J. D. and Harlan P. Banks. 1963. Lycopods of the Devonian of New York. *Palaeontographica Americana* 4(31):219–295.

Banks, Harlan P. 1964. Putative Devonian Ferns. *Mem. Torrey Bot. Club* 21(5):10–25.

Banks, Harlan P. 1964. Upper Devonian plants with gymnospermous anatomy and pteridophytic foliage. Invitation paper. *X Internat. Botanical Congress*. Abstracts, p. 32.

Banks, Harlan P. 1965. Some recent additions to the knowledge of the early land flora. *Phytomorphology* 15(3):235–245.

Banks, Harlan P. 1966. Devonian flora of New York State. *Empire State Geogram* 4(3):10–24.

Banks, Harlan P. 1966. Early land plants and some of their relatives. *Bio Sci* 16(6):432–433.

Carluccio, Leeds M., F. M. Hueber, and Harlan P. Banks. 1966. *Archaeopteris macilenta*, anatomy and morphology of its frond. *Amer. J. Bot.* 53(7):719–730.

Bonamo, P. M. and Harlan P. Banks. 1966. *Calamophyton* in the Middle Devonian of New York State. *Amer. J. Bot.* 53(8):778–791.

Matten, Lawrence C. and Harlan P. Banks. 1966. *Triloboxylon ashlandicum* gen. et sp. nov. from the Upper Devonian of New York. *Amer. J. Bot.* 53:1020–1028.

Hueber, F. M. and Harlan P. Banks. 1967. *Psilophyton princeps:* the search for organic connection. *Taxon* 16:81–85.

Banks, Harlan P. 1967. Plant fossils in central New York. *The Cornell Plantations* 22:55–63.

Matten, Lawrence C. and Harlan P. Banks. 1967. Relationship between the Devonian progymnosperm genera *Sphenoxylon* and *Tetraxylopteris*. *Bull. Torrey Bot. Club* 94:321–333.

Bonamo, P. M. and Harlan P. Banks. 1967. *Tetraxylopteris schmidtii:* Its fertile parts and its relationships within the Aneurophytales. *Amer. J. Bot.* 54:755–768.

Banks et al. 1967. In Chap. 1, Thallophyta, Part 1; Chap. 4, Pteridophyta, Part 1; Chap. 5, Pteridophyta, Part 2. In *The Fossil Record*. W. B. Harland, et al. (eds.). London: Geological Society, pp. 163–180, 219–231, 233–245.

Banks, Harlan P. 1968. The stratigraphic occurrence of early land plants and its bearing on their origin. In *Internat. Symposium on the Devonian System*. D. H. Oswald (ed.) (1967) Vol. 2. Calgary, Canada: Alberta Soc. Petroleum Geologists, pp. 721–730.

Banks, Harlan P. 1968. The early history of land plants. In *Evolution and Environment*. Ellen T. Drake (ed.). New Haven and London: Yale University Press, pp. 73–107.

Banks, Harlan P. and J. D. Grierson. 1968. *Drepanophycus spinaeformis* Goppert in the early Upper Devonian of New York State. *Palaeontographica B.* 123:113–120.

Banks, Harlan P. 1968. Anatomy and affinities of a Devonian *Hostinella*. *Phytomorphology* 17:321–330.

Banks, Harlan P. 1968. Devonian Plants, pp. 1355–1356. In *Developments, trends and outlooks in paleontology*. R. C. Moore (ed.). *J. Paleontol.* 42:1327–1377.

Banks, Harlan P. 1969. Richard Krausel (1890–1966). *Phytomorphology* 18:178–179.

Banks, Harlan P. and M. R. Davis. 1969. *Crenaticaulis*, a new genus of Devonian plants allied to *Zosterophyllum*, and its bearing on the classification of early land plants. *Amer. J. Bot.* 56:436–449.

Matten, Lawrence C. and Harlan P. Banks. 1969. *Stenokoleos bifidus* sp. n. in the Upper Devonian of New York State. *Amer. J. Bot.* 56:880–891.

Banks, Harlan P. 1969. Early Land Plants: Time of Origin and Rate of Change. *J. Paleont.* 43:880. Abstr. No. Amer. Paleont. Conf. Chicago. Sept. 1969.

Banks, Harlan P. Organizer and Chairman. 1970. Symposium on Major Evolutionary Events and the Geological Record of Plants, for XI International Botanical Congress 1969. *Biological Reviews* 45(3):317–454. (Specific authorship: Introduction pp. 317–318; Summary pp. 451–454.)

Banks, Harlan P. 1970. *Evolution and Plants of the Past.* Belmont, California: Wadsworth Pub. Co., Inc. 170 pp.

Banks, Harlan P. 1970. Loren Clifford Petry (1887–1970). *Plant Sci. Bull.* 16(3):10–11.

Banks, Harlan P. 1970. Chipping away at early land plants: of people, places and perturbations. *Plant Sci. Bull.* 16(4):1–6.

Scheckler, S. E. and Harlan P. Banks. 1971. Anatomy and relationships of some Devonian progymnosperms from New York. *Amer. J. Bot.* 58:737–751.

Scheckler, S. E. and Harlan P. Banks. 1971. *Proteokalon* a new genus of progymnosperms from the Devonian of New York State and its bearing on phylogenetic trends in the group. *Amer. J. Bot.* 58:874–884.

Banks, H. P. 1972. The stratigraphic occurrence of early land plants. Palaeontology 15(2):365–377.

Banks, H. P. 1972. The scientific works of Suzanne Leclercq. *Rev. of Palaeobotany & Palynology.* 14:1–5.

Banks, H. P., P. M. Bonamo and J. D. Grierson. 1972. *Leclercqia complexa* gen. et sp. nov., a new lycopod from the Late Middle Devonian of eastern New York. *Rev. of Paleobotany & Palynology* 14:19–40.

Banks, H. P. 1972. Silurian-Devonian Boundaries and early land plants. Twenty-fourth Int. Geol. Congress, Montreal. Abstracts, p. 214.

Skog, J. E. and H. P. Banks. 1973. *Ibyka amphikoma*, gen. et sp. n., a new protoarticulate precursor from the late Middle Devonian of New York State. *Amer. J. Bot.* 60:366–380.

Banks, H. P. 1973. Occurrence of *Cooksonia*, the oldest vascular land plant macrofossil, in the upper Silurian of New York State. *J. Indian Bot. Soc. Golden Jubilee Volume* 50A:227–235.

Scheckler, S. E. and H. P. Banks. 1974. Periderm in some Devonian plants. In *Advances in Plant Morphology. Prof. V. Puri Commem. Vol.* Y. S. Murty, B. M. Johri, H. Y. Mohan Ram, and T. M. Varghese (eds.). Meerut City, U.P. India: Prabhat Press, pp. 58–64.

Banks, H. P., Suzanne Leclercq and F. M. Hueber. 1975. Anatomy and Morphology of *Psilophyton dawsonii* sp. nov. from the Late Lower Devonian of Quebec (Gaspe), and Ontario, Canada. *Palaeontographica Americana* VIII(48):73–127.

Banks, H. P. 1975. The oldest vascular land plants: a note of caution. *Rev. Palaeobot. Palynol.*, 20:13–25.

Banks, H. P. 1975. Palaeogeographic implications of some Silurian-Early Devonian floras. In *Gondwana Geology.* K. S. W. Campbell (ed.). Canberra: Australian

National University Press, pp. 75–97.

Banks, H. P. 1975. Reclassification of Psilophyta. *Taxon* 24:401–413.

Banks, H. P. 1975. Early vascular land plants: Proof and Conjecture. *Bio. Sci.* 25:730–737.

Banks, H. P. 1977. Plant Macrofossils. pp. 298–300. In *The Silurian Devonian Boundary.* A. Martinsson (ed.). I U G S Series A. No. 5. Stuttgart.

Banks, H. P. 1977. Stratigraphic occurrences of Silurian-Devonian megafossils. *J. Paleont.* 51(2 supp.):2.

Stubblefield, Sara and H. P. Banks. 1978. The cuticle of *Drepanophycus spinaeformis* a long-ranging Devonian lycopod from New York and eastern Canada. *Amer. J. Bot.* 65:110–118.

Fairon-Demaret, Murial and H. P. Banks. 1978. Leaves of *Archaeosigillaria vanuxemii.* a Devonian lycopod from New York. *Amer. J. Bot.* 65:246–249.

Serlin, Bruce S. and Harlan P. Banks. 1978. Morphology and anatomy of *Aneurophyton* a progymnosperm from the Late Devonian of New York. *Palaeontographica Americana* 8(51):343–359.

Banks, H. P. 1979. Plant evolution: Study of macrofossil floras. In 1979 Yearbook of Science and Technology. D. N. Lapedes (ed.). New York: McGraw-Hill, pp. 296–298.

Hueber, F. M. and H. P. Banks. 1979. *Serrulacaulis furcatus* gen. et sp. nov., a new zosterophyll from the lower Upper Devonian of New York State. *Rev. Palaeobot. Palynol.* 28:169–189.

Banks, H. P. 1979. Pteridophyta. In *The Encyclopedia of Paleontology.* R. W. Fairbridge and D. Jablonski (eds.). Stroudsburg, Pa: Dowden, Hutchinson and Ross, (886 pp.), pp. 665–669.

Banks, H. P. 1980. The role of *Psilophyton* in the evolution of vascular plants. *Rev. Palaeobot. Palynol.* 29:165–176.

Hartman, Christine M. and Harlan P. Banks. 1980. Pitting in *Psilophyton dawsonii,* and Early Devonian trimerophyte. *Amer. J. Bot.* 67:400–412.

Banks, H. P. (In press). Time of appearance of some plant biocharacters during Siluro-Devonian time. *Bull. Canadian Botanical Association* (supp.).

Banks, H. P. (In press). Floral assemblages in the Siluro-Devonian. In *Biostratigraphy of Fossil Plants: Successional and Palaeoecological Analyses.* D. Dilcher and T. N. Taylor (eds.). Stroudsburg, Pa: Dowden, Hutchinson and Ross.

Banks, H. P. (In press). Peridermal activity (wound repair) in an Early Devonian (Emsian) trimerophyte from the Gaspe Peninsula, Canada. *The Paleobotanist.*

PREFACE

Although it is a truism that the search for the generalized aspects of evolution necessarily involves the perspective gained from diverse areas of specialization and varying levels of abstraction, it is a curiosity that very few paleobiological studies are multidisciplinary. In light of this, a symposium was organized and held at Cornell University on November 16 and 17, 1979. This volume, and its companion, represent 13 of the 14 papers presented and reflect an effort to focus on the nature, direction, and rate of change in evolution.

For various reasons, the symposium emphasized aspects of plant evolution. Perhaps the foremost reason was the initial impetus for its organization—to honor a great paleobotanist, Dr. Harlan P. Banks. In a career spanning five decades, Dr. Banks has contributed much to our understanding of early vascular plant evolution, not merely by his persistent attention to detail, but also because of his innovative approach. A less altruistic but nevertheless equally strong motive for organizing the symposium was the conviction that studies falling under the rubric of paleobotany are undergoing a renaissance and are expanding to include topics in chemistry, genetics, ecology, climatology, and the rigorous quantification of evolutionary theory. This symposium draws attention to an innovation in approach currently seen in fossil plant studies, which, it is hoped, carries on the tradition exemplified by Dr. Banks. The papers collected here demonstrate that the understanding of evolution is a paleo*biological* concern.

The papers are presented in two volumes and are organized in a chronological sequence from the Precambrian to the Tertiary. Volume 1 contains papers dealing with the "formative years" in plant evolution. In addition to contributions relative to specific time periods or relatively circumscribed sets of organisms, some chapters deal with more abstract or theoretical topics. Thus in Volume 1, two chapters on early and late Precambrian biota are followed by discussions on the genesis of fossil soils, biochemical scenarios in early land plant evolution, and the implications of genetic theory on the appearance and longevity of species. With these chapters the stage is set for the diversification of plant life in terrestrial environments, as well as for an appreciation of how many potential species of fossil organisms were never represented in the fossil record. Volume 1 concludes with a treatment of the phyletic implications of *Archaeopteris* and a paleoecologic reconstruction of a Pennsylvanian coal ball swamp. Volume 2 deals predominantly with such megatopics in Mesozoic and Cenozoic plant evolution as the pollination biology of some early seed plants and angiosperms, patterns of land plant diversity, continental drift, and paleoclimatology.

A synthesis of the various chapter topics would be both presumptuous and illogical since their collective scope is evolution. Yet even upon a casual glance, certain leitmotives become evident in the approach taken by the authors. Perhaps the most conspicuous approach is the implicit desire to view fossil organisms as dynamic entities, both in their own life space (in terms of growth, development, and reproduction) and within the context of evolutionary time. A second common approach is the desire to view fossils within their ecosystems—thus, both biotic and abiotic factors are explored in their relationship to functional morphology and evolution. Third, considerable effort is made wherever possible to quantify the data, whether they be morphologic, ecologic, or phyletic. While none of these leitmotives are unique to the literature, they can be placed in apposition to a larger body of earlier studies best described as purely descriptive and lacking in a perspective of temporal change.

The paleobiologist's perception of evolution is most often constrained by the extent to which the morphology of a fossil organism is preserved. There are at least three potentially useful levels of abstraction whereby morphology may be understood. These are the organism's physical properties, its ecologic relationships to other contemporary taxa, and its genotypic and phenotypic potential, which may best be described as "phyletic legacy." Each is reflected in some or all of the papers published here. Various theoretical and applied fields of study bridge the three approaches to understanding a fossil. Developmental patterns, both onto- and epigenetic, link phyletic legacy to the environment and hence to the organisms's ecological role. Functional morphology relates the ecology of a plant or animal to its architecture, while molecular and genetic theory place the biosynthetic capacity of an organism within the context of its evolutionary history and capacity to change. These "bridge concepts" and their respective scientific disciplines require a multidisciplinary approach to evolutionary phenomena and justify the "truism" set forth in the preamble to this introduction.

Like all cooperative efforts, *Paleobotany, Paleoecology, and Evolution* is the product of many people and their hard work. The patience, stamina, and intense motivation of the contributors are the principle reasons for any success that these volumes may enjoy. The College of Agriculture and Life Sciences and the Section of Botany, Genetics and Development of Cornell University are gratefully acknowledged for their financial support, as is The Boyce Thompson Institute for Plant Studies for providing its spacious lecture hall for the formal presentations. We are grateful to President Frank Rhodes, who took time from a busy schedule to welcome all the participants. Mss. Barbara Bernstein, Brenda Colthart, Lois Geesey, Sandra Kisner, Joan Miller, Esther Spielman, and Joan Wilen are to be congratulated and thanked for their assistance and support during and after the symposium. I also wish to thank Dominick Paolillo, Jr., the symposium co-organizer. And, finally, we all wish to thank Dr. Harlan P. Banks, for without the deep love and respect that he engendered during his career such a symposium would not have been possible.

Karl J. Niklas

CONTENTS

PALEOBOTANY, PALEOECOLOGY, AND EVOLUTION

POLLEN AND POLLEN ORGAN EVOLUTION IN EARLY SEED PLANTS

Thomas N. Taylor

In recent years there has been much more emphasis placed in paleobotany on the synthesis and incorporation of biological information in the study of reproductive organs of fossil plants. Examples of this approach might include cone development and pollination biology in *Cycadeoidea* (Crepet 1972, 1974; Crepet and Delevoryas 1972), floral morphology and pollination syndromes of early angiosperms (Crepet 1979; Crepet, Dilcher, and Potter 1975; Dilcher 1979; Dilcher, Potter, and Crepet 1976), embryo development and reproductive strategies among arborescent Carboniferous lycopods (Phillips 1979; DiMichele 1979), and the reproductive biology of early seed plants (Taylor and Millay 1979). To a large degree, all of these studies have relied on the analysis of a large number of specimens, in some cases the utilization of new techniques, and most importantly, the synthesis of a large body of knowledge using extant biological systems as the framework with which to construct working hypotheses. In general, many of these studies have been made possible by detailed anatomical and morphological investigations that have taken place during several decades.

There is probably no flora that has been researched as heavily as the fossil plants of the Carboniferous. In both North America and Europe, the well-known coal ball horizons have been extensively sampled with a considerable amount of structural, morphological, and even developmental information assembled. From these studies, especially in recent years, has come a considerable body of new information about early seed plants, such as the pteridosperms (Pteridospermophyta) and cordaites (Coniferophyta). It is the purpose of this chapter to put into perspective some of the current ideas regarding the evolution of pollen and pollen organs in these plants. In general, the discussion will focus on seed ferns and cordaites of Pennsylvanian age, and will include a consideration of the following orders: Lyginopteridales, Medullosales, Callistophytales, and Cordaitales.

INTRODUCTION

Despite the great amount of anatomical and morphological information presently known about members of the Progymnospermophyta, it is still not possible to identify precisely which plants may represent the Devonian precursors of the seed ferns. At the present time there are two seed-like structures that may be used to establish the presence of seed plants in the Devonian. One of these is *Archaeosperma arnoldii* (Pettitt and Beck 1968), an organ of Famennian age that is represented by two-seeded cupules borne in pairs. The cupule is about 1.5 cm long and consists of elongate segments that are attenuated at their tips. Each seed is approximately 4 mm long and characterized by a spiny outer integument that apically consisted of several free lobes. Unfortunately, nothing is known about the organization of the nucellus, especially the pollen-receiving mechanism; however, each seed did possess a large, sac-like trilete megaspore.

Unlike the radial organization of the *Archaeosperma* seeds, those of the Upper Devonian genus *Spermolithus* were bilaterally symmetrical (Chaloner, Hill, and Lacey 1977). Each seed was about 3.3 mm long with two lobes at the apex separated by a notch believed to be the site of the micropyle. The central body of each seed contained an oval megaspore surrounded by carbonized remnants of the nucellus. In a few specimens of *S. devonicus* the nucellus and integument are separated, suggesting that at least in the micropylar end of the seed, the integument and nucellus were free.

POLLEN ORGAN EVOLUTION

It is far easier to recognize the seed and seed-like structures of early pteridosperms than the pollen organs because the microsporangia of the seed ferns approximate closely the organization found in several Devonian progymnosperms and plants that have been referred to as preferns. It appears quite probable that, like these plants, the earliest seed ferns did not have synangiate microsporangia, but rather possessed aggregations of elongate sporangia borne at the tips of dichotomizing axes. Several Devonian plants, including *Tetraxylopteris schmidtii* (Bonamo and Banks 1967),

Rellimia thomsonii (Bonamo 1977), *Rhacophyton ceratangium* (Andrews and Phillips 1968), and *Psilophyton dawsonii* (Banks, Leclercq, and Hueber 1975) may be used as examples of early stages in the aggregation of sporangia that probably occurred during the evolution of the synangiate microsporangiate pollen organs of the seed ferns. In general, the sporangia of these plants were elongate, often with an apiculate tip, and grouped in clusters of two or more. Most appear to have dehisced longitudinally.

Morphologically indistinguishable from some of these Devonian forms were a number of Lower Carboniferous microsporangiate organs that have been suggested as the pollen-producing organs of seed ferns. It is important to underscore that nothing is known about the structural organization of these microsporangiate organs, nor the plants that produced them; nevertheless, their age and general morphological organization suggest possible affinities with the early seed ferns. One such microsporangiate structure is *Zimmermannitheca cupulaeformis* (Remy and Remy 1959), known from sediments of Lower Namurian A age. In this microsporangiate organ, the number of sporangia per cluster varies from two to seven, with each cluster being borne at the tip of a dichotomizing axis (Figs. 1.1, 1.2). *Simplotheca* is another Lower Namurian A microsporangiate organ that was borne at the ends of dichotomizing axes (Remy and Remy 1955). In *S. silesiaca*, each sporangial cluster is about 2 mm long, with the general morphology suggestive of a synangium (Fig. 1.3). Another Upper Mississippian form that may demonstrate an early type of synangium is *Paracalathiops stachei* (Remy and Remy 1953). Each synangium is approximately 0.7 mm long and thought to consist of several tubular sporangia in turn surrounded by a ring of sterile appendages (Fig. 1.8). Other Upper Devonian-Upper Mississippian probable seed fern microsporangiate organs are described in detail in a recent paper by Millay and Taylor (1979) on this subject. Despite the fact that structural details are unknown for almost all of these compressed aggregations of sporangia making their assignment to the pteridosperms tentative, many possessed pollen; thus, information about potential early pollen types may be considered. Several of the pollen types from these organs will be considered in a later section of this chapter.

In summary, the apparent earliest microsporangiate organs of the seed ferns consisted of terminally borne clusters of elongate sporangia that became longitudinally fused to form synangia. Exactly when the phylogenetic fusion of solitary sporangia took place is not clear, but several probable synangia are present by the early Mississippian. None of these Lower Carboniferous synangiate organs can be shown conclusively to belong to seed ferns; however, in a few instances the pollen is morphologically identical to seed fern pollen, and has also been found in the pollen chambers of several seeds.

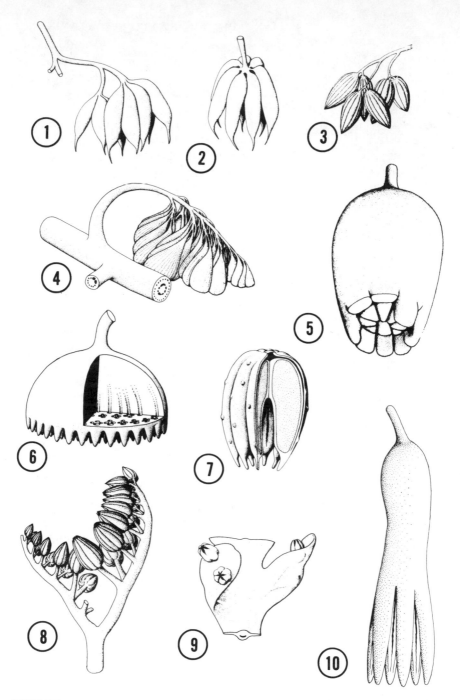

FIGURES 1.1–1.10. Paleozoic pollen organs. **1.1, 1.2,** *Zimmermannitheca cupulaeformis.* **1.3,** *Simplotheca silesiaca.* **1.4,** *Parasporotheca leismanii.* **1.5,** *Sullitheca dactylifera.* **1.6,** *Potoniea illinoiensis.* **1.7,** *Idanothekion glandulosum.* **1.8,** *Paracalathiops stachei.* **1.9,** Foliar segment with *Idanothekion* synangia. **1.10,** *Codonotheca caduca.*

Lyginopteridales

Although the lyginopterid seed ferns represent the group of plants for which the pteridosperms were delimited initially by Oliver and Scott (1904), as entire plants they still remain relatively unknown. Only *Lyginopteris* is known in enough detail so that a characterization of the order may be made. In general, lyginopterid seed ferns were rather small plants, probably less than 2 meters tall, and in habit were vine-like. The earliest recognizable forms of this order had leaves only modified slightly from branching systems; later-appearing members (Pennsylvanian) possessed small, highly dissected leaves that were produced sparingly along the stem. Much of the stem diameter consisted of secondary xylem with tracheids possessing multiseriate oval bordered pits. A broad cortex, often with sclerenchyma plates in a variety of configurations, also contributed to the size of the stems. The seeds of this order were generally small, and produced in multiovulate cupules in some of the earliest members (Taylor and Millay 1980).

Only within the last several years has there been much conclusive information presented about the pollen organs of the lyginopterid seed ferns. Only two (*Telangium* [Fig. 1.38] and *Feraxotheca* [Fig. 1.32]) are known from structurally preserved specimens, with the other taxa used commonly for impression-compression specimens. *Telangiopsis* (Eggert and Taylor 1971) is one such form that is used for Mississippian specimens that consist of a three-dimensional branching system bearing terminal clusters of what have been interpreted as synangia (Fig. 1.42). In 1976, Jennings described a Mississippian specimen of *Telangium* that consists of terminally borne synangia on a dichotomously branching system that appears to replace penultimate pinnae.

Another presumed lyginopterid pollen organ that is known only from compression-impression specimens is *Crossotheca* (Fig. 1.40). In this pollen organ the synangia are borne on the side of an expanded ultimate pinna (Fig. 1.17). In some forms like *C. sagittata*, the synangium is triangular in outline and may have evolved from the fusion of three smaller synangia on a small pinnate frond segment (Fig. 1.16).

One structurally preserved lyginopterid pollen organ of Pennsylvanian age that is frequently associated with the stem genus *Microspermopteris* is *Feraxotheca*. *Feraxotheca culcitaus* (Millay and Taylor 1977) is represented by synangia borne at the ends of ultimate pinnae (Fig. 1.14) on a planated frond segment (Fig. 1.13). There is considerable morphological similarity between *Feraxotheca* and some species of *Crossotheca*, suggesting that the same organ may be represented by differing states of preservation (Millay and Taylor 1978). Each synangium of *Feraxotheca* consists of a parenchymatous pad with six sporangia arising from the flattened lower surface (Figs. 1.32, 1.36). The vascular system of the unit consists of a pinna vascular bundle that spreads out over the surface of the pad, but that does not penetrate the pad or the walls of the individual sporangia.

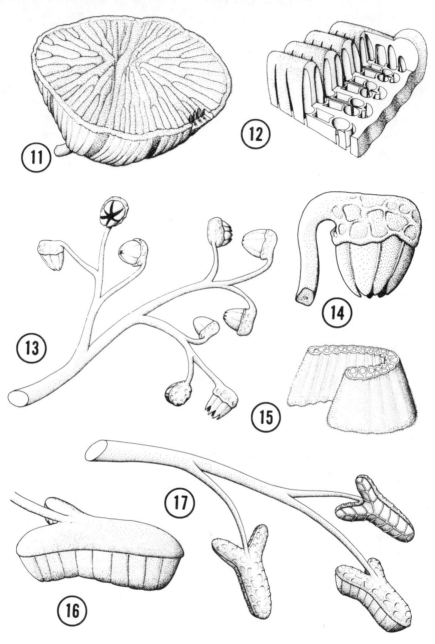

FIGURES 1.11–1.17. Paleozoic pollen organs. **1.11,** Lower surface of *Dolerotheca formosa*. **1.12,** Section from the pollen organ illustrated in Figure 1.11 illustrating organization of the sporangial tubes and lacunae. **1.13,** Fertile pinna of *Feraxotheca culcitaus*. **1.14,** *Feraxotheca culcitaus* synangium. **1.15,** Portion of a synangium of *Parasporotheca leismanii*. **1.16,** *Crossotheca sagittata* synangium. **1.17,** Portion of a fertile pinna of *Crossotheca sagittata*.

The small number of structurally preserved pollen organ taxa that can be assigned to the lyginopterid pteridosperms with any degree of confidence makes deciphering potential evolutionary trends within the group difficult. Nevertheless, within the lyginopterid seed ferns there does appear to be a shift from forms with three-dimensional branching (e.g., *Telangiopsis*) through stages of partial frond planation. This series appears to culminate with partially sterile pinnate frond segments like those of *Feraxotheca* and *Crossotheca*. Lyginopterid synangia may be radial or bilateral, depending on the number of sporangia, with little histological difference among the taxa.

Medullosales

Medullosan pteridosperms were larger plants than the lyginopterids, and judging from the diameter of the stems, were at least 10 m tall. In transverse section the stems contained several segments of vascular tissue (sympodia), each with secondary xylem, which were originally thought to represent the steles of a polystelic plant. These vascular segments are now regarded as continuous units of a highly dissected stele. The leaves of medullosan pteridosperms were massive and consisted of dichotomously branched axes that were regularly pinnate at more distal levels. The seeds were large and radially symmetrical, and in many features closely resembled the seeds of living cycads. In many medullosan seeds both the integument and nucellus contained vascular tissue. Numerous pollen organs have been assigned to this group and include a variety of preservational modes; most of these organs are placed with the medullosan seed ferns because of the presence of a particular pollen type.

In recent years our understanding about the evolution of medullosan pollen organs has increased markedly due to the discovery of numerous structurally preserved forms which have provided an opportunity to homologize features among several taxa. In a recently completed survey of Paleozoic seed fern pollen organs, Millay and Taylor (1979) recognize three basic types among the structurally preserved members of the Medullosales. These include simple, aggregate, and compound synangia.

Simple synangia are constructed typically on a radial plan and may range up to 3 cm in diameter. *Halletheca* (Taylor 1971), a form known from the Upper Pennsylvanian, may be used as an example of this type. Each *H. reticulatus* synangium is constructed of five elongate tubes (sporangia) that surround a central, pentagonally-shaped core of fibers (Fig. 1.39); in the distal end of the organ the center is hollow. Associated with the inner wall of each sporangium is a portion of the central fiber core that split during dehiscence so that pollen was shed into the central hollow of the organ. Each sporangium is vascularized by a single bundle that extends the length of the

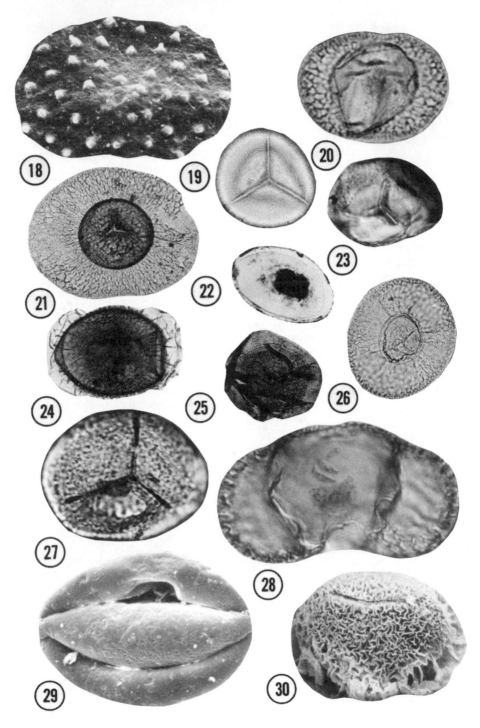

organ through the outer facing wall. It is not known how *Halletheca* was attached to the plant that produced it; however, evidence from a morphologically similar form *(Aulacotheca)* suggests that the synangia may have been produced singly or in small clusters at the ends of delicate branches.

A slightly more complex simple synangium is *Stewartiotheca* (Eggert and Rothwell 1979). In *S. warrenae*, about 80 elongate sporangial tubes are arranged in a uniseriate ring that has been internally folded or plicated (Fig. 1.35). A solid core of fibers is present near the peduncle of the organ, becoming hollow in the direction of the distal surface. Sporangial tubes occur in pairs, with the longitudinal dehiscence slits positioned oppositely. Each sporangial tube is vascularized by a single bundle on the dorsal surface; between each pair of tubes is an elongate hollow area. The internal folding (plication) of a uniseriate ring of sporangial tubes is also a feature of *Sullitheca dactylifera* (Stidd, Leisman, and Phillips 1977). The center of this pollen organ contains an H-shaped fibrous structure (Fig. 1.31), which is considered homologous with the five-sided sclerenchyma core of *Halletheca*. The individual sporangial tubes alternate with lacunae and appear to dehisce through slits that develop on the surface of facing sporangia. At the distal end of the organ is a ring of solid finger-like projections (Fig. 1.5), which may have functioned to prevent desiccation during development.

There are two basic types of aggregate synangia known for the medullosan seed ferns. In one type, represented by *Rhetinotheca tetrasolenata* (Leisman and Peters 1970), each synangium consists of four small radial sporangial tubes (3.6 mm long) positioned around a reduced core of fibers. In the distal part of each synangium the fiber core is absent. Like all of the Paleozoic seed fern pollen organs that have been studied in detail, the sporangial contents were shed into the central hollow by means of longitudinal dehiscence slits. Extending from the epidermis of each synangium were numerous peg-like trichomes, each with a bulbous apex, which interdigitated with the trichomes of neighboring synangia to form a massive reproductive structure. A similar system of interlocking trichomes appears to have been the method used to hold together the simple synangia of *Parasporotheca leismanii* (Dennis and Eggert 1978). This medullosan pollen organ is especially interesting because the individual synangia are bilaterally

FIGURES 1.18–1.30. Paleozoic pollen types. **1.18,** *Nanoxanthiopollenites mcmurrayii* (distal). × 600. **1.19,** *Crossotheca* cf. *mcluckiei* (proximal). × 480. **1.20,** *Idanothekion (Vesicaspora) glandulosum.* (proximal). × 960. **1.21,** *Felixipollenites macroreticulata* (proximal). × 250. **1.22,** *Monoletes* grain with two sperm. × 65. **1.23,** *Potoniea* sp. (proximal). × 800. **1.24,** *Parasporotheca (Parasporites) leismanii.* (proximal). × 150. **1.25,** *Rhabdosporites langii.* (proximal). × 200. **1.26,** *Cordaianthus (Florinites) concinnus.* (proximal). × 400. **1.27,** *Feraxotheca culcitaus.* (proximal). × 1000. **1.28,** *Idanothekion (Vesicaspora) glandulosum.* (lateral). ×·1500. **1.29,** *Dolerotheca (Monoletes) formosa.* (diatal). × 200. **1.30,** *Parasporotheca (Parasporites) leismanii.* (proximal). × 225.

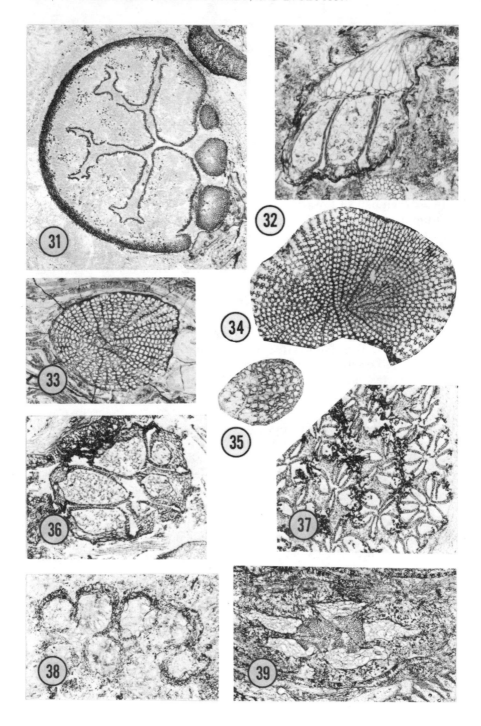

symmetrical (Figs. 1.4, 1.41), whereas all of the other medullosan pollen organs described to date possess radial symmetry. Each scoop-like synangium consists of up to 20 elongate tubes that alternate with lacunae in the ground tissue (Fig. 1.15). A vascular strand is present in the dorsal wall of each sporangium.

The third basic type of medullosan pollen organ is compound in its organization. The best known example is *Dolerotheca*, a taxon that was rather common throughout the Carboniferous and includes both structurally preserved (Fig. 1.33) and compression specimens. Some specimens of *D. formosa* are known that were at least 4 cm in diameter and consisted of what appear as radiating rows of sporangia (Fig. 1.34). The internal organization of this synangium has been regarded as facing pairs of elongate sporangia that longitudinally dehisced into a common slit-like hollow area. Like the organization in *Parasporotheca*, *Stewartiotheca* and several others, lacunae alternate with the elongate sporangial tubes in *Dolerotheca* (Fig. 1.12). A transverse section made through the distal face of a *Dolerotheca* pollen organ resembles superficially a section of *Stewartiotheca*, except that in *Dolerotheca* there are basically four components to the campanulum (Fig. 1.11). Each unit represents a single synangium that has been internally symmetrically plicated similar to the organization in *Stewartiotheca* and *Sullitheca*.

An alternative view has been expressed recently regarding the possible way in which the *Dolerotheca* pollen organ may have evolved (Stidd 1980). According to this idea, the *Dolerotheca* campanulum has evolved from a whorl or tight spiral of dichotomously branched axes bearing pinnately arranged elongate tubular sporangia along the branches. Stidd accounts for the paired sporangia in *Dolerotheca* as the result of adjacent sporangia fusing from different branches. One of the major drawbacks regarding this hypothesis is that the basic pollen organ type used to derive the *Dolerotheca* organ is the ill-understood impression-compression taxon *Codonotheca* (Fig. 1.10), which may be far more complex than was originally interpreted.

FIGURES 1.31–1.39. Structurally preserved Paleozoic pollen organs. **1.31**, *Sullitheca dactylifera*. Transverse section near the distal end of the organ showing a portion of the H-shaped sclerenchyma zone and three of the finger-like projections. × 6.5. **1.32**, *Feraxotheca culcitaus*. Longitudinal section of synangium showing parenchymatous pad and three sporangia. × 25. **1.33**, *Dolerotheca schopfii*. Transverse section through distal end of organ showing the paired sporangial tubes and highly plicated unit. × 1. **1.34**, *Dolerotheca formosa*. Transverse section through distal end of organ showing folded nature of the four component synangia. × 1. **1.35**, *Stewartiotheca warrane*. Transverse section near distal surface. × 2.1. **1.36**, *Feraxotheca culcitaus*. Transverse section of synangium showing thin inner facing sporangial walls. × 25. **1.37**, *Potoniea illinoiensis*. Transverse section of organ showing several rows of simple synangia. × 12. **1.38**, *Telangium scotti*. Transverse section of synangium. × 14. **1.39**, *Halletheca reticulatus*. Transverse section of organ showing central sclerenchyma core surrounded by five pollen sacs. × 10.

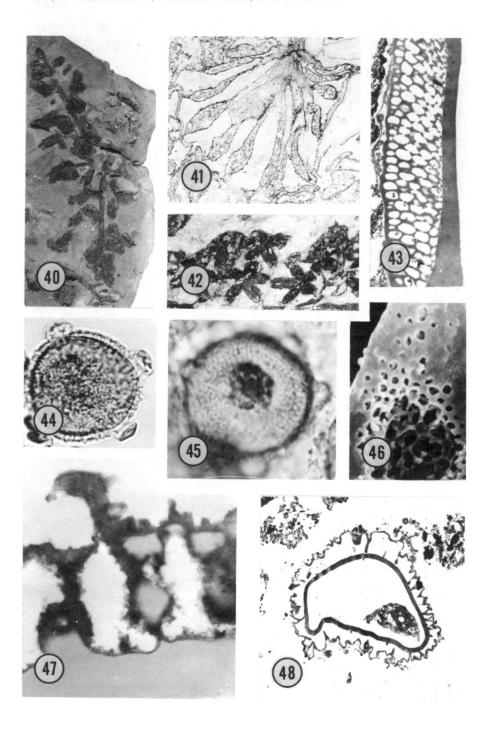

Another compound synangium believed by some to be associated with the medullosan seed ferns, but structurally different from *Dolerotheca*, is *Potoniea*. Until recently this pollen organ was known only from compressed specimens that because of their preservation were impossible to interpret. The recent discovery of *P. illinoiensis* (Stidd 1978) is especially important in that it provides for the first time information about the structural complexity of this organ, as well as providing a means whereby features of this pollen organ can be homologized with other structurally preserved seed fern synangia. *Potoniea illinoiensis* is a bell-shaped organ, approximately 1 cm in diameter, that contains elongate pollen sacs embedded in a cortical ground tissue. On the distal surface of the campanulum, the sporangial tubes were free for about 1 mm (Fig. 1.6). A transverse section of the unit indicates that the basic organization consists of clusters of simple sporangia arranged in five concentric rings (Fig. 1.37). Each sporangial cluster consists of four to six sporangia arranged around a central core, much like the simple synangia of *Halletheca*. In *Potoniea*, sporangial dehiscence appears to have taken place toward the center of each simple, synangial cluster.

Thus, within the Medullosales, the evolution among the pollen organs appears to have taken place along several distinct lines in the direction of increasing the number of sporangial tubes in each reproductive unit. According to this interpretation the simple radial uniseriate ring of sporangia represents the basic organizational type. Increasing the number of sporangia, while keeping the surface area of the organ relatively small, appears to have taken place through the process of plication or internal folding. This internal folding of a uniseriate ring of sporangia that alternate with lacunae is present to a limited extent in medullosan organs like *Sullitheca*, and is particularly prominent in simply synangia, such as *Stewartiotheca*.

The clustering of individual simple synangia into a massive pollen-producing unit also appears to have taken place within the order through the aggregation of individual synangia held together by epidermal trichomes. This type of synangial clustering is illustrated by *Rhetinotheca* and *Parasporotheca*, and includes both radial and bilateral, simple synangia.

FIGURES 1.40–1.48. Paleozoic pollen and pollen organs. **1.40**, *Crossotheca* cf. *mcluckiei*. ×1. **1.41**, Several synangia of *Parasporotheca leismanii*. ×8. **1.42**, Several synangia of *Telangiopsis* sp. ×6. **1.43**, Section through the sporoderm of *Monoletes* pollen grain showing the inner (right) homogeneous nexine and alveolate sexine. ×2700. **1.44**, Equatorial view of *Lasiostrobus polysacci* pollen grain showing several sacci. ×1200. **1.45**, *Lasiostrobus polysacci* pollen grain showing nucleus and cytoplasmic materials. ×1700. **1.46**, Fractured surface of a spine of *Nanoxanthiopollenites mcmurrayii* showing alveolate sexine. ×10,000. **1.47**, Ultrathin section of *Lasiostrobus polysacci* pollen grain wall showing columellate nature of exine. ×29,000. **1.48**, *Lasiostrobus polysacci* grain showing organization of the sporoderm and central inclusion. ×2400.

Lastly, at least two types of medullosan pollen organs appear to have evolved in the direction of compound units. In *Dolerotheca* the evidence appears overwhelming that the basic campanulum has evolved from the fusion of four highly plicated uniseriate rings of tubular sporangia, each structurally homologous to a simple synangium like *Stewartiotheca*. It is quite probable that detailed studies of some *Dolerotheca* specimens may indicate that two, and even three plicated synangia make up the basic bell-shaped unit.

In the case of *Potoniea*, the compound nature of the organ is far more difficult to interpret. If the interpretation is accurate that each sporangial cluster is fundamentally a simple synangium, then the entire pollen organ appears to have evolved through the telescoping and fusion of a branching system bearing small radial synangia. One further point that needs to be noted concerns the pollen produced by *Potoniea*, which is morphologically and ultrastructurally different from the pollen produced by all other medullosan pollen organs. This strongly suggests that *Potoniea* belongs to some other group of seed ferns, such as the Lyginopteridales, and that the unusual structural organization when compared with all other medullosan pollen organs merely reflects the inaccurate systematic assignment of the taxon.

Callistophytales

The Callistophytales is represented by a small group of seed ferns that range from the Pennsylvanian into the Permian. Despite the fact that this order has been known for a relatively short period of time, there can be little question that the plants within the order constitute the best known members of the pteridosperms. Callistophytaleans were profusely branched small plants that were probably shrub-like in their general habit. They produced pinnately compound fern-like leaves with multilobed pinnules. Seeds were small and bilaterally symmetrical, and are believed to have been produced on the lower surface of pinnules. The pollen organs consisted of sessile radial synangia also borne on the abaxial surface of pinnules (Fig. 1.9).

The pollen organs of the callistophytalean pteridosperms were described initially by the generic names *Idanothekion* (Millay and Eggert 1970) and *Callandrium* (Stidd and Hall 1970); however, recent studies suggest that the two genera probably represent different preservational and/or developmental stages of the same taxon (Rothwell 1972b). In general these synangia consist of a ring of five to nine elongate, exannulate sporangia that surround a parenchymatous core in the base of the organ (Fig. 1.7). Distally, the central region becomes hollow with the attenuated tips of the sporangia curved toward the center of the synangium. Dehiscence appears to have taken place toward the center of the organ through the thinner, inner-facing walls of the sporangia. Superficially at least, the pinnule-borne synangia of the Callistophytales appear to be most closely related to some members of the lyginopterid pteridosperms. Both have relatively small radial synangia that appear morphologically identical to some of the early forms of *Telangium*.

POLLEN EVOLUTION

During the last several years, there has been a considerable amount of interest directed at the pollen grains and spores extracted from the reproductive organs of various Carboniferous seed plants. These studies have included several broadly defined areas of investigation, such as ornamentation, sporoderm ultrastructure, and the significance of various haptotypic features (Millay and Taylor 1976). The study of *in situ* pollen has been especially rewarding in providing clues concerning the origin and subsequent evolution of several different lines of Paleozoic seed plant pollen.

Before considering the pollen grains of several groups of Carboniferous seed plants it is important to discuss briefly the microspores produced by various plants that are included presently within the Progymnospermophyta. Typically, the spores of these plants were quite large with most falling in the 100- to 200-μm size range. Many show some separation of the sporoderm into a structure termed a pseudosaccus, which surrounds the central body of the grain. One of the most common *sporae dispersae* grain types found in the sporangia of members of the Progymnospermophyta is *Rhabdosporites* (Richardson 1960). Spores of this type are generally oval, trilete, and possess a central body that is completely enclosed by a layer of the sporoderm (monosaccus or pseudosaccus) (Fig. 1.25). Attachment between the pseudosaccus and central body occurs only in the region of the proximal surface. Closely packed rods with truncated tips form a uniform ornamentation over the entire grain surface; no ornamentation has been described on the inner surface of the pseudosaccus. Within individual sporangia of several progymnosperms there is often an extensive size range in spore diameter (e.g., *Tetraxylopteris* 73 to 176 μm; *Rellimia* 91 to 191 μm), which suggests a potential shift in the direction of sporangial heterospory. The recently completed study by Brauer (1979) on the Devonian genus *Barinophyton* also suggests the presence of sporangial heterospory.

Morphologically, grains of the *Rhabdosporites*-type appear to suggest dispersal via wind and gravity; however, the presence of an encircling pseudosaccus except in the proximal region where germination took place, may have also functioned as an early insulating mechanism. Although ultrastructural data is desperately needed for several important progymnosperm spore taxa, there appears to have been a marked reduction in the thickness of the central body wall concomitantly with the evolution of the saccus in many Carboniferous grain types. Although the *Rhabdosporites* grain type may be traced to early pollen types that characterize the Carboniferous cordaites and some conifers, recognizing pre-Carboniferous pteridosperm pollen types has not met with much success.

Lyginopteridales

When compared with the spores of other plants, the pollen of lyginopterid seed ferns bears a striking morphological and ultrastructural similarity to the grains produced by many homosporous ferns. The pollen of the lyginopterids is rather uniform, and includes oval grains that rarely exceed 80 μm in diameter, with most of the *in situ* grains reported within the 40- to 50-μm size range. All are trilete (Figs. 1.19, 1.27), except for one report in which both monolete and trilete grains were described as occurring in the same sporangium (Jennings 1976). Pollen grains extracted from *Crossotheca*, *Feraxotheca*, *Telangium*, and *Telangiopsis* are all characterized by well-developed trilete sutures on the proximal surface with the laesurae extending to near the amb of the spore. Preliminary ultrastructural information about the sporoderm of *Crossotheca* indicates that the exine is about 2 μm thick, except in the region of the suture where the wall thickens to about 4 μm. The innermost (nexine) portion of the sporoderm is thin (0.2 μm) in comparison to the thicker sexine. Ornamentation on the surface of lyginopterid pollen is generally uniform on all surfaces and ranges from granulate to conate.

Trilete grains have also been reported from several problematical lyginopterid pollen organs, including *Alcicornopteris*, *Geminitheca*, *Staphlotheca*, and *Zimmermannitheca*, while monosaccate grains have been extracted from the sporangia of *Paracalithiops* and *Simplotheca*.

Medullosales

There are three basic morphological types of pollen that are included within the medullosan pteridosperms. By far the most common type, and the one that has been used to associate various isolated pollen organs of varying preservational states with the medullosan seed ferns, is *Monoletes*. Of the several dozen pollen organs attributed to the medullosan seed ferns, grains assignable to the *sporae dispersae* genus *Monoletes* have been identified from eleven. Grains of this morphotype are bilaterally symmetrical and range from 100 to over 500 μm in length. On the proximal surface is a single suture that is often slightly bent near the middle; two longitudinal grooves ornament the distal surface (Fig. 1.29). The sporoderm is thick (approximately 10 μm in most mature grains) and divided into a homogeneous nexine and characteristically honey-combed (alveolate) sexine (Figs. 1.43, 1.46). In some *Monoletes* grains up to 30 distinct lamellae have been noted in the nexine (Taylor 1978).

Two dense structures that morphologically resemble the sperm of some extant cycads found in a *Monoletes* grain (Fig. 1.22) provide the only information regarding the nature of the microgametophyte in these seed ferns (Stewart 1951).

A second type of pollen grain that is known from at least one medullosan pollen organ is *Parasporites*. This grain type initially described from coal macerates, is now known to have been produced by the aggregate synangiate pollen organ *Parasporotheca leismanii* (Dennis and Eggert 1978). One unusual feature of this grain is the presence of a reduced saccus at each end of a large slightly elliptical central body (Fig. 1.24). The maximum length of the grain, including the sacci, is 273 μm; the width ranges up to 187 μm. On the proximal surface is a monolete suture that possesses the same median deflection as does the suture of *Monoletes*. The ornamentation of the corpus of this bisaccate grain consists of closely spaced folds that give the surface a very rugged appearance (Fig. 1.30). The sacci, although greatly reduced and probably vestigal, lack external ornamentation. The exine infrastructure of *Parasporites* is similar to the alveolate sexine of *Monoletes* (Millay, Eggert, and Dennis 1978). Another Carboniferous pollen grain with an alveolate sexine that probably belongs to the Medullosales is *Nanoxanthiopollenites* (Fig. 1.18). Specimens of *N. mcmurrayii* were described initially as *sporae dispersae* specimens recovered from sediments of the Upper Pennsylvania (Clendening and Nygreen 1976). Unlike *Monoletes and Parasporites*, the surface of *Nanoxanthiopollenites* is ornamented by several large conical spines up to 12 μm long (Taylor 1979).

Earlier it was noted that some specimens of *Potoniea* (Fig. 1.23) may be the pollen-producing structures of a pteridosperm group like the Lyginopteridales instead of the medullosan seed ferns. This suggestion is based on the morphological and ultrastructural organization of *P. illinoiensis* pollen that is unlike any known pollen in the Medullosales. The pollen of *P. illinoiensis* is small (39 to 46 μm), spherical, and distinctly trilete. The proximally positioned laesurae are short; on the distal surface the grains are often collapsed or invaginated. Unlike the other medullosan pollen grains, the exine of *Potoniea* is homogeneous and comparable ultrastructurally to the sporoderm of all lyginopterid pollen types that have been investigated to date. One interesting feature of *P. illinoiensis* pollen concerns the presence of a delicate, outer separable layer that has been suggested as being homologous with a perine.

Callistophytales

The pollen produced by the Callistophytales is monosaccate and in many features the grains closely conform morphologically to the pollen of many extant conifers. Pollen of the *Vesicaspora*-type ranges up to 54 μm in length and is further characterized by a proximal cappus lacking any distinct haptotypic features. On the distal surface is a well-defined functional germinal aperture (Fig. 1.28). Internally, the saccus wall is ornamented by numerous endoreticulations that are believed to provide structural support to the thin-walled encircling saccus (Fig. 1.20). Well-preserved stages in the development of the microgametophyte have been described in detail by Millay and Eggert (1974), who have demonstrated them to consist of an axial row of at least three prothallial cells in the cappus region of the grain (Fig. 1.28). These authors have also identified what have been interpreted as an embryonal and tube cell in one grain; however, subsequent stages in the development of the microgametophyte remain to be discovered. Rothwell's (1972a) exciting discovery of a *Vesicaspora* pollen grain producing a pollen tube from the distal surface represents the only conclusive evidence for the existence of pollen tubes in any Carboniferous seed plants to date.

In discussing the evolution of early seed plant pollen, two additional groups of plants need to be briefly considered. One of these is the Cordaitales, a group of Carboniferous and Permian seed plants that have been important in tracing the early stages in the evolution of the conifer cone scale. The second pollen type is known only from an isolated reproductive organ (*Lasiostrobus polysacci*), which has been suggested as having affinities with either the cycads or conifers (Taylor 1970). In the case of the cordaites, there is little documented evidence to suggest how the pollen-producing organs may have evolved, nor is there any evidence in this regard concerning the pollen cone *Lasiostrobus*.

Cordaitales

Suggested stages in the evolution of cordaite pollen have been described elsewhere and will only be summarized here (Millay and Taylor 1974, 1976). One grain that has been regarded as a primitive cordaite pollen type is the Lower Pennsylvanian genus *Felixipollenites* (Millay and Taylor 1974). Grains of this morphotype macerated from the cordaitean reproductive organ *Gothania* (Daghlian and Taylor 1979) are monosaccate and up to 180 μm in diameter (Fig. 1.21). Both radially symmetrical forms with trilete sutures and bilaterally symmetrical types with monolete sutures have been recovered from the same pollen sacs. In addition, approximately 70 percent of the grains had suture types that were intermediate between the trilete and monolete forms. The organization of the suture in *Felixipollenites* is complex and consists of a narrow depression that supports a median, slightly elevated ridge. Transversely oriented delicate ribs situated at right angles to the long axis of the laesurae support the ridge. Ornamentation includes uneven muri that cover the proximal surface of the grain. Saccus-corpus attachment was described initially as occurring at both the proximal and distal poles; however, recent studies detailing the development of this grain indicate that the large monosaccus was attached only to the proximal surface of the corpus (Taylor and Daghlian 1979).

Monosaccate grains of a slightly smaller size (70 μm) may be used to illustrate a more specialized type of cordaitean pollen. Specimens of *Sullisaccites* are oval and possess a prominent trilete suture on the proximal surface. Like the majority of *Felixipollenites* grains, the suture in this type is asymmetrical, with one arm of the trilete slightly longer. The saccus is attached to the central body at both the proximal and distal poles, with the attachment more pronounced in the proximal region. Internally, the saccus is ornamented by numerous endoreticulations; externally the exine is granular.

The slightly younger (Middle Pennsylvania) alete pollen of the *Florinites*-type is representative of the third cordaitean pollen morphotype. These monosaccate grains are generally elliptical and range from 60 to 75 μm in diameter (Fig. 1.26). Endoreticulations are a prominent feature of the saccus that is attached to both the proximal and distal poles of the corpus. Ultrathin sections in the proximal region show the sporoderm appreciably thickened, while on the distal surface the exine is thinner and in some *Florinites* grains there is evidence of a preformed germination site. Based on the ultrastructure of the exine and the absence of proximally positioned haptotypic features, it appears certain that germination in *Florinites* took place through the distal surface.

Using these three examples of *in situ* pollen types from a reasonably well-defined taxonomic group (Cordaitales), it is possible to suggest several stages in the evolution of one gymnospermous pollen type. In this series the shift from large monosaccate, proximally germinating grains has involved a reduction of grain size and sporoderm thickness (corpus) that was accompanied by a gradual loss of haptotypic features and ornamentation. In addition, the site of germination has shifted from an elaborate trilete suture on the proximal pole to a thin slit-like aperture on the distal grain face. Although purely speculative at this time, saccus attachment to both poles of the corpus constitutes an early stage in the evolution of bisaccate grains in which the distal inclination of the sacci cover the thin distal suture, thus preventing excessive water loss.

Cordaitean microgametophytes have been described from *Florinites* grains extracted from pollen cones of the *Cordaianthus*-type (Florin 1936). Although the antheridial jacket originally described in the corpus of these grains appears to represent folds of the sporoderm, the grains do contain up to four prothallial cells and an embryonal cell that closely compare to the organization of the microgametophyte in the Callistophytales.

Lasiostrobus

The pollen of *Lasiostrobus* is an unusual grain type that appears to demonstrate a rather advanced suite of features. The pollen of *L. polysacci* is circular to subcircular in outline and ranges up to about 30 μm in diameter. Extending from the surface of the grain are numerous (three to eight) sexinous inflations (sacci) that are positioned typically near the equator of the grain (Fig. 1.44). The majority of grains examined are inaperturate; however, a few have been observed that possess a faint trilete suture on the proximal surface. What is perhaps most unusual about the pollen of *Lasiostrobus* is the organization and infrastructure of the sporoderm. Ultrathin sections show the exine constructed of two distinct layers, an inner homogeneous nexine, about 0.5 μm in thickness, and a three-parted sexine. The innermost portion of the sexine is represented by a delicate layer that is contiguous with the nexine. Arising from this layer are numerous regularly spaced columns up to 1.2 μm high (Fig. 1.47). A thin sporoderm component fused to the tips of the columns represents the outermost layer of the exine. Structurally, there is a close correspondence between the pollen of *Lasiostrobus* and the tectate pollen structure that is characteristic of most angiosperms.

The pollen of *Lasiostrobus* is of further interest in certain features associated with the development of the microgametophyte. Many of the grains contain spherical, dense structures that represent the shrunken remains of the microspore cytoplasm and nucleus (Fig. 1.45). By far the majority of grains containing inclusions are those that were present in dehisced pollen sacs or were free among the sporophylls of the cone (Taylor and Millay 1977). None of the grains examined showed any evidence of multicellular microgametophytes like those that have been reported in the Cordaitales and Callistophytales, suggesting that in *Lasiostrobus* the central inclusion constitutes the remnant of a single cell (Fig. 1.48). The fact that these grains are present in dehisced sporangia suggests that the pollen of *Lasiostrobus* was shed in the microspore stage of development, a feature that is regarded as being advanced evolutionarily among extant gymnosperms (Sterling 1963). It should be noted that the possibility exists that prothallial cell development may have been greatly delayed in *Lasiostrobus*; however, comparable examples are not known to exist among living gymnosperms. In the case of *Lasiostrobus* pollen, both structural features of the sporoderm (inaperturate, tectate-like exine) and functional characters associated with the formation of a microgametophyte suggest a grain considerably more advanced than any known Paleozoic pollen type.

CONCLUSIONS

Information about the pollination systems of early seed plants has been relatively slow to accumulate. In many instances, all that is known about a critical stage in the life history of an entire order has been determined from a single specimen. This is especially true for stages of the microgametophyte. In general, the microgametophytes of Paleozoic seed plants are morphologically similar to those of several extant gymnosperm groups, and do not demonstrate any of the primitive features that have been postulated for the microgametophyte during the evolution of the seed habit.

Many of the pollen grains produced by these early seed plants possess features that clearly indicate an anemophilous pollination syndrome, and that appear similar to the large monosaccate microspores of some members of the Progymnospermophyta. Almost all of the earliest seed plants appear to have large grains that germinated from the proximal surface. Within the Cordaitales grains became progressively smaller with germination shifting to the distal face. *Vesicaspora* (Callistophytales) might be used as an example of a specialized wind pollinated grain type in which a pollen tube was produced from the distal surface. Just what factors initially influenced the evolution of saccate and pseudosaccate grains can only be speculated on; however, two possible pressures in addition to grain buoyancy might include the insulation of the central body contents and a mechanism against predation. Information about the pollen in some extant gymnosperms suggests that the principal function of sacci in bisaccate grains is to prevent desiccation from the thin distal aperture. It is interesting to note that the saccate grain morphotype was also present in some nonseed plants during the Carboniferous (e.g., several groups of lycopods). The pollen organs of many early seed plants also suggest that wind pollination was the dominant early syndrome. For example, in many living plants in which wind pollination is the primary method there is some form of arresting mechanism that insures that all of the grains are not shed at the same time. In many of the Paleozoic seed plants this mechanism may have involved sporangial dehiscence toward the center of the pollen organ that impeded large numbers of grains from being released at one time.

In addition to the numerous early seed plant saccate grain types that suggest wind pollination, there are a few pollen types that may suggest early stages in the evolution of insect pollination. Notable among these is *Monoletes*, a grain that because of its size was probably not carried much of a distance via wind. In this grain the alveolate exine structure represented one method whereby the buoyancy of the grain could be increased, and later, in association with the distal surface grooves, may have functioned in a harmomegathic manner. It has been suggested that such an exine infrastructure may have also been selected as a site for wall-held materials involved in germination and also recognition substances associated with insect pollination (Taylor 1978). A shift away from anemophily in the medullosan seed ferns may also be inferred from the organization of the pollen types *Parasporites* and *Nanoxanthiopollenites*. Although both possess the alveolate sporoderm structure, and are smaller than *Monoletes*, the sacci of *Parasporites* are clearly vestigial, while the spines that ornament the surface of *Nanoxanthiopollenites* are more typical of ornamentation patterns found on entomophilious grains. Further attesting to the probability that some of the early seed plants were pollinated by other than solely abiotic means is the increasing frequency of reports of pollen-rich coprolites (Scott 1977; Baxendale 1979). In several instances one coprolite may contain a single pollen type, possibly suggesting some early selectivity regarding foraging behavior. What was responsible for the initial indiscriminate foraging between pollen organs and ovules is not known; however, the fact that pollen is a primary attractant in some syndromes because it is a rich source of food, especially proteins, may have been an important factor. It is interesting to note that cycad pollen is a primary source of food for beetles and that the Cycadales as a group appear to have insects as secondary pollinators. This is especially noteworthy since the cycads are regarded by most as evolutionarily related to the medullosan seed ferns.

While a discussion of the pollination syndromes and reproductive biology of early seed plants still remains highly speculative, there can be little doubt that the fossil record can contribute to a more complete picture of the mechanisms associated with these systems. There are many structural features of the seeds including cupules, glands, and pollen-receiving devices that no doubt played an important role in various stages of the pollination in these early seed plants. An analysis of these features, together with a synthesis of information about the pollen organs and pollen, will not only be important in understanding the entire biology of a particular fossil organism, but holds great promise for providing a method of interpreting some of the more biological dimensions of paleobotany.

ACKNOWLEDGMENT

This investigation was supported in part by funds from the National Science Foundation through grant DEB77-02239.

REFERENCES

Andrews, H. N., and T. L. Phillips. 1968. *Rhacophyton* from the Upper Devonian of West Virginia. *J. Linn. Soc. Bot.* 61:37–64.

Banks, H. P., S. Leclercq, and F. M. Hueber. 1975. Anatomy and morphology of *Psilophyton dawsonii*, sp. n. from the Late Lower Devonian of Quebec (Gaspé), and Ontario, Canada. *Palaeontogr. Amer.* 8:77-127.

Baxendale, R. W. 1979. Plant-bearing coprolites from North American Pennsylvanian coal balls. *Palaeontology* 22:537-548.

Bonamo, P. M. 1977. *Rellimia thomsonii* (Progymnospermopsida) from the Middle Devonian of New York State. *Amer. J. Bot.* 64:1272-1285.

Bonamo, P. M., and H. P. Banks. 1967. *Tetraxylopteris schmiatii*: its fertile parts and its relationships within the Aneurophytales. *Amer. J. Bot.* 54:755-768.

Brauer, D. F. 1979. Barinophytacean plants from the Upper Devonian Catskill Formation of northern Pennsylvania. Ph.D. Dissertation. SUNY at Binghamton.

Chaloner, W. G., A. J. Hill, and W. S. Lacey. 1977. First Devonian platyspermic seed and its implications in gymnosperm evolution. *Nature* 265:233-235.

Clendening, J. A., and P. W. Nygreen. 1976. *Nanoxanthiopollenites*, a new monolete miospore from the Upper Pennsylvanian of Kansas. *Geoscience Man* 15:125-127.

Crepet, W. L. 1972. Investigations of North American cycadeoids: pollination mechanisms in *Cycadeoidea*. *Amer. J. Bot.* 59:1048-1056.

———. 1974. Investigations of North American cycadeoids: the reproductive biology of *Cycadeoidea*. *Palaeontographica* 148B:144-169.

———. 1979. Some aspects of the pollination biology of Middle Eocene angiosperms. *Rev. Palaeobot. Palynol.* 27:213-238.

Crepet, W. L., and T. Delevoryas. 1972. Investigations of North American cycadeoids: early ovule ontogeny. *Amer. J. Bot.* 59:209-215.

Crepet, W. L., D. L. Dilcher, and F. W. Potter. 1975. Investigations of angiosperms from the Eocene of North America: a catkin with Juglandaceous affinities. *Amer. J. Bot.* 62:813-823.

Daghlian, C. P., and T. N. Taylor. 1979. A new structurally preserved Pennsylvanian cordaitean pollen organ. *Amer. J. Bot.* 66:290-300.

Dennis, R. L., and D. A. Eggert. 1978. *Parasporotheca*, gen. nov., and its bearing on the interpretation of the morphology of permineralized medullosan pollen organs. *Bot. Gaz.* 139:117-139.

Dilcher, D. L. 1979. Early angiosperm reproduction: an introductory report. *Rev. Palaeobot. Palynol.* 27:291-328.

Dilcher, D. L., F. W. Potter, Jr., and W. L. Crepet. 1976. Investigations of angiosperms from the Eocene of North America: Juglandaceous winged fruits. *Amer. J. Bot.* 63:532-544.

DiMichele, W. A. 1979. Arborescent lycopods of Pennsylvanian age coals: *Lepidophloios*. *Palaeontographica* 171B:57-77.

Eggert, D. A., and G. W. Rothwell. 1979. *Stewartiotheca* gen. n. and the nature and origin of complex permineralized medullosan pollen organs. *Amer. J. Bot.* 66:851-866.

Eggert, D. A., and T. N. Taylor. 1971. *Telangiopsis* gen. nov., and Upper Mississippian pollen organ from Arkansas. *Bot. Gaz.* 132:30-37.

Florin, R. 1936. On the structure of the pollen-grains in the Cordaitales. *Svensk. Bot. Tidskr.*, 30:624-651.

Jennings, J. R. 1976. The morphology and relationships of *Rhodea, Telangium, Telangiopsis*, and *Heterangium. Amer. J. Bot.* 63:1119–1133.

Leisman, G. A., and J. S. Peters. 1970. A new pteridosperm male fructification from the Middle Pennsylvanian of Illinois. *Amer. J. Bot.* 57:867–873.

Millay, M. A., and D. A. Eggert. 1970. *Idanothekion* gen. n., a synangiate pollen organ with saccate pollen from the Middle Pennsylvanian of Illinois. *Amer. J. Bot.* 57:50–61.

———. 1974. Microgametophyte development in the Paleozoic seed fern family Callistophytaceae. *Amer. J. Bot.* 61:1067–1075.

Millay, M. A., D. A. Eggert, and R. L. Dennis. 1978. Morphology and ultrastructure of four Pennsylvanian prepollen types. *Micropaleontology* 24:303–315.

Millay, M. A. and T. N. Taylor. 1974. Morphological studies of Paleozoic saccate pollen. *Palaeontographica* 147B:75–99.

———. 1976. Evolutionary trends in fossil gymnosperm pollen. *Rev. Palaeobot. Palynol.* 21:65–91.

———. 1977. *Feraxotheca* gen. n., a lyginopterid pollen organ from the Pennsylvanian of North America. *Amer. J. Bot.* 64:177–185.

———. 1978. Fertile and sterile frond segments of the lyginopterid seed fern *Feraxotheca. Rev. Palaeobot. Palynol.* 25:151–162.

———. 1979. Paleozoic seed fern pollen organs. *Bot. Rev.* 45:301–375.

Oliver, F. W., and D. H. Scott. 1904. On the structure of the Palaeozoic seed *Lagenostoma lomaxi* with a statement of the evidence upon which it is referred to *Lyginodendron. Phil. Trans. Roy. Soc.* (Lond.) 197B:193–247.

Pettitt, J. M., and C. B. Beck. 1968. *Archaeosperma arnoldii*—a cupulate seed from the Upper Devonian of North America. *Cont. Mus. Paleo. Univ. Mich.* 22:139–154.

Phillips, T. L. 1979. Reproduction of heterosporous arborescent lycopods in the Mississippian-Pennsylvanian of Euramerica. *Rev. Palaeobot. Palynol.* 27:239–289.

Remy, R., and W. Remy. 1955. *Simplotheca silesiaca* n. gen. st sp. *Abh. Dtsch. Akad. Wiss. Berlin Kl. Chem. Geol. Biol.* 2:1–7.

Remy, W., and R. Remy. 1953. Untersuchungen über einige Fruktifikationen von Farnen und Pteridospermen aus dem Mitteleuropäischen Karbon und Perm. *Abh. Dtsch. Akad. Wiss. Berlin Kl. Math. Allg. Naturwiss.* 2:4–38.

———. 1959. *Zimmermannitheca cupulaeformis* n. gen. n. sp. *Monatsber. Dtsch. Akad. Wiss. Berlin* 1:767–776.

Richardson, J. B. 1960. Spores from the Middle Old Red Sandstone of Cromarty. Scotland. *Palaeontology* 3:45–63.

Rothwell, G. W. 1972a. Evidence of pollen tubes in Paleozoic pteridosperms. *Science* 175:772–774.

Rothwell, G. W. 1972b. Pollen organs of the Pennsylvanian Callistophytaceae (Pteridospermopsida). *Amer. J. Bot.* 59:993–999.

Scott, A. C. 1977. Coprolites containing plant material from the Carboniferous of Britain. *Palaeontology* 20:59–68.

Sterling, C. 1963. Structure of the male gametophyte in gymnosperms. *Biol. Rev.* 38:167–203.

Stewart, W. N. 1951. A new *Pachytesta* from the Berryville locality of southeastern Illinois. *Amer. Midl. Nat.* 46:717–742.

Stidd, B. M. 1978. An anatomically preserved *Potoniea* with *in situ* spores from the Pennsylvanian of Illinois. *Amer. J. Bot.* 65:677–683.

———. 1980. The current status of medullosan seed ferns. *Rev. Palaeobot. Palynol.* (in press).

Stidd, B. M., and J. W. Hall. 1970. *Callandrium callistophytoides*, gen. et sp. nov., the probable pollen-bearing organ of the seed fern, *Callistophyton. Amer. J. Bot.* 57:394–403.

Stidd, B. M., G. A. Leisman, and T. L. Phillips. 1977. *Sullitheca dactylifera* gen. et sp. n.: a new medullosan pollen organ and its evolutionary significance. *Amer. J. Bot.* 64:994–1002.

Taylor, T. N. 1970. *Lasiostrobus* gen. n., a staminate strobilus of gymnospermous affinity from the Pennsylvanian of North America. *Amer. J. Bot.* 57:670–690.

———. 1971. *Halletheca reticulatus* gen. et sp. n.: a synangiate Pennsylvanian pteridosperm pollen organ. *Amer. J. Bot.* 58:300–308.

———. 1978. The ultrastructure and reproductive significance of *Monoletes* (Pteridospermales) pollen. *Can. J. Bot.* 56:3105–3118.

———. 1979. Ultrastructural studies of pteridosperm pollen: *Nanoxanthiopollenites. Rev. Palaeobot. Palynol.* 56:3105–3118.

Taylor, T. N., and C. P. Daghlian. 1979. The morphology and ultrastructure of *Gothania* (Cordaitales) pollen. *Rev. Palaeobot. Palynol.* 29:1–14.

Taylor, T. N., and M. A. Millay. 1977. The ultrastructure and reproductive significance of *Lasiostrobus* microspores. *Rev. Palaeobot. Palynol.* 23:129–137.

———. 1979. Pollination biology and reproduction in early seed plants. *Rev. Paleobot. Palynol.* 27:329–355.

———. 1980. Morphologic variability of Pennsylvanian lyginopterid seed ferns. *Rev. Palaeobot. Palynol.* (in press).

2

A COASTAL HYPOTHESIS FOR THE DISPERSAL AND RISE TO DOMINANCE OF FLOWERING PLANTS

Greg Retallack and David L. Dilcher

INTRODUCTION

Few subjects in natural history have attracted so many diverse speculations as the origin of flowering plants. According to the most popular concept of early angiosperms, they were tropical, *Magnolia*-like, evergreen trees, pollinated by insects, their large seeds dispersed by animal ingestion (Takhtajan 1969; Regal 1977). The apparent lack of any fossil evidence for this concept was explained by Axelrod (1952, 1966), who postulated that angiosperms differentiated in tropical uplands, perhaps as long ago as 250 million years (during the Permian period) before invading lowland depositional environments about 110 million years ago (during the early Cretaceous period). Building on some aspects of Axelrod's work and also on a general observation made by Seward (1926, 1933), Raven (1977) suggested that the poleward spread of insect-pollinated tropical angiosperms was aided by climatic warming of coastal areas as a consequence of the dramatic mid-Cretaceous expansion of epeiric seas. Like most other views of early angiosperms, these are based largely on inferences drawn from the comparative anatomy, morphology, and distribution of living plants.

The most encouraging effect of the recent upsurge in research on angiosperm origins is the gathering recognition that the fossil record may furnish critical evidence for the origin and diversification of flowering plants, as it has for understanding the early history of amphibians, reptiles, birds, mammals, and humans. For example, Doyle and Hickey (1976; Hickey and Doyle 1977) have shown that the oldest angiosperm-like leaves and pollen are not as advanced as Axelrod assumed. They have also found much support for Stebbins's (1974) hypothesis that some early angiosperms were small-leaved shrubs of seasonally arid environments which migrated into disturbed habitats of mesic regions as streamside weed trees.

In our own studies of the diverse and well-preserved fossil plants, soils, and sedimentary environments of the mid-Cretaceous Dakota Formation in Russell County, Kansas, we have been able to extend our observations to early angiosperm fructifications and to a variety of coastal sedimentary environments, in addition to coastal stream deposits. We agree with the general concept of pioneer early angiosperms, as presented by Stebbins, Doyle, and Hickey, but also believe that this view can no longer be reconciled with the traditional concept of upland, *Magnolia*-like ancestral angiosperms. Accordingly, we propose the following new hypothesis for the dispersal and rise to dominance of flowering plants, based primarily on early angiosperm fossils and their ancient environments.

Some of the earliest angiosperms were probably woody, microphyllous plants of the rift valley system joining Africa and South America during the Early Cretaceous. Some of these plants were evidently adapted to pioneering disturbed coastal environments and so were exceptionally widespread during the remarkable transgressions and regressions of epeiric seas during the mid-Cretaceous. In the North American interior, the first angiosperms penetrated from the Gulf Coast to as far north as Alberta, Canada, with the Late Albian marine transgression and sea temperature maximum. Angiosperms became dominant rapidly in temperate paleolatitudes and first appeared in Alaska during the latest Albian regression. As angiosperms became more abundant, characteristic Early Cretaceous mangroves (*Weichselia, Frenelopsis, Pseudofrenelopsis*), ferns of freshwater coastal swamps and marshes (*Tempskya* and other ferns), and levee plants of coastal streams (*Cycadeoidea*) became extinct, but conifers of floodplains and uplands were less affected. Globally dispersed populations of these pioneering angiosperms were increasingly isolated during the Late Cretaceous by continued continental drift and realignment, by the general retreat of the epeiric seas, by local dispersal of angiosperms into disturbed and depositional environments further inland, and by increasingly specialized coadaptation with local animal pollinators and seed dispersers.

PALEOECOLOGY OF THE DAKOTA FORMATION

The upper Dakota Formation in Russell County, Kansas (Fig. 2.1) was deposited about 92 to 94 million years ago (Obradovich and Cobban 1975) during the early Cenomanian age of the Cretaceous. These are sediments of a complex lagoonal and deltaic coastal plain disconformably overlying the

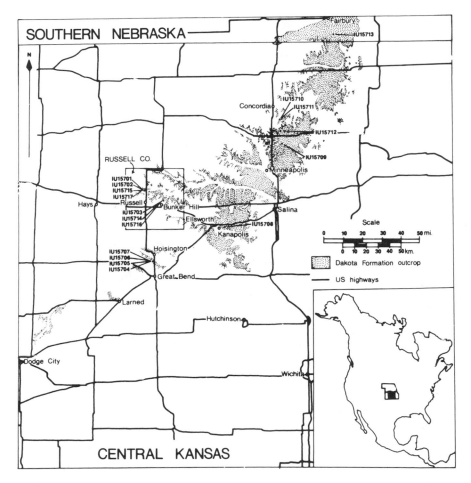

FIGURE 2.1. Mid-Cretaceous fossil plant localities (numbers prefixed by IU) in the Dakota Formation (stipple) in central Kansas and southern Nebraska.

Late Albian marine Kiowa Shale and older Paleozoic rocks, and formed as shallow mid-Cretaceous seas expanded over central North America. As discussed in more detail later in this chapter, angiosperms had only migrated into the interior of the United States a few million years before, during the Late Albian. This was also a time when the initial evolutionary radiation of early angiosperms was just beginning to wane (Doyle and Hickey 1976). Some of the oldest angiosperm leaf and pollen types, many of which became extinct soon after, were still present along with more modern types (Hickey, in press). Angiosperm fossils from the Dakota Formation are thus evidence of a critical phase in the early evolution and dispersal of flowering plants.

Perhaps more important, however, is the excellent preservation and abundance of the fossil plants, including common fructifications, and also the diversity of sedimentary paleoenvironments in the upper Dakota Formation (Table 2.1), which is not equaled by any older localities for early angiosperm fossils yet reported.

Shallow Marine Sands and Muds

The contact between the Dakota Formation and the overlying marine Graneros Shale can be arbitrarily drawn at the top of the highest sandstone, but represents a transition from nonmarine to marine sedimentation through a complex of coastal paleoenvironments. The uppermost sandstones of the Dakota Formation often contain marine fossils (Hattin 1967). One of these north of Russell (Figs. 2.2, 2.3) is a bed of ferruginized, bioturbated sandstone with molds and casts of disarticulated marine bivalves, probably deposited in a barrier or off-shore bar (in the sense of Shepard 1952). The invertebrate fauna of the lower Graneros Shale includes oysters (*Exogyra columbella*), venerid bivalves (*Aphrondina lamarensis*), brachiopods (*Lingula*), and arenaceous foraminifera. The fauna is sparse compared to that found in overlying marine sediments and was probably adapted to more brackish seawater near deltaic distributaries (Hattin 1965). Features such as these indicate that the Dakota Formation and Graneros Shale were produced by continuous sedimentation as the interior seaway expanded eastward into Minnesota.

Angiosperm Debris in Coastal Lagoons

The most diverse assemblages of fossil plants in the upper Dakota Formation are found in the shales of coastal lagoons. Fairly complete impressions of leaves and fructifications are common, along with trails of more thoroughly comminuted plant debris. These plant remains all appear to have been transported some distance from where the plants lived originally and are probably mixed from several former plant communities. *Sequoia*-like conifer shoots are common, but are far outnumbered by angiosperm remains of almost every taxon known from the Dakota Formation.

Remains of this kind are found at Indiana University plant locality IU15702, along the Saline River, near U.S. Highway 281, north of Russell (Figs. 2.2, 2.3). Sandy flaser and linsel interbeds within these lagoonal shales show opposed cross-bedding, indicating some tidal influence. Hattin (1965) found arenaceous foraminifera at two levels in this same unit only 100 m along strike to the north of the most productive fossil plant locality. Siemers (1971) found disarticulated valves of mussels (*Brachidontes arlingtonanus*) at the base of the shaly unit near the fossil plant locality. Living species of *Brachidontes* are found in nearshore or lagoonal waters of brackish and fluctuating salinity (Siemers 1976; Scott and Taylor 1977), as also are arenaceous foraminifera (Hattin 1965).

At another locality (IU15707) in the clay pits south of Hoisington (Fig. 2.1), fossil plant fragments are scattered through shales near the top of the Dakota Formation with flaser and linsel interbeds, and also a variety of trace fossils. In the same shaly unit, some 150 m south along strike from this is a remarkable accumulation of diverse and complete impressions of plants (IU15706) in the C horizon of a very weakly differentiated paleosol. This fossil soil has only a few roots, scattered ferric mottles, and much relict bedding. The plant fossils were probably a mixed accumulation of plant debris near a lagoonal shoreline only intermittently and sparsely vegetated as the shoreline fluctuated.

As in modern lagoonal systems, salinity probably varied dramatically in time as well as in space, from fully marine salinities more usual near the opening into the sea to completely fresh water common in the innermost reaches of the lagoons (Phleger 1969). This is especially apparent from the paleoecology of invertebrate fossils of the upper Dakota Formation (Hattin 1967; Siemers 1976; Hattin, Siemers, and Stewart 1978). Low-diversity assemblages of bivalves (*Brachidontes* and *Ostrea*) and serpulid worm tubes, probably lived in brackish water, typical of the more inland portions of coastal lagoons. Other assemblages include brackish elements (*Brachidontes*, *Crassostrea*), and also a variety of more typically marine bivalves (such as *Breviarca*, *Cymbophora*, *Geltina*, *Laternula*, *Parmicorbula*, *Tellina*, and *Volsella*), which indicate normal marine salinity common in the outer (seaward) parts of lagoons.

Distributary Sands

The nature of the sandstones of the upper Dakota Formation is the key to understanding the mid-Cretaceous coastal geomorphology of Kansas, and its attendant suite of subenvironments.

The Rocktown channel sandstone has a distinctive sinuous outcrop pattern across Russell County (Fig. 2.2A). This quartz-sandstone lacks fossils and is conspicuously cross-bedded, with foresets dipping mainly to the west (Siemers 1976). It was probably deposited by a meandering coastal stream (Siemers 1976; Karl 1976).

Planar-bedded sandstones are often associated with this and other channel sandstones in the upper Dakota Formation and form a prominent marker horizon throughout Russell County, Kansas (Figs. 2.2B,C, 2.3). Paleocurrents deduced from low-angle cross-bedding and ripple marks in these sandstones are bimodal or polymodal (Siemers 1976), indicating

TABLE 2.1: Reconstructed Vegetation as Evidenced by Fossil Plant Assemblages, Soils, and Sedimentary Environments of the Mid-Cretaceous, Upper Dakota Formation in Kansas and Nebraska.

	floodplain forest	lake margin scrub	levee scrub of coastal streams	swamp woodlands of lagoonal margins and interdistributary areas	mangroves of tidally influenced distributaries	plant debris drifted into coastal lagoons
Reconstructed vegetation						
Sedimentary facies	redbeds (superimposed paleosols)	laminated shale with scattered plants and insects	epsilon crossbeds with scour-and-fill and ferruginized sandstones	coal and carbonaceous shales	massive gray mudstones and carbonaceous shale adjacent to channel sandstones	linsel and flaser bedded shales with diverse trace fossils, arenaceous foraminifera and brackish to marine molluscs

TABLE 2.1 (continued): Reconstructed Vegatation as Evidenced by Fossil Plant Assemblages, Soils, and Sedimentary Environments of the Mid-Cretaceous, Upper Dakota Formation in Kansas and Nebraska.

Paleosols	none	weakly differentiated clayey	coal-bearing (thick organic horizon)	weakly differentiated sandy	weakly differentiated sandy	well differentiated with reddish B horizon
Characteristic fossil plants	very diverse, fragments of nearly all plant taxa found in the Dakota Formation	*"Acerites multiformis"* of Lesquereux 1892	*Magnoliaephyllum* sp., *Sapindopsis* sp., *Liriophyllum* sp.	*Araliopsoides cretacea*	*Platanus*-like leaves and *Sequoia*-like conifer shoots	none preserved (likely habitat of conifers in drifted plant beds and regional pollen rain)
Representative localities (Fig. 2.1)	Hoisington (IU15706, IU15707), Saline River (IU15702)	Rose Creek (IU15713) Saline River (IU15717, IU15715)	Linnenbergers's Ranch (IU15703, IU15714, IU15716), Saline River (IU15701)	Hoisington (IU15704 IU15705), Kanapolis (IU15708)	Braun's Ranch (IU15709)	Saline River, Kanapolis, (areas where fossil soils have been found)

Source: Compiled by the authors.

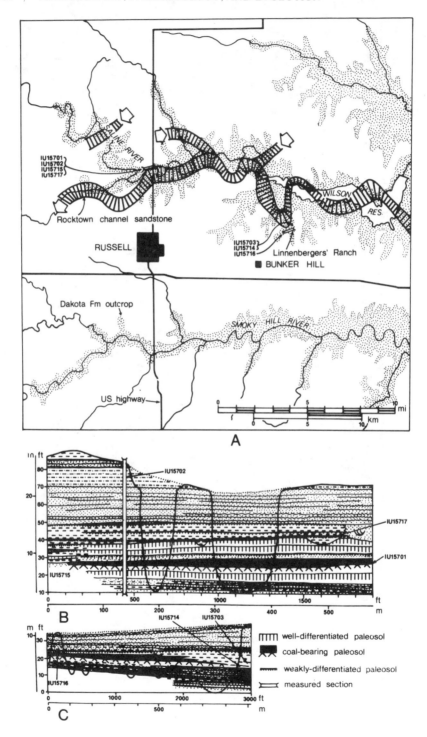

A

B

C

well-differentiated paleosol

coal-bearing paleosol

weakly-differentiated paleosol

measured section

considerable tidal influence. Most of the sandstone is more or less bioturbated, with trace fossils such as *Planolites*, *Skolithus*, *Arenicolites*, and *Chondrites*. They also contain brackish-adapted fossil bivalves (*Corbicula*) and serpulid worm tubes (Hattin, Siemers, and Stewart 1978).

These marine-influenced sandstones were probably deposited in the wide shoaling mouths of distributaries of a tide-dominated delta, like the modern Klang Langat (Coleman, Gagliano, and Smith 1970), Ganges-Brahmaputra, and Mekong deltas (Morgan 1970). This is indicated especially by the close relationship of the channel and marine-influenced sandstones in Russell County, by the westward thickening of the marine-influenced sandstone at the expense of the channel sandstone, and by the restriction of marine fossils in the base of the marine-influenced sandstones to localities where the channel sandstone is also present (Siemers 1971, 1976; Hattin, Siemers, and Stewart 1978). The term *tide-dominated delta* is used here rather than estuary, to indicate that these sediments were part of a depositional landscape and sequence rather than lying on an older erosional land surface. Franks (1980) has interpreted the Longford Member of the Kiowa Formation (underlying the Dakota Formation in Kansas) as an estuarine coast. Such tide-dominated sedimentation was probably widespread in the stable low-relief coasts of the eastern margin of Cretaceous seaways in the interior of the United States.

Tide-dominated deltas may support different plant communities than river-dominated or digitate deltas, like the Mississippi delta (Shepard 1956; Frazier and Osanik 1969) or wave-dominated arcuate deltas, like the Grijalva-Usumacinta delta of Mexico (Thom 1967). The seaward flaring distributaries of tide-dominated deltas are marine-influenced for some distance inland and their margins are usually muddy and well vegetated. The various subenvironments of the tide-dominated deltas of India and Malaya are forested largely by different and diverse mangal communities (Blasco 1977; Chapman 1977b).

Mangal Paleosols of Distributary Margins

Weakly differentiated gray clayey paleosols are common in the uppermost Dakota Formation, associated with marine-influenced and channel sandstones. Many of these soils were probably marine influenced. Some

FIGURE 2.2. Geology of mid-Cretaceous fossil plant localities in the Dakota Formation, in Russell County, Kansas. A. Outcrop of the Dakota Formation and the Rocktown Channel Sandstone in Russell County. Paleocurrents are indicated by arrows. B and C. Geological cross-sections with considerable vertical exaggeration. Present topography indicated by heavy line. Lithological symbols are mostly as in Figure 2.3. B. North of the Saline River, near U.S. Highway 281, north of Russell. C. On Linnenbergers's Ranch, northeast of Bunker Hill.

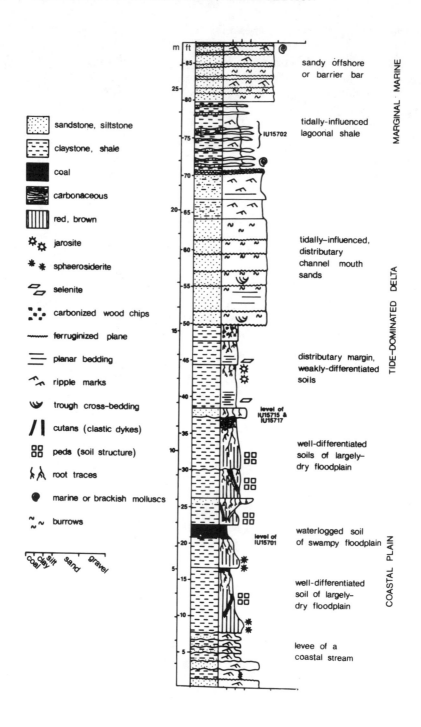

sandstone, siltstone

claystone, shale

coal

carbonaceous

red, brown

jarosite

sphaerosiderite

selenite

carbonized wood chips

ferruginized plane

planar bedding

ripple marks

trough cross-bedding

cutans (clastic dykes)

peds (soil structure)

root traces

marine or brackish molluscs

burrows

coal clay silt sand gravel

m ft

sandy offshore
or barrier bar

MARGINAL · MARINE

tidally-influenced
lagoonal shale

IU15702

tidally-influenced,
distributary
channel mouth
sands

TIDE-DOMINATED DELTA

distributary margin,
weakly-differentiated
soils

level of
IU15715 &
IU15717

well-differentiated
soils of largely-
dry floodplain

waterlogged soil
of swampy floodplain

level of
IU15701

well-differentiated
soil of largely-
dry floodplain

levee of a
coastal stream

COASTAL PLAIN

even contain brackish-adapted mussels in life position. Judging from the fossil roots and leaf litter found in these paleosols, they were vegetated by scrubby angiosperms, characterized by "*Acerites multiformis*" (of Lesquereux 1892; Figs. 2.4, 2.22K).

The clearest example of a mangal paleosol was found near Rose Creek, south of Fairbury, Nebraska (locality IU 15713). Within this paleosol, Dr. J. F. Basinger found recently a specimen of *Brachidontes*, of normal size (25.5 mm long) and preserved articulated and closed, in life position (Fig. 2.5). This epibyssate bivalve evidently lived for some time in brackish water bathing mudflats near or within a stand of mangrove-like plants. Work is still in progress on the plant remains found in the leaf litter of this paleosol. Leaves of "*Acerites multiformis*" (Figs. 2.4, 2.22K) and radially symmetrical perfect flowers (Figs. 2.6, 2.7) are the most common fossils. A limited diversity of other angiosperm leaves was also found. Fern and conifer fragments are rare.

Comparable fossil soils are also found immediately underlying the marine influenced sandstones north of the Saline River and on Linnenbergers's Ranch, near Bunker Hill in Russell County, Kansas (Figs. 2.2, 2.3). The leaf litter of one of these (IU 15717) contains small *Araliopsoides* leaves, leaflets of *Sapindopsis*, less common *Liriodendron*-like leaves, fragments of fern and cycadophyte leaves, and angiosperm fructifications. No brackish invertebrates have yet been found in these other paleosols, but like the Rose Creek paleosol, they too have conspicuous jarosite blooms, probably from the surficial weathering of pyrite and marcasite (Battey 1972). The apparent abundance of these sulfides in the leaf litter horizons of these fossil soils is comparable with that of marine-influenced coals (Mackowsky 1975; Carruccio and Geidel 1979). Large selenite crystals up to 7 cm long are also common in the B horizons of these and the Rose Creek paleosol, but, probably, do not indicate Cretaceous marine influence. Pyrite and marcasite have often been reported from drill cores and cuttings through the subsurface Dakota Formation, but gypsum has never been found in them (Rubey and Bass 1925; Landes 1930; Landes and Keroher 1938, 1939; Swineford and Williams 1945; Schoewe 1952; Merriam et al. 1959; Mack 1962; Siemers 1971). The selenite is probably not an original marine evaporite, but more likely precipitated in the present outcrops by ground water acidified by the weathering of iron sulfides (Hattin 1965).

Like the modern mangal, the vegetation of these clayey Cretaceous soils was apparently woody and adapted to fluctuating salinities of marine-influenced distributary margins. The analogy with modern mangroves

FIGURE 2.3. Stratigraphical section and interpreted paleoenvironments of the upper Dakota Formation at fossil plant locality IU 15702, north of the Saline River, near U.S. Highway 281, north of Russell.

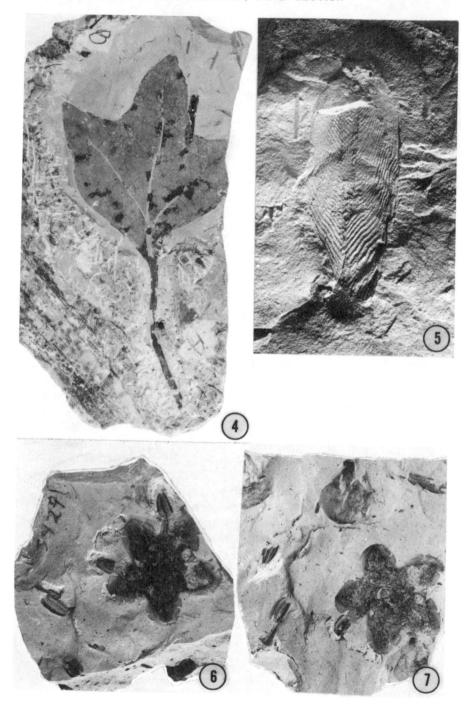

should not, however, be taken too literally. There is no evidence in these Cretaceous fossils for such peculiarities as prop roots, air roots, or vivipary, found in a few of the more common genera of modern mangroves. There is no fossil record of any modern mangrove genera in rocks older than Eocene (Tralau 1964; Muller 1970; Churchill 1973). Before the Eocene, woody intertidal plants were very different taxonomically, and in their reproduction and adaptations to salinity, from modern mangroves. Extinct mangroves probably belonged to such diverse groups as lycopods (Retallack 1975), true ferns (Daber 1968), seed ferns (Harris 1966; Retallack 1977b), cordaites (Cridland 1964; Eggert et al. 1979), and conifers (Jung 1974).

Angiosperm Woodlands of Interdistributary Swamps

Many of the coals of the upper Dakota Formation contain large fossil roots, also extending into the underclay. Some of these swamp woodland paleosols also show evidence of abundant pyrite, probably indicating limited marine influence. They evidently formed in interdistributary depressions, in part fringing coastal lagoons and sheltered bays. From plant debris and pollen associated with the coals and from more complete larger remains in nearby subautochthonous accumulations, these swamps appear to have been wooded largely by angiosperms, commonly with leaves of *Magnoliaephyllum*, *Liriophyllum*, and *Sapindopsis*. There is also good evidence that these swamp woodlands had an extensive ground cover of true ferns.

On Linnenbergers's Ranch, near Bunker Hill in Russell County, Kansas (Figs. 2.1, 2.2), coal was mined as recently as 1927, but most of the old adits are now covered by slumps. Our excavations of the productive coal seam revealed numerous large carbonaceous roots in the underclay and also common jarosite blooms. Only indeterminate angiosperm leaf fragments were collected from the underclay (locality IU 15716). Palynological preparations made recently from the coal by Michael Zavada (University of Connecticut, Storrs) contain abundant fern spores, common angiosperm pollen, and rare gymnosperm pollen. About 300 m northeast along strike from where the coal was mined, the coal passes laterally into fossiliferous shale filling a swale within levee deposits adjacent to the Rocktown channel sandstone and overlain by marine-influenced sandstone (localities IU 15703, IU 15714, Fig. 2.2). The fossil plants in this shale appear to have accumulated close to where the plants lived. This is indicated by the low taxonomic

---

FIGURES 2.4–2.7. Fossils from near Rose Creek, Nebraska (locality IU 15713). **2.4.** *"Acerites multiformis"* of Lesquereux 1892, a very common leaf at this locality. IU 15713-3317. × 1.1. **2.5.** *Brachidontes*, a brackish water bivalve preserved in life position alongside plant fossils shown in Figures 2.4, 2.6, and 2.7. IU 15713-3303. × 2.9. **2.6** and **2.7.** Remains of perfect flowers showing calyces, stamens, and a detached petal. Figure 2.6. IU 15713-3429'. × 1.5. Figure 2.7. IU 15713-3429. × 1.5.

diversity in our large fossil plant collection from here, by the excellent preservation of organic material in these remains, and by the mixture of leaves, fructifications, and logs, organs which would be sorted quickly during transport by wind and water. Angiosperm remains are most abundant, particularly leaves of *Magnoliaephyllum* (Figs. 2.9, 2.22J), *Liriophyllum* (Figs. 2.14, 2.22I), and *Sapindopsis* (Fig. 2.8) and a variety of fructifications which, on anatomical grounds, can be attributed to these same extinct angiosperms (Figs. 2.10, 2.11, 2.15). Ferns and *Brachyphyllum*-like conifer shoots are rare. These remains were probably derived from the vegetation of the adjacent coal-bearing paleosols. The composition of this megafossil assemblage is in broad agreement with the palynology of the coal. Many elements of the megafossil assemblage are also common in drifted plant accumulations of coastal lagoons, but few are common in levee vegetation better known from ferruginized sandstones of the Dakota Formation.

Another paleosol with a coaly organic horizon was found with associated angiosperm debris along the Saline River, north of Russell (IU15701, Figs. 2.2, 2.3). Similar paleosols are widespread in the Dakota Formation of Kansas and in other mid-Cretaceous rocks of the North American interior. Coals with autochthonous palynofloras dominated by angiosperms and ferns, and with rare conifers and microplankton have been found in mid-Cretaceous rocks of Arizona, Utah, Oklahoma, and Minnesota (Pierce 1961; Hedlund 1966; Agasie 1969; May and Traverse 1973; Romans 1975). The ferny understory of one of these mid-Cretaceous swamp woodlands has been remarkably well preserved by volcanic ash fall in a coal of the Dakota Formation in Utah (Rushforth 1971). As with coastal peats accumulating in modern deltas (Frazier, Osanik, and Elsik 1978), these swamp woodlands, probably vegetated interdistributary areas with fresh to brackish groundwater.

Sediments and Angiosperms of Coastal Streamsides

As in Cretaceous rocks of the eastern United States (Doyle and Hickey 1976; Hickey and Doyle 1977), there is good evidence in Kansas that scrubby pioneer angiosperms colonized levees and point bars of the freshwater reaches of coastal streams. From the relationships of sedimentary facies exposed in large clay pits, it seems that the leaf impressions of ferruginized sandy flagstones of Kansas, made famous by the pioneering monographs of Lesquereux (1874, 1883, 1892), are the remains of levee vegetation. The regional flora of these sandstones is very diverse, although few species were found at any one locality, perhaps an indication that the remains accumulated close to the living plants. Overall, *Araliopsoides cretacea* (Fig. 2.22G) is the most characteristic species of these angiosperm-dominated assemblages.

The sedimentary setting of these ferruginized sandstones is best seen in the clay pits south of Hoisington (Fig. 2.1). Underlying the lagoonal shales there are well-exposed deposits of a stream channel and its levee, overlying and partly eroding a well-differentiated floodplain paleosol. At the northern end of the northern clay pit is a trough cross-bedded, quartz sandstone, with numerous fossil logs. This passes southward along strike into a unit of finely interbedded sandstone and shale, cross-bedded at a very low angle and disrupted by several large scour-and-fill structures also containing finely-interbedded sandstone and shale. The interbedded unit is an epsilon cross bed (as defined by Allen 1963), typical of levees of meandering streams (Retallack 1977a). This levee deposit includes beds of ferruginized sandstone (locality IU15705) with numerous impressions of *Araliopsoides* and other angiosperm leaves. Some of the shale beds are also fossiliferous (IU15704) with impressions of aquatic leaves, such as *Nelumbites* (Fig. 2.22H). These probably represent plants growing on levee tops and swales, respectively. Fossil soils are never well differentiated within these levee deposits, seldom showing more than small root traces and surficial ferruginization. The angiosperm-dominated levee vegetation was probably scrubby and (geologically) impermanent.

Paleosols, Sediments, and Angiosperms of Lake Margins

Shales of freshwater coastal lakes also contain abundant angiosperms, particularly *Platanus*-like leaves (Figs. 2.12, 2.22F) and their likely fructifications (Fig. 2.13), but also common *Sequoia*-like conifer shoots. Such shales are exposed for a distance of about a mile in gullies and a small quarry on Braun's Ranch, southeast of the intersection of highways 81 and 24, Kansas (IU15709, Fig. 2.1). Unlike the lagoonal shales of the upper Dakota Formation, there is no indication of tidal influence or marine fauna in these shales. Angiosperm leaves and conifer shoots are well preserved in the even shale partings along with fructifications and insects. No rocks were seen to overlie the shales here, but they are probably somewhere in the middle of the Dakota Formation, judging from their location in the middle of the outcrop area of this westerly dipping formation. Underlying the shales are sandstones with carbonized wood chips, low angle cross bedding, and ferruginized bedding planes, probably, the deposit of a lakeside beach. The sandstones overlie a well-differentiated reddish paleosol, formed in a largely dry floodplain. Fossil plants in the shale were probably derived from lake margin communities, as well as the floodplain forests.

Conifer Forested Floodplains

No fossil plant remains other than root traces are preserved in or associated with the thick well-differentiated reddish paleosols characteristic

of the middle Dakota Formation (Figs. 2.2, 2.3). Their clear differentiation into a gray or yellowish A horizon and a reddish brown B horizon indicates that they were forested soils of floodplains in which the water table was more or less permanently at or below about a meter from the surface. As conifer remains are abundant in mixed debris of coastal lagoons and lakes and dominated the regional pollen rain, they were probably prominent in these floodplain forests. The most common megafossil remains of conifers in the Dakota Formation of Kansas are *Sequoia*-like shoots. A number of other studies (Pierce 1961; Hedlund 1966; Agasie 1969; Burgess 1971; May and Traverse 1973; Romans 1975; Hickey and Doyle 1977) also indicate that conifer dominance of inland and floodplain vegetation of North America persisted well into the Late Cretaceous.

Overview

The mid-Cretaceous coastal plain of Kansas consisted of a variety of environments, including offshore and barrier bars, large coastal lagoons, tide-dominated deltas, lakes, and largely dry floodplains. Within this varied landscape, scrubby angiospermous mangal, characterized by "*Acerites multiformis*," evidently colonized distributary margins of tide-dominated deltas. Swamp woodlands extending inland from lagoons and interdistributary bay margins were also dominated by angiospermous trees, with leaves of *Magnoliaephyllum*, *Liriophyllum* and *Sapindopsis*, shading a ferny understory. Scrubby angiosperms such as *Araliopsoides*, and aquatic angiosperms such as *Nelumbites*, vegetated levees and swales of freshwater coastal streamsides. Angiosperms with *Platanus*-like leaves were common around freshwater lakes. However, the largely dry floodplains were forested mainly by conifers.

Angiosperms were most abundant in a variety of unstable depositional environments. Many were probably early successional plants. These conclusions are also supported by the nature of early angiosperm reproductive organs, discussed in the following section.

EARLY ANGIOSPERM REPRODUCTION

Structurally preserved early angiosperm fructifications are few. Our material from the Dakota Formation (Dilcher et al. 1976; Dilcher 1979) is

FIGURES 2.8–2.11. Leaves and unisexual reproductive organs of two mid-Cretaceous plants common at both Linnenbergers's Ranch (IU15703) and Hoisington (IU15706). These organs are attributed to the same original plants on the basis of anatomical evidence and their occurrence together. **2.8.** Leaflets of *Sapindopsis* sp. IU15703-2494. × 1.1. **2.9.** Leaf referred to as *Magnoliaephyllum* sp. IU15703-2455. × 0.6. **2.10.** Pollen-bearing florets of *Sapindopsis* sp. IU15703-2751. × 1.75. **2.11.** Raceme of multifollicular axes of *Magnoliaephyllum*. IU15706-3402. × 0.7.

among the oldest. Vachrameev and Krassilov (1979) have also described a compressed fructification from Middle Albian rocks of western Kazakhstan. These few well-preserved fructifications allow a better understanding of a variety of poorly preserved fructifications from Early and mid-Cretaceous rocks. As our investigations into these remains continue, we are forming a concept of early angiosperm reproduction which can no longer be reconciled easily with traditional views.

Pollination

The interpretation of pollination syndromes from the morphology of dispersed pollen grains is difficult. Hickey and Doyle (1977) regarded the exine structure of early angiosperm-like pollen as "strong evidence that the plants which produced them were insect pollinated." On the other hand, Walker (1976) compared this same pollen to that of the living Chlorantha-ceae, which he regarded as wind pollinated. This is probably not a true dilemma. Both are likely, but only to a degree.

The oldest angiosperm-like pollen to appear in most regions have features which are not optimal for either wind or insect pollination. Plants producing these pollen were probably generalists, pollinated both by wind and by a variety of insects, like living *Salix* and *Rhizophora* (Faegri and van der Pijl 1966; Muller and Caratini 1977; Tomlinson, Primack, and Bunt 1979). This kind of pollination is best suited to early successional plants and communities (Proctor 1978). Fossil angiosperm-like pollen appearing at slightly higher stratigraphic levels show a dramatic adaptive radiation of pollen into additional types adapted more clearly to either wind or insect pollination. Much of the later evolution of angiosperms was, probably, a result of increasingly specialized methods of pollination.

The oldest angiosperm-like pollen appearing in both southern England and the eastern United States (*Clavatipollenites hughesii*) during the Barremian may well have been produced by a generalist. Its size (12 to 30 μm, according to Kemp 1968) is on the boundary of the optimum range for wind pollination (20 to 40 μm, according to Whitehead 1969). The exine of *Clavatipollenites* is tectate-perforate to semitectate, so it is unclear if these are true sculptural elements or whether the depressions between them were filled with substances involved in compatibility mechanisms, in adhering to the stigma or in preventing water loss, as in modern angiosperms (Heslop-

FIGURES 2.12–2.15. Leaves and reproductive organs of mid-Cretaceous plants common at Linnenbergers's Ranch (IU 15703), Hoisington (IU 15704), and Braun's Ranch (IU 15709). These organs are attributed to the same original plants on the basis of anatomical evidence and their occurrence together. **2.12.** *Platanus*-like leaf. IU 15709-3135. ×0.5. **2.13.** *Platanus*-like reproductive axis. IU 15709-3165. × 0.34. **2.14.** *Liriophyllum* sp. leaf. IU 15706-3188. × 0.40. **2.15.** Reproductive axis of *Liriophyllum*. IU 15706-3084. × 0.73.

Harrison 1976). Even if regarded as sculpture, it is much finer than that usual in the modern pollen dispersed by insects studied by Skvarla and Larson (1965). If *Clavatipollenites* were dispersed in clumps (also argued by Hickey and Doyle 1977), then these remain to be found. Finally, the monosulcus, as in *Clavatipollenites*, is a feature which has developed independently in pteridosperms, *Ginkgo*, cycads, cycadeoids, and angiosperms (Townrow 1960; Wodehouse 1935). As argued by Heslop-Harrison (1976), the monosulcus is largely a mechanism to prevent desiccation, which is more critical for wind pollination than for specialized insect pollination.

Soon after the appearance of the first generalist angiosperm-like pollen in Barremian rocks of England and the United States, angiosperm-like pollen diversified to include types better suited either to wind or insect pollination. *C. rotundus*, first found in Lower Albian rocks of England, is similar to *C. hughesii*, but larger (20 to 32 μm), with a more conspicuous sulcus and more rounded outline (Kemp 1968). These may well have been dispersed by wind. Larger monosulcate grains with a more prominent stellate exine sculpture, but only 29 to 39 μm long, appear in rocks of latest Barremian to Early Aptian age in England (Hughes, Drewry, and Laing 1979). These may indicate a tendency toward more frequent or more specialized insect pollination. The large size (36 to 73 μm) and prominent stellate exine sculpture of *Stellatopollis barghoornii*, first appearing in Middle to Late Albian rocks of the eastern United States (Doyle, van Campo, and Lugardon 1975) are fully comparable with the pollen of modern insect-pollinated angiosperms. Very small (9 to 17 μm), finely reticulate pollen also first appear in Middle to Late Albian rocks of southern England (Kemp 1968) and the eastern United States (Doyle et al. 1975; Hickey and Doyle 1977). Pollen of this size would be deflected easily around the stigma by boundary layers of air flow, and was probably also dispersed by insects.

The first angiosperm-like pollen appeared in Australia during the Middle Albian (Dettmann 1973), and may also have been produced by generalists. The most common of these, *Phimopollenites pannosus*, is tricolporoidate, prolate to subspheroidal, 13 to 28 μm in size and microreticulate, appearing almost smooth in the light microscope. The other of the oldest angiosperm-like pollen, *Rousea georgensis*, is tricolpate, 15 to 24 μm in size and also microreticulate. During the Late Albian there was a greater diversity of angiosperm-like pollen, including likely generalists (*Clavatipollenites* sp., 17 to 26 μm), as well as pollen possibly dispersed by wind (*Phimopollenites augathellaensis*, 22 to 40 μm) and by insects (*Tricolpites minutus*, 10 to 16 μm, and also found in dispersed tetrads; Dettmann 1973).

A greater variety of pollination syndromes may be represented by the earliest angiosperm-like pollen found in Gabon, west Africa. These include small, weakly-sculptured *Clavatipollenites*, medium-sized moderately sculptured *Retimonocolpites*, and larger stellately sculptured *Stellatopollis* (Doyle et al. 1977). If this is indeed near the center of origin for angiosperms (Hickey and Doyle 1977), there may have been a variety of pollination syndromes there. The generalist pollination of early angiosperms best applies to those which emigrated to achieve one of the most dramatic revolutions of the world flora in the geological history of plants. Unfortunately, further details of these very ancient angiosperm-like pollen have not yet been published.

The generalist pollination of early angiosperms is supported only in an indirect way by known early angiosperm reproductive organs, as most of these are fruits rather than flowers. The oldest structurally preserved angiosperm fructification is a branched catkin-like aggregate of multifollicles (*Caspiocarpus paniculiger* Vachrameev and Krassilov 1979) in the axils of leaves (*Cissites* sp. cf. *C. parvifolius*). Although Vachrameev and Krassilov refer this Middle Albian fossil to the Ranunculales, there is no evidence of any stamens, perianth parts, or scars from their previous attachment. We have studied a comparable raceme of multifollicles from the Dakota Formation (Fig. 2.11), for which we have about 100 specimens, many of them beautifully preserved and with a complete covering of cuticle. These also show no evidence of stamens, petals, sepals, or nectaries. Apparently apetalous unisexual multifollicles are the most common Early Cretaceous angiosperm fructifications. These include Late Barremian to Aptian remains referred to as "*Carpolithus*" *virginiensis*, "*C.*" *geminatus*, "*C*" *sessilis*, and "capsules sp." of Fontaine (1889) from the eastern United States (Dilcher 1979), Early Albian remains referred to as "*Carpolithus*" *ternatus* and "*C.*" *fascicularis* by Fontaine (1889) from eastern North America (geological ages after Doyle and Hickey 1976), and Middle Albian remains referred to as "*Carpolithus*" *karatcheensis* by Vachrameev (1952) from western Kazakhstan (age discussed by Vachrameev and Krassilov 1979). Similar fructifications, apparently also derived from unisexual flowers, are common in Late Cretaceous and Tertiary floras of the North Hemisphere. These have been called *reproductive axes* by Dilcher et al. (1976; Dilcher 1979), "*Carpolithus arcticus*" by Hickey (1977), *Trochodendrocarpus* by Krassilov (1973a, 1977), *Jenkinsella* by Chandler (1961), and *Cercidiphyllum* by Brown (1939, 1962), Becker (1961, 1969, 1973), Chandrasekharam (1974), and Crane (1978). The absence of any evidence for features usually associated with insect pollination, such as nectaries, stamens, petals, and sepals in these fructifications may be because they were pollinated by wind or were generalists.

Other Early Cretaceous angiosperm-like fructifications also lack evidence of nectaries, stamens, petals, and sepals. These are terminal clusters of free follicles, including the Late Barremian to Aptian *Callitris* sp. and the Early Albian *Carpolithus conjugatus* and "indet. plant e" of Fontaine (1889) from eastern North America (Dilcher 1979; geological age from Doyle and

Hickey 1976) and the Albian *Ranunculaecarpus quinquecarpellatus* from the far eastern U.S.S.R. (Samylina 1968; Takhtajan 1969). As in modern flowers (van der Pijl 1972), the radial symmetry of these reproductive organs would have been recognized more easily by animals. This was unlikely to have been a specialized animal interaction, as the most similar modern flower is that of *Cercidiphyllum*, which is wind pollinated (Lawrence 1951).

A variety of angiosperm fructifications has been found in mid-Cretaceous rocks, although diversity is still low. In the Dakota Formation in Kansas, these include elongate multifollicles, catkin-like inflorescences, globular *Platanus*-like fructifications, and five-lobed flower calyxes (Figs. 2.6, 2.7, 2.10, 2.11, 2.13, 2.15; Dilcher et al. 1976; Dilcher 1979). In the Woodbridge Clay Member of the Raritan Formation in the eastern United States, there are terminal clusters of multifollicles, catkin-like inflorescences, and five-lobed flower calyxes (Newberry 1895; geology after Doyle and Robbins 1977). In the Elk Neck Beds of the upper Potomac Group, globular *Platanus*-like fructifications are found (Hickey and Doyle 1977; Doyle and Robbins 1977). In Cenomanian rocks of Czechoslovakia, there are also globular *Platanus*-like fructifications, arranged in spikes and in the axils of leafy shoots, and also spikes of five-lobed flower calyxes (Velenovsky 1889). Some of the catkin-like inflorescences may have been generalists or wind pollinated. Recent studies of one of the five-lobed flower calyxes from the Dakota Formation (Dilcher and Basinger work in progress) have shown that these are bisexual flowers with conspicuous petals and stamens containing very small, weakly-ornamented pollen. These were probably pollinated by insects.

In summary, many early angiosperms, including the first to migrate into many regions, were probably generalists, pollinated by wind and a variety of insects, like many modern early successional plants. Once these pioneering plants were established, their adaptive radiation very soon included species pollinated exclusively by wind or by insects, as well as persistent generalists.

Seed Dispersal

A variety of Early Cretaceous and Jurassic fossil plant propagules have been identified as angiospermous, largely on the basis of their various adaptations for animal dispersal, such as strong spines and pitted or ribbed, thick stony layers. These have been called *Nyssidium*, *Kenella*, *Ievlevia*, *Onoana*, *Lappacarpus*, and *Tyrmocarpus* (by Chandler and Axelrod 1961; Krassilov 1967, 1973a, 1973b; Samylina 1968; Douglas 1969). As also indicated by Wolfe, Doyle, and Page (1975) and by Hughes (1976b), none of these fossils have any definitive angiospermous character. They could equally be considered propagules of extinct gymnosperms, so they need not be considered further here.

Spines and stony and fleshy layers are not found in any of the Cenomanian and older angiosperm fructifications already discussed. Instead, these appear to have been without exception, dry, dehiscent follicles with numerous small seeds. The Middle Albian *Caspiocarpus paniculiger* is a catkin-like aggregation of hundreds of follicles, each about 1 mm long, and containing about six ovate seeds about 0.8 x 0.5 mm in size (Vachrameev and Krassilov 1979). The follicles are fully cutinized inside and out and appear neither fleshy nor woody. Individual follicles from a Cenomanian fructification in Kansas are similar (Fig. 2.11). This fructification is a raceme of at least four multifollicles, each with 50 to 90 helically arranged follicles. These follicles average 2.0 x 2.7 mm and each contains two to six seeds, averaging 0.8 x 0.5 mm in size. The lax catkin-like form of these and similar fructifications would have been a poor visual attractant to animals. Neither the seeds nor the follicles have stony layers to withstand ingestion by animals. Nor do they have well-developed wings, like the modern winged fruits of *Engelhardia* and *Acer*, adapted to wind dispersal. The large number of small seeds could probably float or be blown considerable distances. They were probably dispersed in several ways, mainly by wind and water.

The apparent generalist pollination, large number of small seeds, wind and water dispersal, and lack of any evident interdependence on specific animal pollinators or dispersers are a syndrome of features found today largely in weeds and other early successional plants (Rorison 1973; Heinrich 1976). Early angiosperms appear to have had a potential for colonizing clearings, open woodland, fluvial and deltaic levees and crevasse splays, and tidal flats.

VEGETATIVE FEATURES OF EARLY ANGIOSPERMS

The woody and small-leaved nature of the oldest angiosperms has long been proposed from studies of the comparative morphology of living plants (Cronquist 1968; Takhtajan 1969). Both these features are also well in evidence from the fossil record of early angiosperms. These additional constraints serve to refine our concept of the habit and habitat of early angiosperms.

Woody Early Angiosperms

Comparison of woody and herbaceous forms within different orders, families, and genera of living plants, has almost universally led to the conclusion that herbaceous angiosperms were derived from woody ancestors

(Takhtajan 1969). The woody nature of most early angiosperms is also evident from their fossils. Most of the early angiosperm fructifications already discussed are attached to woody peduncles. In the Early Cenomanian Dakota Formation of Kansas, abscission scars have been found on several twigs and angiosperm leaves. This indicates that these plants were woody perennials and shed their leaves naturally. It is not yet certain, however, whether leaf fall was seasonal. Fossil soils and root traces in the Dakota Formation also indicate that much of the angiosperm-dominated vegetation consisted of shrubs and trees.

Some possible herbaceous early angiosperm fossils are notable for their antiquity: an Aptian or latest Barremian leafy shoot of *Acaciaephyllum spatulatum* from the eastern United States (Doyle 1973) and a Middle Albian fertile leafy shoot bearing catkins of *Caspiocarpus paniculiger* and leaves of *Cissites* sp. cf. *C. parvifolius* (Vachrameev and Krassilov 1979). Perhaps some early angiosperms were indeed herbaceous or perhaps these are merely young shoots. More detailed studies of these older plants and associated sediments and soils are needed. Herbaceous aquatic angiosperms, such as "*Menispermites*" and *Vitiphyllum*, are common in Albian and younger rocks (Doyle and Hickey 1976; Hickey and Doyle 1977).

As we have demonstrated, the reproduction of early angiosperms was most like that of modern early successional plants, most of which are now herbaceous weedy plants. Perhaps a better comparison is with modern larger woody plants of a similar early successional nature, such as the river colonizing species of *Platanus* (Moore 1972), and *Casuarina* (Beadle, Evans, and Carolin 1972) and the mangrove *Rhizophora* (West 1977).

Microphyllous Ancestors

The earliest angiosperm-like leaves in many regions are smaller than later angiosperm-like leaves in the same region (Vachrameev 1952; Samylina 1968; Krassilov 1973). The weakly organized venation of some of the oldest angiosperm-like leaves has been interpreted by Doyle and Hickey (1976) as evidence that their ancestors had even smaller leaves. They attribute the rapid size increase of early angiosperm leaves to the Early Cretaceous innovation of intercalary meristem in angiosperm leaves, thus enabling the expansion of formerly microphyllous leaves before the regularity of higher venation was determined genetically. It is also likely that most of the blade of parallel-veined monocotyledonous leaves was derived from a phyllode of ancestral plants (Kaplan 1973). Microphylly was probably only one of a number of features developed in stressful environments during the early evolution of angiosperms. Other features include the closed carpel, reduced gametophyte stages, double fertilization, shortened reproductive cycle, and long dormancy of the seeds (Stebbins 1974).

The selective pressure producing such an hypothetical microphyllous ancestral angiosperm has been thought to have been a locally or regionally arid habitat (Stebbins 1974; Hickey and Doyle 1977). On botanical grounds, this is at least as likely to have been an adaptation to the physiological aridity of tidally-influenced mudflats and hypersaline lagoons. Considering all the existing geological evidence, discussed in the next section of this chapter, the so-called *xeromorphic bottleneck* in angiosperm evolution (Hickey and Doyle 1977) was perhaps more likely induced by near marine rather than arid environments.

EARLY DISPERSAL OF ANGIOSPERMS

The Early Cretaceous dispersal and rise to dominance of angiosperms was a remarkable phenomenon. In only ten million years they had spread over coastal areas of the earth from the equator to the poles, apparently immune to oceanic barriers, different soils and climates, and competition from preexisting local vegetation, all of which prevented dispersal of many other plants. It is probably not a coincidence that the widespread dispersal of angiosperms coincides with maximum oscillations of epicontinental seas during the Aptian, Albian, and Cenomanian. Widespread changes in sea level would have presented a unique opportunity for pioneering coastal angiosperms with generalized methods of pollination and dispersal. This is indicated especially by the geological record of early angiosperm dispersal.

Evidence of Angiosperm-like Pollen

Palynological research provides a good overview of the early migrations of angiosperms and is in broad agreement with the appearance of angiosperm-like megafossils (as discussed by Axelrod 1959; Teslenko, Golbert, and Poliakova 1966). According to Hickey and Doyle (1977), the earliest angiosperm-like monosulcate pollen dispersed "instantaneously" around tropical Tethyan coasts during the Barremian, as it is found in deposits of roughly similar age in the eastern United States, western Africa, England, Israel, and Patagonia. The first angiosperm-like tricolpate pollen appeared in Brazil and western Africa during the Aptian and may be as old as the Barremian in Israel. It evidently reached England, the eastern United States, and Siberia by the Early Albian, penetrated the North American interior as far north as Alberta, Canada, and also south and east to Australia by the Middle Albian, and finally reached Alaska by the Cenomanian. From the precocious appearance and diversity of tricolpate and other angiosperm-like pollen in Early Cretaceous rocks of Brazil and western Africa, Hickey and Doyle also suggest that this region of west Gondwanaland was the center from which angiosperms dispersed.

These migrations would not be so remarkable if angiosperms were polyphyletically derived from a number of gymnospermous groups, as Krassilov (1977) has argued. However, the oldest angiosperm-like pollen is similar everywhere, morphologically conservative and low in diversity. This and the orderly evolutionary radiation of angiosperm-like pollen in younger rocks are additional support for the existing floral, morphological, and embryological evidence that angiosperms are monophyletic (Doyle 1978). As a group, these plants seem to have been undeterred by oceanic barriers, such as the deep equatorial Tethyan ocean and several meridional epicontinental seas and embayments. The successful crossing of the Tethys sea by two successive waves of angiosperm-like pollen was most remarkable, as considerations of plate tectonics (Dietz and Holden 1970; Dewey et al. 1973; Spencer 1974), marine faunal interchange (Ager 1967; Dilley 1973; Kauffman 1973; Stevens 1973), and likely oceanic circulation (Luydendyk, Forsyth, and Phillips 1972; Gordon 1973), indicate continuous oceanic separation of the northern from the southern continents during the Early Cretaceous. These early angiosperms also appear to have been little affected by climatic zonation which restricted the distribution of preexisting flora and fauna (Bergquist 1971; Sohl 1971; Vachrameev 1978). Angiosperms probably migrated largely in coastal regions, where the climate was more uniform.

Although the oldest cosmopolitan angiosperm-like pollen are of limited diversity, they became increasingly provincial and diverse during the Late Cretaceous (Brenner 1976). By the end of the Cretaceous, the rudiments of modern floristic provinces and higher taxonomic groups of angiosperms, including several modern families, had already appeared (Muller 1970; Raven and Axelrod 1974; Pacltova 1978; Martin 1977). Even the marine angiospermous seagrasses seem to have originated during the Late Cretaceous (Brasier 1975).

According to our hypothesis, regional differences in the widely dispersed early angiosperms would have been initiated as their range extended along the world's coastlines. Once dispersed, these pioneer angiosperms appear to have adapted to other unstable environments, such as river levees (Hickey and Doyle 1977). These environments may have served as secondary dispersal routes to other disturbed habitats further inland. Continued continental drift culminated in the increased separation of continents by deep meridional oceans (Dietz and Holden 1970), a geographic configuration which has induced provinciality even in modern mangroves (van Steenis 1962). The general retreat of epeiric seas by the end of the Cretaceous (Kauffman 1977a) may have left isolated populations of coastal angiosperms in inland areas. Finally, coevolution of angiosperms with local animal pollinators and dispersers has proceeded to extraordinary levels in modern angiosperms and evidently has been a major factor in the speciation of both angiosperms and animals (Faegri and van der Pijl 1966; van der Pijl 1972).

Early Angiosperms in the Rift Valleys of West Gondwanaland

Several authors (Brenner 1976; Doyle et al. 1977) have argued, from the record of angiosperm-like pollen in the Early Cretaceous rift valley sediments which connected Africa and South America, that regional aridity was a stimulus to angiosperm evolution. However, what evidence there is for regional aridity in west Gondwanaland is not only too late to explain the origin of the oldest angiosperm-like pollen there, but also inextricably linked with marine transgression. The main evidence of aridity is the widespread Aptian evaporites which accumulated in restricted embayments and coastal lagoons as marine transgression proceeded into the rift valley system (Reyre 1966; Reyre et al. 1966; Brognon and Verrier 1966; Hourcq 1966; Reyment and Tait 1972). Angiosperm-like pollen first appear in sediments stratigraphically below the evaporites at levels where there is little evidence of aridity. Just below the evaporites, bisaccate pollen and fern spores become much less abundant at the expense of *Classopollis* and *Ephedripites* (Jardiné, Kieser, and Reyre 1974; Doyle et al., 1977). In Gabon, angiosperm-like pollen becomes spectacularly diverse in rocks underlying and interbedded with the salt (Doyle et al. 1977). It is difficult to determine whether these changes are due to increasing aridity, salinity, or both. At least one genus of the bizarre extinct pollen found in Early Cretaceous rocks of this region, *Classopollis*, is known to have been produced by halophytic conifers whose shoots are referred to as *Frenelopsis* (Figs. 2.18, 2.19) and *Pseudofrenelopsis* (Oldham 1976; Daghlian and Person 1977; Hluštík and Konzalová 1976; Alvin, Spicer, and Watson 1978). Further studies of the palynofloras and further work on the Early Cretaceous megafossil floras (additional to that of Axelrod and Raven 1978; Batton 1965; Plumstead 1969; Berry 1939, 1945) of west Gondwanaland are needed, particularly in view of the possibility outlined by Hickey and Doyle (1977) that this was the center of origin for the angiosperms.

English Coastal Lagoons

Angiosperm-like pollen first appears in Barremian (middle Early Cretaceous) rocks of England within the Weald Clay (Hughes 1976a, 1977; Hughes et al. 1979). At this level, Wealden rocks accumulated largely in estuaries and coastal lagoons of laterally variable and temporally fluctuating salinity, with brackish water fossils, redeposited marine fossils, and interbeds of glauconitic sandstone (Kilenyi and Allen 1963; Allen 1976; Allen et al. 1973; Kennedy 1978). Angiosperm-like megafossils have not yet been recognized in Wealden rocks, although megafossil plants are common on many lower horizons, nor have angiosperm-like pollen been found in older Wealden palynofloras (Hughes 1976a, 1977; Hughes and Moody-Stuart

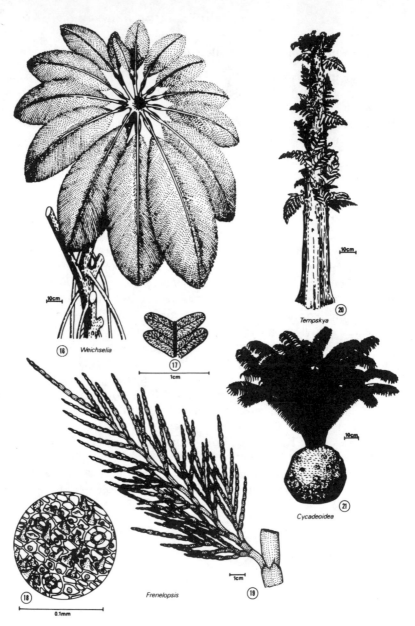

FIGURES 2.16–2.21. Characteristic Early Cretaceous warm temperate to tropical, coastal plants which became extinct as angiosperms became abundant. **2.16, 2.17.** *Weichselia reticulata* (Stokes et Webb) Fontaine in Ward 1899. **2.16.** Reconstruction (after Alvin 1971, with permission). **2.17.** Detail of pinnules, with characteristic reticulate venation (after Fontaine in Ward 1899). **2.18, 2.19.** *Frenelopsis ramosissima* Fontaine 1889. **2.18.** Partially reconstructed shoot (after Fontaine 1889). **2.19.** Strongly papillate cuticle with thickly cutinized and sunken stomata (after Berry 1910). **2.20.** *Tempskya*, a reconstruction. (*Source:* Andrews, H. N., and Kern, E. M. *Ann. Miss. Bot. Gard.* 34:119–186, 1947, with permission.) **2.21.** *Cycadeoidea*, a reconstruction (*Source:* Delevoryas, T., *Proc. Nth Am. Paleont. Conv.* 1L:1672, 1971, with permission.)

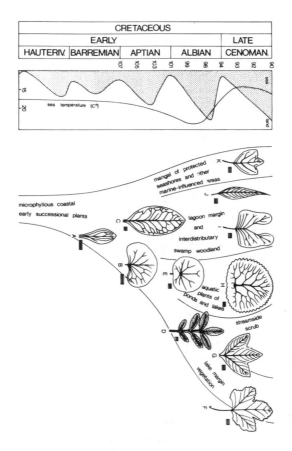

FIGURE 2.22. A coastal hypothesis for the dispersal and rise to dominance of angiosperms, based largely on the geological record of early angiosperms in later Early and earliest Late Cretaceous rocks of North America. The radiometric time scale, marine transgressions, and regressions are based largely on studies on the North American interior (Kauffman 1977a, 1977b). Figured fossil angiosperm leaves are; A, *Acaciaephyllum spatulatum* Fontaine 1889 (after Doyle and Hickey 1979); B, *Proteaephyllum reniforme* Fontaine 1889 (after Doyle and Hickey 1976); C, *Ficophyllum crassinerve* Fontaine 1889 (after Doyle and Hickey 1976); D, *Sapindopsis belviderensis* Berry 1922b (after Berry 1922b); E, *Nelumbites virginiensis* (Fontaine) Berry 1911 (after Doyle and Hickey 1976); F, *Platanus*-like leaf (Indiana University collections, Dakota Formation); G, *Araliopsoides cretacea* (Newberry) Berry 1916b (after Lesquereux 1874); H, *Nelumbites* sp. (Indiana University Collection); I, *Liriophyllum* sp. (Indiana University Collection); J, *Magnoliaephyllum* sp. (Indiana University Collection); K, "*Acerites multiformis*" of Lesquereux 1892 (Indiana University Collection). Scale bars are all 1 cm.

1967). Angiosperm-like pollen reaches its highest concentrations (up to 10 percent in the shallow marine and prodeltaic Atherfield Clay of the overlying Greensand and persists in concentrations somewhat less than 1 percent in overlying marine rocks of Aptian to Albian age (Casey 1961; Kemp 1968;

Hughes 1976b; Middlemiss 1976). It is notable that angiosperm-like pollen first appears rarely in the marine-influenced upper portion of a terrestrial sequence well-known for fossil plant remains, is most common in shallow marine prodeltaic sediments, and persists in younger marine rocks. Considering this timing and observations of Churchill (1973) and Hughes (1976b) on distinguishing coastal vegetation from regional pollen rain into the sea, it is likely that these early angiosperms were most abundant along the coast, but less prominent, if present at all, further inland.

Into the Interior of North America

The coincidence between sea level changes and the appearance and rise to dominance of angiosperms is seen most strikingly in the geological record of early flowering plants in North America.

The oldest generally accepted angiosperm-like remains in North America have been found in the Aptian or perhaps latest Barremian portion of the lower Potomac Group of the eastern United States (Doyle and Hickey 1976; Hickey and Doyle 1977). As the fossiliferous outcrops are about 100 km west of laterally equivalent marine rocks intersected in deep boreholes (Doyle and Robbins 1977), only a limited number of largely fluvial environments are preserved in the outcropping Potomac Group. As in the Dakota Formation in Kansas, the floodplains of the Potomac Group were forested largely by conifers, while angiosperms were most common on fluvial levees (Hickey and Doyle 1977). Even though inland, some of the major changes in the appearance and abundance of fossil angiosperms correlate with sea level changes better documented in deep drill cores of equivalent rocks to the east (Vokes 1948; Stephenson 1948; Doyle and Robbins 1977). Angiosperm-like fossils appear within the Patuxent Formation, at a time of regression after at least two earlier marine transgressions. They became more common and diverse above the contact between the Arundel Clay and the Patapsco Formation, at a time of marked local regression. They became abundant in the Elk Neck Beds and Raritan Formation, during the most pronounced fluctuations of sea level. There are also indications that some large coastal estuaries, like those of the modern mid-Atlantic coast of North America, may have penetrated even as far inland as the outcropping Potomac Group during the Cretaceous. Such an irregular topography would be expected, considering the Early Cretaceous intrusions and underlying rift valley systems of basement rocks revealed by recent deep drilling (Schlee et al. 1976, 1977; Scholle 1977). More direct evidence includes the estuarine grain size distribution of some outcropping sandstones of the Potomac Group (Groot 1952; Glaser 1969) and estuarine molluscan faunas in outcrops of the overlying Raritan Formation (Wolfe and Pakiser 1971). This may also explain the local abundance of *Frenelopsis* and *Pseudofrenelopsis* at several fossil plant localities and stratigraphic levels in the Potomac Group, such as Trents Reach, Fredericksburg, Fort Washington, and Chinkapin Hollow of Fontaine (1889), Knowlton (1919), and Doyle and Hickey (1976). Almost monodominant assemblages of these species in other parts of the world have been interpreted usually as the remains of estuarine gallery or lagoonal margin mangal (Oldham 1976; Daghlian and Person 1977; Doludenko 1978). It is likely that *Pseudofrenelopsis*, *Frenelopsis*, and early angiosperms dispersed inland along estuaries and coastal streams which deeply dissected vegetation dominated by ferns, cycadeoids, and conifers.

In marine rocks of the northwestern Great Valley of California, fossil plant remains indicate that angiosperms had dispersed right around the southern shoreline and up the west coast of the United States by the Albian (when most of Central America was not yet attached to North America). Most of the supposed angiosperms from these marine rocks (reported by Fontaine in Ward 1905; Diller 1908; Chandler and Axelrod 1961) cannot be considered definitively angiospermous (Wolfe, Doyle, and Page 1975; Hughes 1976b; examination by Dilcher). However, the fragments reported by Fontaine (in Ward 1905) from Elder Creek, west of Red Bluff (locality 23 of Ward 1905; interval 9500–10,000' of Diller and Stanton 1894) probably are angiospermous, although unidentifiable. These are in the Albian or Aptian portion of the succession. Angiosperms are again found more commonly in conglomerates of Late Albian or Cenomanian age in the same region (Diller and Stanton 1894; Murphy 1965; Popenoe, Imlay, and Murphy 1960).

Angiosperms do not appear to have penetrated the interior of North America until the great meridional epeiric sea was established, linking the western Tethyan and the Arctic oceans during the Middle to Late Albian stand of high sea level. The Early Cenomanian and Early Turonian high sea level stands were of successively greater extent and were followed by lesser oscillations of sea level until the sea retreated for the last time from the North American interior by the latest Cretaceous (Kauffman 1977a).

These epeiric seas were characterized by dramatic sea level changes and probably also significant tidal fluctuations. Active uplift and volcanism on the hilly western margin contrasted with the stable coastal plains and low hills of the eastern margin. Kauffman (1977a) has argued that, during times of high sea level, warm currents from the south increased the salinity and temperature of the interior seaway. During low stands the waters were not only colder, but probably also had a low-density layer of brackish water on the surface and near the coasts. The Late Albian transgression and temperature maximum, during which marine temperatures at midlatitudes were raised as much as 5°C, appear to have been the most critical time for wide dispersal of angiosperms into the North American interior.

In Texas and Oklahoma, along what was then the Gulf Coast, a Middle Albian fossil flora of conifers and cycads, and the likely mangrove, *Frenelopsis*, has been found in the Glen Rose Limestone (Daghlian and Person 1977; Perkins, Langston, and Stone 1979). A single loose nodule containing another likely extinct mangrove, *Weichselia* (Figs. 2.6A,B) (Daber 1968; Alvin 1971; Batten 1975) may have come from this or a younger Albian limestone (Berry, 1928). The oldest angiosperm fossils in the region are found in the later Middle Albian, marginal-marine Paluxy Sandstone, and equivalents (Ball 1937; Hedlund and Norris 1968). During the latest Middle to earliest Late Albian, angiosperms appear in fossil floras as rare elements throughout the interior as far north as Alberta, Canada. In both the Beaver Mines Formation in Alberta and in the upper Lakota Formation of the Black Hills of western South Dakota and eastern Wyoming, these pioneer angiosperms, represented largely by *Sapindopsis* leaf fossils, are rare elements of the fossil flora and occur in successions with older Albian fossil floras lacking angiosperms. In the Black Hills, angiosperms appear with *Weichselia* (Ward 1899; geology also discussed by Cobban and Reeside 1952; Gookey et al. 1972; Hickey and Doyle 1977). In Alberta, angiosperms appear soon after a minor marine transgression, evidenced by fossil foraminifera in the lowest Beaver Mines Formation (Bell 1956; Mellon 1967; Stott 1974; Jeletsky 1978). During the Late Albian, angiosperms became common in fossil floras of the Cheyenne Sandstone in Kansas (Berry 1922b; Scott and Taylor 1977), the Fall River Sandstone in the Black Hills (Ward 1899), and the Mill Creek Formation in Alberta (Bell 1956). Only by latest Albian times do the first rare angiosperms appear in the Chandler Formation in the northern foothills of the Brooks Range, Alaska (Scott and Smiley 1979). The Chandler Formation has also yielded older Albian nonangiospermous fossil floras in its lower part and was apparently deposited by a series of large river-dominated marine deltas (Ahlbrandt et al. 1979). Following widespread earliest Cenomanian regression, angiosperms came to dominate coastal depositional environments rapidly, as preserved in the Woodbine Formation of Texas and Oklahoma (Berry 1922a; MacNeal 1958; Kauffman, Hattin, and Power 1977) and in the Dakota Formation in central Kansas (Dilcher et al. 1976; Dilcher, Potter, and Reynolds 1978; Dilcher 1979). Similar dominance was not achieved in Alaska until the Turonian, when the marginal marine Seabee Formation was deposited (Smiley 1969).

Two features of the fossil record of early angiosperms in North America should be noted. Firstly, nonangiospermous mangroves, prominent in Albian and older fossil floras, become rare or extinct as angiosperms become more abundant. Secondly, the introduction and rapid increase in abundance of angiosperms are coincident with episodes of marine influence. This is most striking in areas where there is a good fossil record of preexisting inland nonangiospermous floras.

Early Cretaceous Coasts of the Far Southeastern U.S.S.R.

Judging from Krassilov's (1973c) description of floral changes in the far southeastern U.S.S.R., the appearance of angiosperms there was in many ways similar to that of the North American interior. The first rare angiosperms appear in the Sutschan Basin at a level where oyster fossils occur for the first time in a largely nonmarine sequence with abundant older nonangiospermous fossil floras. Higher in the sequence, a more pronounced marine transgression is recorded by deposition of the "*Trigonia* sandstones" and overlying, marginal-marine black shales. Plant fossils, including more common angiosperms, "occur occasionally in the *Trigonia* beds as well as in the black shales. More numerous and better preserved specimens come from the fine-grained coaliferous sandstones, 70 m thick, which are transitional between marine and lacustrinal deposits" (Krassilov 1973c).

Along the Seashores of Central Australia

In Australia, angiosperm-like pollen first appears in coastal and marine sediments of Late Middle Albian age. As in Early Cretaceous rocks of England, angiosperm-like pollen are more abundant (up to 60 percent of the total palynological assemblage) in nearshore and lagoonal deposits than in marine rocks further offshore (Dettmann 1973). The transgression of Albian epeiric seas into Australia also introduced several cosmopolitan marine mollusks into endemic faunas remaining from older epeiric seas (Day 1969), and coincides with a diversification of newly introduced gleicheniaceous and schizaeaceous fern spores (Dettmann and Playford 1969) and perhaps also, with rising marine temperatures (Stevens 1971). This was a particularly appropriate time for the immigration of tropical coastal vegetation.

The oldest angiosperm-like leaves in Australia are found in Albian sediments of coastal basins of Victoria (Medwell 1954) and Queensland (Walkom 1919). Angiosperm-like pollen have not yet been found in these sediments (Dettman 1963; Dettmann and Playford 1969), but palynological preparations from several interbedded horizons contain abundant microplankton (de Jersey 1960; Douglas 1969). Other Early Cretaceous angiosperm-like fossils (as reported by Douglas 1963, 1965, 1969) cannot be regarded as convincing records of angiosperms.

For the later history of early angiosperms in Australia, both palynological (Dettmann 1973) and megafossil (Berry 1916a; Douglas 1969; Senior, Mond, and Harrison 1978) records are in agreement. Angiosperms become rapidly dominant following withdrawal of epeiric seas during the Cenomanian.

COMPETITIVE REPLACEMENT OF PREEXISTING VEGETATION

The key elements of the Cretaceous floral revolution were the appearance and rise to dominance of angiosperms and a synchronous decline and even extinction of characteristic earlier Cretaceous true ferns, seed ferns, cycadeoids, and ginko-like plants (Delevoryas 1971; Scott and Smiley 1979). The nonangiospermous plants affected most by these changes were those most prolific in lowland and coastal environments. However, conifers which were abundant in upland and floodplain environments (Pierce 1961; May and Traverse 1973), were relatively unaffected in either abundance or diversity by Early Cretaceous invasions of angiosperms (Delevoryas 1971; Miller 1977). The more dramatic overturn of other kinds of plants could be due to competitive replacement of preexisting lowland and coastal plants by angiosperms adapted better to pioneering similar disturbed coastal environments. Such an interpretation is particularly evident from the case histories of four prominent and characteristic Early Cretaceous plants. Each of these became extinct just as angiosperms were becoming common, and each vegetated specific coastal environments.

A Mangrove Fern

Weichselia (Figs. 2.16, 2.17) was one of the most common and characteristic fossil ferns along Tethyan and Proto-Atlantic seashores, from North America to Japan on the north shore and from Peru to north central India on the south shore. It is most common and widespread in rocks of Early Cretaceous age, and a useful index fossil for rocks of that age, although there are a few doubtful records of fragments of leaves and of similar petrified axes from Jurassic and Late Cretaceous rocks (Sahni 1936; Berry 1945; Batton 1965; Jongmans and Dijkstra 1965; Alvin 1971). *Weichselia* is strikingly different from other matoniaceous ferns: its leptosporangia are enclosed tightly by interlocking peltate indusia which form tight soral clusters; its cuticle is exceptionally thick for a fern; its pinnules are strongly recurved; and its many-branched stems bear numerous aerophores and probably formed tangled thickets supported by numerous prop roots (Fig. 2.16). *Weichselia* is found commonly in monospecific fossil associations in marine or near-marine sediments. Batten (1975) suggested that it may have been either a coastal dune-binder or a mangrove. Considering its occurrence in an evidently waterlogged humic paleosol (Daber 1968), the latter interpretation seems more likely. Thus, *Weichselia* probably formed a pantropical mangal during the Early Cretaceous.

Conifer Mangroves

Frenelopsis (Figs. 2.18, 2.19) and *Pseudofrenelopsis* are fossil genera for distinctive jointed shoots with reduced sheathing leaves. They were evidently woody trees or shrubs (Watson 1977), which thrived along the northern coasts of the Tethys and Proto-Atlantic oceans from Texas to China and also in northern Africa (Boureau 1953; Batton 1965; Jongmans and Dijkstra 1973; Alvin et al. 1978; Vachrameev 1978; Doludenko 1978). They were most common during the Early Cretaceous, but persisted in a few places into the Late Cretaceous (Jongmans and Dijkstra 1973; Alvin 1977; Pons and Broutin 1978). The Late Cretaceous decline of these conifers is seen more clearly from the pollen produced by the same plants, *Classopollis* (Hluštık and Konzalová 1976; Alvin et al. 1978) and *Classoidites* (which differs only in detail from *Classopollis*; Pons and Broutin 1978). As this pollen has also been found in several cheirolepidiaceous conifer cones (Srivastava 1976; Miller 1977), its decline may indicate the decreasing importance of a whole group of plants, as well as *Frenelopsis* and *Pseudofrenelopsis*. *Classopollis* wanes dramatically in abundance in North America after the Aptian (Brenner 1976), when angiosperms first appear. It is extinct almost everywhere by the Turonian (Srivastava 1976), when angiosperms appear to have dominated coastal vegetation. *Classopollis* appears only to have persisted into the later Cretaceous in south central Asia (Vachrameev 1978).

The shoots of *Frenelopsis* and *Pseudofrenelopsis* are strongly vascularized, so evidently belonged to woody shrubs or trees (Watson 1977). They also have a distinctly succulent appearance, compared by several paleobotanists (Zeiller 1882; Reymanówna and Watson 1976) with the extant chenopodiaceous angiosperm *Salicornia*, which is a dominant plant of many modern salt marshes and hypersaline mudflats (Chapman 1977a). The geological occurrence of these shoots also indicates that these plants were coastal halophytes. Their shoots and dispersed cuticles commonly form monodominant fossil associations in near-channel and prodelta sediments. Oldham (1976) has suggested that they may have been comparable to the almost monospecific gallery mangals lining seashores, estuaries, and coastal streams in Florida today. Low-diversity assemblages of *Frenelopsis* or *Pseudofrenelopsis* have also been found in hypersaline lagoonal deposits (Daghlian and Person 1977) and with a variety of marine trace fossils and pectinid bivalves of a coastal lagoon or estuary (Doludenko 1978). A small branch, about 3 cm in diameter, has been found in an almost monodominant assemblage of *Pseudofrenelopsis* shoots, with a perfectly symmetrical encrustation of serpulid worm tubes (Perkins, Langston, and Stone 1979). This branch was evidently vertical, and probably living, when bathed in brackish water. *Frenelopsis* and *Pseudofrenelopsis* appear to have been the shoots of woody shrubs or trees which colonized estuarine shorelines, and tidally-influenced and hypersaline mudflats throughout the tropical coasts of southern Laurasia and parts of northern Africa.

Ferns of Freshwater Coastal Swamplands

Tempskya is a characteristic and distinctive fern false stem, representative of a number of widespread ferns of Early Cretaceous coastal regions (Fig. 2.20). It is found north of the Tethys and Proto-Atlantic oceans, from Oregon through much of the United States and Europe to southern central U.S.S.R. and Japan (Jongmans and Dijkstra 1965; Banks et al. 1967; Endo 1926). In North America, *Tempskya* was extinct by the end of the Albian (Read and Ash 1961). It evidently persisted in parts of Europe, at least until the middle Late Cretaceous (Jongmans and Dijkstra 1965). *Tempskya* has been found preserved in growth position in pyritic-carbonaceous shale which accumulated in a freshwater swamp (Tidwell, Thayn, and Roth 1976). Petrified *Tempskya* false stems penetrated by angiosperm stems, could also be interpreted as evidence of direct competition (Tidwell et al. 1977). Considering the paleogeography of Cretaceous epeiric seas and Early Cretaceous tectonics (particularly of Oregon and Idaho; Jones, Silberling, and Hillhouse 1977), *Tempskya* appears to have been restricted to coastal plain areas. A widespread and diverse assemblage of ferns evidently thrived in Early Cretaceous freshwater peat swamps of these coastal plains (see Oishi 1940; Pierce 1961; Hedlund 1966; Agasie 1969; Rushforth 1970, 1971; May and Traverse 1973; Hughes 1976a). *Tempskya* was evidently an integral part of these Early Cretaceous fern brakes throughout freshwater coastal swamps of southern Laurasia.

Cycadeoids of Coastal Streamsides

Cycadeoidea (Fig. 2.21) was one of the last of a very important Mesozoic group of gymnosperms. These distinctive squat, silicified fertile trunks are found in Early Cretaceous rocks throughout North America and Europe, from California to Czechoslovakia (Jongmans and Dijkstra 1958). *Cycadeoidea* probably did not survive into the Late Cretaceous. Records of Late Cretaceous *Cycadeoidea* are based on outdated ideas of the geological age of sediments (Andrews and Kern 1947; Read and Ash 1961); on small fragments found loose on the surface, often in areas of marine rocks (Wieland 1906, p. 19, 1916, pp. 121, 122, 1934, p. 94; Leriche 1909); on fragments lacking the diagnostic cones of *Cycadeoidea* (Chrysler 1932; Endo 1953; Baikovskaya 1956; Wieland 1928b); or on specimens whose locality is uncertain, such as those gained from Navajo Indians and archaeological excavations in New Mexico (Wieland 1928a, 1934; Delevoryas 1959). *Cycadeoidea*, apparently, became extinct just as angiosperms were becoming common in North America and Europe.

Only in a few places has *Cycadeoidea* been collected in place and reported by trained geologists. These are in the Lakota Formation of the Black Hills, eastern Wyoming and western South Dakota (Ward 1899; Wieland 1906, p. 23), in the Cedar Mesa Sandstone near Moab, Utah (Tidwell, Thayn, and Roth 1976) and in the uppermost Patuxent Formation in Maryland (Ward 1905, p. 405). All these were in streamside or levee deposits. This can be seen especially well from the better known occurrences from the Black Hills. Near Piedmont, South Dakota, on the eastern margin of the Black Hills, Ward (1899, p, 564) found cycadeoids in place in a soft yellowish sandstone and sandy shale, with occasional interbeds of reddish clay. Near Matias Peak ("Matties Peak" of Ward, 1899, p. 555) on the southern margin of the Black Hills, the unit (bed 12 of Ward) containing numerous upright cycadeoids is now known to be gray claystone, with numerous interbeds of red and orange-yellow, very fine grained sandstone (from stratigraphic drilling by Silver King Mines, Edgemont, South Dakota, courtesy of Mr. E. Faulkner). Both these occurrences are probably in deposits of levees, unlike the nearby cross-bedded sandstones of in-channel bars, including common large silicified conifer logs (Ward 1899). Unfortunately, the roots of these cycadeoids have never been found to prove that they were in growth position (Wieland 1906, p. 208). It is likely that the clayey material seen under some trunks (Wieland 1906, p. 224) was impervious to silicifying solutions which preserved the rest of the plant. Other evidence that they may have been in place is the rapid silicification of delicate young ovules and embryos in some of these trunks (Crepet and Delevoryas 1972; Crepet 1974). Like modern pachycaul trunks, such as the Australian *Macrozamia* and *Xanthorrhea*, *Cycadeoidea* would also have disintegrated if transported any distance by streams. However, the fossil cycadeoids are still intact, even though some trunks are in advanced stages of decay, indicated by internal disorganization of tissue, conspicuous insect borings (Crepet 1974), and collapsed crowns ("birds nest" form described by Wieland 1906, pp. 22, 23). Also compatible with a levee habitat for *Cycadeoidea*, is the likelihood that they were self-pollinating (Crepet 1974). This is a common adaptation among pioneer plants of disturbed habitats (Stebbins 1977, p. 55), such as streamsides. *Cycadeoidea* was evidently a prominent plant of coastal streamsides and levees in North America and Europe, which also became extinct as angiosperms were becoming common in similar habitats.

A SCENARIO

The conclusions discussed in previous sections are summarized here into a possible scenario for the dispersal and rise to dominance of angiosperms (Fig. 2.22).

A variety of plants related to the ancestral complex of angiosperms may have flourished in the rift valley systems of west Gondwanaland during the Mesozoic. Some of these plants were evidently microphyllous and adapted to coastal environments. These plants were probably woody, both wind and insect pollinated and their numerous small seeds dispersed largely by wind and water. Such generalized pollination and dispersal strategies, independent of local animal and plant communities, made these plants ideal for long-range dispersal by pioneering fresh sedimentary surfaces of coastal deltas, lagoons, and tidal flats.

As the extent of marine transgression and regression began to increase during the Early Cretaceous, these pioneering early angiosperms or angiosperm ancestors began to disperse along coastlines out of the rift valley system of west Gondwanaland. By the Aptian, these plants had dispersed to a limited extent into tropical Tethyan and Proto-Atlantic coasts. Their opportunist reproduction enabled them to exploit rapidly a variety of unstable coastal depositional environments, such as the shores of estuaries, river levees, coastal lagoons, and swamps. Initial morphological diversification of these plants went hand in hand with their early exploitation of a variety of coastal environments.

The most extensive dispersal of these early angiosperms was largely a result of the remarkable transgression and regression of epeiric seas during the Late Albian and earliest Cenomanian. In the interior of North America, the first rare angiosperms were distributed from the Gulf Coast to as far north as Alberta, Canada, during the Late Albian marine transgression and sea temperature maximum. During the latest Albian regression, they became dominant in the central interior and first reached Alaska. By the end of the Albian they had dispersed widely to Siberia and Australia.

At this time, many common and characteristic Early Cretaceous plants of mangrove (*Weichselia, Frenelopsis, Pseudofrenelopsis*), freshwater coastal swamp (*Tempskya*), and fluvial levee environments (*Cycadeoidea*) began to wane in abundance and even become extinct as increasing numbers of angiosperms competed more and more effectively for the same coastal depositional environments.

Regional differences in the far-flung initial complex of angiosperms must have been initiated as their range extended to the far corners of the world's coasts. By the end of the Albian, diversification of this once-conservative group of plants was also encouraged by several other factors. Angiosperms were then globally dispersed and living in a variety of different local environments and climates. Angiosperm dominance of coastal streamsides was well established and this served as an additional dispersal route to other disturbed and depositional environments further inland. By this time also many angiosperms were becoming increasingly coadapted for pollination and dispersal by local animals. Diversification was further enhanced during the Late Cretaceous by continental breakup and realignment due to continued sea-floor spreading. The general retreat of epeiric seas may have allowed the persistence of coastal vegetation in more inland locations. By the end of the Cretaceous most lowland vegetation was dominated by angiosperms and many major taxonomic groups and floral provinces (evident in modern angiosperms) were established.

ACKNOWLEDGMENTS

The authors appreciate helpful discussions and correspondence with Drs. W. L. Crepet (University of Connecticut, Storrs), J. A. Doyle and G. L. Stebbins (University of California, Davis), L. J. Hickey and E. G. Kauffman (U.S. National Museum, Washington, D.C.), G. J. Gastony and D. E. Hattin (University of Indiana, Bloomington), W. D. Tidwell (Brigham Young University, Provo, Utah), T. Delevoryas (University of Texas, Austin), P. R. Crane (University of Reading, England), and W. A. Cobban (U.S. Geological Survey, Denver). Few of these colleagues would agree with all the interpretations presented here, for which we assume complete responsibility. Dr. L. J. Hickey loaned material and showed us localities in the Potomac Group. Dr. J. F. Basinger helped with the preliminary information on fossil plants from the Dakota Formation at the Rose Creek, Nebraska locality. Mr. M. Zavada provided information on the palynology of the upper Dakota Formation in Kansas. Mr. E. Faulkner (Silver King Mines, Edgemont, South Dakota) provided geological data from drilling in the southern Black Hills. Help during fieldwork and collecting in Kansas was provided by Mr. Karl Longstreth, Mr. Thomas Halter, Mr. Charles Beeker, and Ms. Carolyn Diefenbach (University of Indiana, Bloomington); Drs. W. L. Crepet (University of Connecticut, Storrs); F. W. Potter, Jr. and H. C. Reynolds (Fort Hays State University, Hays, Kansas); and Roger Pabian (University of Nebraska, Lincoln, Nebraska). Typing and final preparation of the manuscript was done by Ms. Virginia Flack, Indiana University. This research was funded by National Science Foundation grants BMS 75-02268, DEB 77-04846, and DEB 79-10720 to D. L. Dilcher.

REFERENCES

Agasie, J. M. 1969. Late Cretaceous palynomorphs from northeastern Arizona. *Micropalaeontology* 15:13–30.

Ager, D. V. 1967. Some Mesozoic brachiopods in the Tethys region. In *Aspects of Tethyan Biogeography* (C. G. Adams and D. V. Ager, eds.). *Publ. Syst. Assoc.* (Lond.) 7:135–151.

Ahlbrandt, T. S., A. C. Huffman, J. E. Fox, and I. Pasternack, 1979. Depositional framework and reservoir-quality studies of selected Nanushuk Group outcrops, North Slope, Alaska. *Circ. U.S. Geol. Surv.* 794:14–31.

Allen, J. R. L. 1963. The classification of cross-stratified units, with notes on their origin. *Sedimentology* 2:93–114.

Allen, P. 1976. Wealden of the Weald: a new model. *Proc. Geol. Assoc.* 86:389–436.

Allen, P., M. L. Keith, F. C. Tan, and P. Dines, 1973. Isotopic ratios and Wealden environments. *Palaeontology* 16:607–621.

Alvin, K. L. 1971. *Weichselia reticulata* (Stokes et Webb) Fontaine, from the Wealden of Belgium. *Mem. Inst. Royal. Sci. Nat. Belgique* 166:33.

———. 1977. The conifers *Frenelopsis* and *Manica* in the Cretaceous of Portugal. *Palaeontology* 20:387–404.

Alvin, K. L., R. A. Spicer, and J. Watson. 1978. A *Classopollis*-containing male cone associated with *Pseudofrenelopsis*. *Palaeontology* 21:849–856.

Andrews, H. N., and E. M. Kern. 1947. The Idaho *Tempskyas* and associated fossil plants. *Ann. Miss. Bot. Gard.* 34:119–186.

Axelrod, D. I. 1952. A theory of angiosperm evolution. *Evolution* 6:29–60.

———. 1959. Poleward migration of early angiosperm flora. *Science* 130:203–207.

———. 1966. Origin of deciduous and evergreen habits in temperate forests. *Evolution* 20:1–15.

Axelrod, D. I., and P. H. Raven. 1978. Late Cretaceous and Tertiary vegetation history of Africa. In *Biogeography and Ecology of Southern Africa* (M. J. A. Werger and A. C. van Bruggen eds.). The Hague: Junk, pp. 77–130.

Baikovskaya, T. N. 1956. Melovie flora severnoi azii (Cretaceous flora of northern Asia). *Acta Komarov Bot. Inst. Paleobotanika* 2:103.

Ball, O. M. 1937. Flora of the Trinity Group. *J. Geol.* 45:528–537.

Banks, H. P., M. G. Gollett, F. R. Gnauk, and N. F. Hughes. 1967. Pteridophyta-2. In *The Fossil Record* (W. B. Harland, C. H. Holland, M. R. House, N. F. Hughes, A. B. Reynolds, M. J. S. Rudwick, G. E. Satterthwaite, L. B. H. Tarlo, and E. C. Willey, eds.). London: Geological Society of London, pp. 233–245.

Batten, D. J. 1975. Wealden palaeoecology from the distribution of plant fossils. *Proc. Geol. Assoc.* 85:433–458.

Battey, M. H. 1972. *Mineralogy for Students*. London: Oliver and Boyd.

Batton, G. 1965. Contribution a l'étude anatomique et biostratigraphique de la flore du continental intercalaire Saharien. *Publ. Centr. Nat. Rech. Scient. Paris, Ser. Geol.* 6:8–95.

Beadle, N. C. W., O. D. Evans, and R. C. Carolin. 1972. *Flora of the Sydney Region*. Wellington: A. H. and A. W. Reed.

Becker, H. F. 1961. Oligocene plants from the upper Ruby River Basin, southwestern Montana. *Mem. Geol. Soc. Am.* 82:127.

———. 1969. Fossil plants of the Tertiary Beaverhead Basins in southwestern Montana. *Palaeontographica j1B127:1–142*.

———. 1973. The York Ranch flora of the upper Ruby River Basin, southwestern Montana. *Palaeontographica* B143:18–93.

Bell, W. A. 1956. Lower Cretaceous floras of western Canada. *Mem. Geol. Surv. Canada* 285:331.

Bergquist, H. R. 1971. Biogeographical review of the Cretaceous foraminifera of the western hemisphere. *Proc. Nth Am. Paleont. Conv.* 1L:1565–1609.

Berry, E. W. 1910. The epidermal characters of *Frenelopsis ramosissima*. *Bot. Gaz.* 50:305–309.

———. 1911. Systematic paleontology, Lower Cretaceous Dicotyledonae. In *Lower Cretaceous* (W. B. Clark, A. B. Bibbins, and E. W. Berry, eds.) Baltimore: Maryland Geological Survey and Johns Hopkins University Press, pp. 214–508.

———. 1916a. The Upper Cretaceous floras of the world. In *Upper Cretaceous* (W. B. Clark, ed.). Baltimore: Maryland Geological Survey and Johns Hopkins University Press, pp. 183–313.

———. 1916b. Systematic paleontology, Upper Cretaceous, Angiospermophyta. In *Upper Cretaceous* (W. B. Clark, ed.). Baltimore: Maryland Geological Survey and Johns Hopkins University Press, pp. 946–987.

———. 1922a. The flora of the Woodbine Sand at Arthurs Bluff, Texas. *Prof. Pap. U.S. Geol. Surv.* 129G:153–180.

———. 1922b. The flora of the Cheyenne Sandstone of Kansas. *Prof. Pap. U.S. Geol. Surv.* 129I:199–225.

———. 1928. *Weichselia* from the Lower Cretaceous of Texas. *J. Wash. Acad. Sci.* 18:1–5.

———. 1939. The fossil plants from Huallanca, Peru. *Stud. Geol. Johns Hopkins Univ.* 13:73–93.

———. 1945. The *Weichselia* Stage in the Andean Geosyncline. *Stud. Geol. Johns Hopkins Univ.* 14:151–170.

Blasco, F. 1977. Outlines of the ecology, botany and forestry of the mangals of the Indian subcontinent. In *Wet Coastal Ecosystems* (V. J. Chapman, ed.). Amsterdam: Elsevier, pp. 241–260.

Boureau, E. 1953. Sur quelques plantes fossiles du Tchad et du Nord-Cameroun. *Bull. Serv. Mines Terr. Cameroun* 1:121–132.

Brasier, M. D. 1975. An outline of the history of seagrass communities. *Palaeontology* 18:691–702.

Brenner, G. J. 1976. Middle Cretaceous floral provinces and early migration of angiosperms. In *Origin and Early Evolution of Angiosperms* (C. B. Beck, ed.). New York: Columbia University Press, pp. 23–47.

Brognon, G., and G. Verrier. 1966. Tectonique et sédimentation dans le bassin du Cuanza (Angola). In *Sedimentary Basins of the African Coasts. Part 1, Atlantic Coast* (D. Reyre, ed.). Paris: Association of African Geological Surveys, pp. 207–252.

Brown, R. W. 1939. Fossil leaves, fruits and seeds of *Cercidiphyllum J. Paleont.* 13:485–499.

———. 1962. Paleocene flora of the Rocky Mountains and Great Plains. *Prof. Pap. U.S. Geol. Surv.* 375:119.

Burgess, J. D. 1971. Palynological interpretation of Frontier environments in central Wyoming. *Geoscience and Man* 3:69–81.

Carruccio, F. T., and G. Geidel. 1979. Using the paleoenvironment of strata to characterize mine drainage quality. In *Carboniferous Depositional Environ-*

ments in the Appalachian Region: Collected Papers and Field Guide (C. Ferm and J. C. Horne, eds.). Columbia: Department of Geology, University of South Carolina, pp. 587–596.

Casey, R. 1961. The stratigraphical palaeontology of the Lower Greensand. Palaeontology 3:487–621.

Chandler, M. E. J. 1961. The Lower Tertiary Floras of Southern England. 1. Palaeocene Floras, London Clay Flora (Supplement). London: British Museum of Natural History.

Chandler, M. E. J., and D. I. Axelrod. 1961. An early Cretaceous (Hauterivian) angiosperm fruit from California. Am. J. Sci. 259:441–446.

Chandrasekharam, A. 1974. Megafossil flora from the Genesee locality, Alberta, Canada. Palaeontographica B147:1–41.

Chapman, V. J. 1977a. Introduction. In Wet Coastal Ecosystems (V. J. Chapman, ed.) Amsterdam: Elsevier, pp. 1–29.

――――. 1977b. Wet coastal formations of Indo-Malesia and Papua-New Guinea. In Wet Coastal Ecosystems (V. J. Chapman, ed.). Amsterdam: Elsevier, pp. 261–270.

Chrysler, M. A. 1932. A new cycadeoid from New Jersey. Am. J. Bot. 19:679–692.

Churchill, D. M. 1973. The ecological significance of tropical mangroves in the early Tertiary floras of southern Australia. Spec. Publs. Geol. Soc. Aust. 4:79–86.

Cobban, W. A., and J. R. Reeside. 1952. Correlation of the Cretaceous formations of the western interior of the United States. Bull. Geol. Soc. Am. 63:1011–1044.

Coleman, J. M., S. M. Gagliano, and W. G. Smith. 1970. Sedimentation in a Malaysian high tide tropical delta. In Sedimentation, Modern and Ancient (J. P. Morgan, ed.). Spec. Publ. Soc. Econ. Paleont. Miner. Tulsa 15:185–197.

Crane, P. R. 1978. Angiosperm leaves from the lower Tertiary of southern England. Cour. Forsch.-Inst. Senckenberg 30:126–132.

Crepet, W. L. 1974. Investigations of North American cycadeoids: the reproductive biology of Cycadeoidea. Palaeontographica B148:144–169.

Crepet, W. L., and T. Delevoryas. 1972. Investigations of North American cycadeoids: early ovule ontogeny. Am. J. Bot. 59:209–215.

Cridland, A. A. 1964. Amyelon in American coal balls. Palaeontology 7:186–209.

Cronquist, A. 1968. The Evolution and Classification of Flowering Plants. Boston: Houghton Mifflin.

Daber, R. 1968. A Weichselia-Stiehleria-Matoniaceae community within the Quedlinburg Estuary of Lower Cretaceous age. J. Linn. Soc. Bot. 61:75–85.

Daghlian, C. P., and C. P. Person. 1977. The cuticular anatomy of Frenelopsis varians from the Lower Cretaceous of central Texas. Am. J. Bot. 64:564–569.

Day, R. W. 1969. The Lower Cretaceous of the Great Artesian Basin. In Stratigraphy and Palaeontology, Essays in Honour of Dorothy Hill (K. S. W. Campbell, ed.). Canberra: Australian National University Press, pp. 140–173.

de Jersey, N. J. 1960. The Styx Coal Measures. In The Geology of Queensland (D. Hill and A. K. Demmead, eds.). J. Geol. Soc. Aust. 7:330–333.

Delevoryas, T. 1959. Investigations of North American cycadeoids: Monanthesia. Am. J. Bot. 46:657–666.

――――. 1971. Biotic provinces and the Jurassic-Cretaceous floral transition. Proc. Nth Am. Paleont. Conv. 1L:1660–1674.

Dettmann, M. E. 1963. Upper Mesozoic microfloras from southeastern Australia. *Proc. R. Soc. Vict.* 77:1–148.

———. 1973. Angiospermous pollen from Albian to Turonian sediments of eastern Australia. *Spec. Publ. Geol. Soc. Aust.* 4:3–34.

Dettmann, M. E., and G. Playford. 1969. Palynology of the Australian Cretaceous: a review. In *Stratigraphy and Palaeontology, Essays in Honour of Dorothy Hill* (K. S. W. Campbell, ed.). Canberra: Australian National University Press, pp. 174–210.

Dewey, J. F., W. C. Pitman, B. F. Ryan, and J. Bonnin, 1973. Plate tectonics and the evolution of the Alpine System. *Bull. Geol. Soc. Am.* 84:3137–3180.

Dietz, R. S., and J. C. Holden. 1970. Reconstruction of Pangaea: breakup and dispersion of continents, Permian to present. *J. Geophys. Res.* 75:4939–4956.

Dilcher, D. L. 1979. Early angiosperm reproduction: an introductory report. In *Plant Reproduction in the Fossil Record* (T. N. Taylor, D. L. Dilcher, and T. Delevoryas, eds.). *Rev. Palaeobot. Palynol.* 27:291–328.

Dilcher, D. L., W. L. Crepet, C. D. Beeker, and H. C. Reynolds. 1976. Reproductive and vegetative morphology of a Cretaceous angiosperm. *Science* 191:854–856.

Dilcher, D. L., F. Potter, and H. C. Reynolds. 1978. Preliminary account of middle Cretaceous angiosperm remains from the interior of North America. *Cour. Forsch.-Inst. Senckenberg* 30:9–15.

Diller, J. S. 1908. Strata containing the Jurassic flora of Oregon. *Bull. Geol. Soc. Am.* 19:367–402.

Diller, J. S., and T. W. Stanton. 1894. The Shasta-Chico Series. *Bull. Geol. Soc. Am.* 5:435–464.

Dilley, F. C. 1973. Cretaceous larger foraminifera. In *Atlas of Palaeobiogeography* (A. Hallam, ed.). Amsterdam: Elsevier, pp. 403–419.

Doludenko, M. P. 1978. The genus *Frenelopsis* (Coniferales) and its occurrence in the Cretaceous of the U.S.S.R. *Paleont. J.* 3:384–398.

Douglas, J. G. 1963. Nut-like impressions attributed to aquatic dicotyledons from Victorian Mesozoic sediments. *Proc. R. Soc. Vict.* 76:23–28.

———. 1965. A Mesozoic dicotyledonous leaf from Yangery no. 1 bore, Koroit, Victoria. *Min. Geol. J. Vict.* 6:64–67.

———. 1969. The Mesozoic flora of Victoria. *Mem. Geol. Surv. Vict.* 28:310.

Doyle, J. A. 1973. Fossil evidence on the early evolution of monocotyledons. *Q. Rev. Biol.* 48:399–413.

———. 1978. Origin of angiosperms. *Ann. Rev. Ecol. Syst.* 9:365–392.

Doyle, J. A., P. Biens, and A. Doerenkamp, and S. Jardiné. 1977. Angiosperm pollen from the pre-Albian Lower Cretaceous of equatorial Africa. *Bull. Cent. Rech. Explor. Prod. Elf-Aquitaine* 1:451–473.

Doyle, J. A., and L. J. Hickey. 1976. Pollen and leaves from the mid-Cretaceous Potomac Group and their bearing on early angiosperm evolution. In *Origin and Early Evolution of Angiosperms* (C. B. Beck, ed.). New York: Columbia University Press, pp. 136–206.

Doyle, J. A., and E. I. Robbins. 1977. Angiosperm pollen zonation of continental Cretaceous of the Atlantic coastal plain and its application to deep wells in the Salisbury Embayment. *Palynology* 1:43–78.

Doyle, J. A., M. van Campo, and B. Lugardon. 1975. Observations on exine structure of *Eucommiidites* and Lower Cretaceous angiosperm pollen. *Pollen et Spores* 17:429–486.

Eggert, D. L., T. L. Phillips, W. A. Dimichele, P. R. Johnson, and R. A. Peppers. 1979. *Guidebook to Environments of Plant Deposition—Coal Balls, Paper Coals, and Gray Shale Floras, in Fountain and Parke Counties, Indiana.* Urbana: Ninth International Congress of Carboniferous Stratigraphy and Geology.

Endo, S. 1926. On the genus *Tempskya* from the vicinity of Yuasa, Kii. *Chikyû* 6:5–9.

——. 1953. A new *Cycadeoidea* from S. Sakhalin. *Kumamoto J. Sci.* B2:1–7.

Faegri, K., and L. van der Pijl. 1966. *The Principles of Pollination Biology.* Oxford: Pergamon Press.

Fontaine, W. M. 1889. The Potomac or younger Mesozoic flora. *Monogr. U.S. Geol. Surv.* 15:377.

Franks, P. C. 1980. Models of marine transgression—example from Lower Cretaceous fluvial and paralic deposits, north-central Kansas. *Geology* 8:56–61.

Frazier, D. E., and A. Osanik. 1969. Recent peat deposits—Louisiana coastal plain. In *Environments of Coal Deposition* (E. C. Dapples and M. E. Hopkins, eds.). *Spec. Pap. Geol. Soc. Am.* 114:63–85.

Frazier, D. E., A. Osanik, and W. C. Elsik. 1978. Environments of peat accumulation—coastal Louisiana. In *Proceedings of the Gulf Coast Conference: Geology, Utilization and Environmental Aspects, 1976* (W. R. Kaiser, ed.). *Rept. Invest. Bur. Econ. Geol. Univ. Texas Austin* 90:5–20.

Glaser, J. D. 1969. Petrology and origin of Potomac and Magothy (Cretaceous) sediments, middle Atlantic coastal plain. *Rept. Invest. Geol. Surv. Maryland* 11:102.

Gookey, D. P., J. D. Haun, L. A. Hale, H. G. Goodell, D. G. McCubbin, R. J. Weimer, and G. R. Wulf. 1972. Cretaceous System. In *Geologic Atlas of the Rocky Mountains* (W. W. Mallory, ed.) Denver: Rocky Mountain Association of Geologists, pp. 190–228.

Gordon, W. A. 1973. Marine life and ocean surface currents in the Cretaceous. *J. Geol.* 81:269–284.

Groot, J. J. 1955. Sedimentary petrology of the Cretaceous sediments of northern Delaware. *Bull. Geol. Surv. Delaware* 5:157.

Harris, T. M. 1966. Dispersed cuticles. *Palaeobotanist* 14:102–105.

Hattin, D. E. 1965. Stratigraphy of the Graneros Shale (Upper Cretaceous) in central Kansas. *Bull. Geol. Surv. Kansas* 178:83.

——. 1967. Stratigraphy and paleoecologic significance of macroinvertebrate fossils in the Dakota Formation (Upper Cretaceous) of Kansas. In *Essays in Paleontology and Stratigraphy, Raymond C. Moore Commemorative Volume* (C. Teichert and E. L. Yochelson, eds.). *Spec. Publ. Dept. Geol. Univ. Kansas* 2:570–589.

Hattin, D. E., C. T. Seimers, and G. F. Stewart. 1978. Guidebook: Upper Cretaceous stratigraphy and depositional environments of western Kansas. *Guidebook Geol. Surv. Kansas* 3:102.

Hedlund, R. W. 1966. Palynology of the Red Branch Member of the Woodbine Formation (Cenomanian), Bryan County, Oklahoma. *Bull. Geol. Surv. Oklahoma* 112:69.

Hedlund, R. W., and G. Norris. 1968. Spores and pollen grains from Fredericks-burgian (Albian) strata, Marshall County, Oklahoma. *Pollen et Spores* 10:129–159.

Heinrich, B. 1976. Flowering phenologies: bog, woodland and disturbed habitats. *Ecology* 57:890–899.

Heslop-Harrison, J. 1976. The adaptive significance of the exine. In *The Evolutionary Significance of the Exine* (I. K. Ferguson and J. Muller, eds.). London: Academic Press, pp. 27–37.

Hickey, L. J. 1977. Stratigraphy and paleobotany of the Golden Valley Formation (early Tertiary) of western North Dakota. *Mem. Geol. Soc. Am.* 150:183.

———. in press. Changes in the angiosperm flora across the Cretaceous-Tertiary boundary. In *The New Uniformitarianism* (W. A. Berggren and J. van Couvering, eds.). Princeton: Princeton University Press.

Hickey, L. J., and J. A. Doyle. 1977. Early Cretaceous fossil evidence for angiosperm evolution. *Bot. Rev.* 43:3–104.

Hluštík, A., and M. Konzalová. 1976. Polliniferous cones of *Frenelopsis alata* (K. Feistm.) Knobloch, from the Cenomanian of Czechoslovakia. *Vest. Ustred. Ust. Geol.* 51:37–45.

Hourcq, V. 1966. Les grands traits de la geologie des bassins côtiers du Groupe Equatorial. In *Sedimentary Basins of the African Coasts. Part I, Atlantic Coast* (D. Reyre, ed.). Paris: Association of African Geological Surveys, pp. 163–170.

Hughes, N. F. 1976a. Plant succession in the English Wealden strata. *Proc. Geol. Assoc.* 86:439–455.

———. 1976b. *Palaeobiology of Angiosperm Origins.* Cambridge: Cambridge University Press.

———. 1977. Palaeo-succession of earliest angiosperm evolution. *Bot. Rev.* 43:105–127.

Hughes, N. F., G. E. Drewry, and J. F. Laing. 1979. Barremian earliest angiosperm pollen. *Palaeontology* 22:513–535.

Hughes, N. F., and J. C. Moody-Stuart. 1967. Palynological facies and correlation in the English Wealden. *Rev. Palaeobot. Palynol.* 1:259–268.

Jardiné, S., G. Kieser, and D. Reyre. 1974. L'individualization progressive du continent africain vue a travers les donnees palynologiques de l'ere secondaire. *Bull. Sci. Geol. Strasbourg* 27:69–85.

Jeletzky, J. A. 1978. Causes of Cretaceous oscillations of sea level in western and Arctic Canada and some general geotectonic implications. *Pap. Geol. Surv. Canada* 77–18:44.

Jones, D. L., N. J. Silberling, and J. Hillhouse. 1977. Wrangellia—a displaced terrane in northwestern North America. *Canad. J. Earth Sci.* 14:2565–2577.

Jongmans, W., and S. J. Dijkstra. 1958. *Fossilium Catalogus. II Plantae Pars 34. Filicales, Pteridospermae, Cycadales.* Gravenhage: Junk.

———. 1965. *Fossilium Catalogus. II Plantae Pars 63. Filicales, Pteridospermae, Cycadales.* Gravenhage: Junk.

———. 1973. *Fossilium Catalogus. II Plantae Pars 82. Gymnospermae (Ginkgophyta & Coniferae).* Gravenhage: Junk.

Jung, W. 1974. Die Konifere *Brachyphyllum nepos* Saporta aus den Solnhofer Plattenkalken (unteres Untertithon), ein Halophyt. *Mitte. Bayer. Statesamml.*

Paläont. Hist. Geol. 14:49–58.

Kaplan, D. R. 1973. The problem of leaf morphology and evolution in the monocotyledons. *Q. Rev. Biol.* 48:437–457.

Karl, H. A. 1976. Depositional history of Dakota Formation (Cretaceous) Sandstones, southeastern Nebraska. *J. Sedim. Petrol.* 46:124–131.

Kauffman, E. G. 1973. Cretaceous bivalvia. In *Atlas of Palaeobiogeography* (A. Hallam, ed.). Amsterdam: Elsevier, pp. 353–383.

———. 1977I. Geological and biological overview, western interior Cretaceous Basin. *Mountain Geologist* 14:75–99.

———. 1977b. Evolutionary rates and biostratigraphy. In *Concepts and Methods of Biostratigraphy* (E. G. Kauffman and J. E. Hazel, eds.). Stroudsburg, Pennsylvania: Dowden, Hutchinson and Ross, pp. 109–141.

Kauffman, E. G., D. E. Hattin, and J. D. Powell. 1977. Stratigraphic, paleontologic and paleoenvironmental analysis of the Upper Cretaceous rocks of Cimarron County, northwestern Oklahoma. *Mem. Geol. Soc. Am.* 149:150

Kemp, E. M. 1968. Probable angiosperm pollen from British Barremian to Albian strata. *Palaeontology* 11:421–434.

Kennedy, W. J. 1978. Cretaceous. In *The Ecology of Fossils* (W. S. McKerrow, ed.). Cambridge: Massachusetts Institute of Technology Press, pp. 280–322.

Kilenyi, T. I., and N. W. Allen. 1968. Marine-brackish bands and their microfauna from the lower part of the Weald Clay of Sussex and Surrey. *Palaeontology* 11:141–162.

Knowlton, F. H. 1919. A catalogue of the Mesozoic and Cenozoic plants of North America. *Bull. U.S. Geol. Surv.* 696:815.

Krassilov, V. A. 1967. *Rannemelovaya Flora Yznogo Primoriya i ee Znachenie Dlaya Stratigafii (Early Cretaceous Flora from Southern Primorie and Its Significance for Stratigraphy).* Moscow: Nauka.

———. 1973a. Mesozoic plants and the problem of angiosperm ancestry. *Lethaia* 6:163–178.

———. 1973b. The Jurassic disseminules with pappus and their bearing on the problem of angiosperm ancestry. *Geophytology* 3:1–4.

———. 1973c. Climatic changes in eastern Asia as indicated by fossil floras. I. Early Cretaceous. *Palaeogeogr. Palaeoclim. Palaeoec.* 13:261–273.

———. 1977. The origin of angiosperms. *Bot. Rev.* 43:143–176.

Landes, K. K. 1930. The geology of Mitchell and Osborne Counties, Kansas. *Bull. Geol. Surv. Kansas* 16:1–55.

Landes, K. K., and R. P. Keroher. 1938. Oil and gas resources of Rush County, Kansas. *Min. Res. Circ. Geol. Surv. Kansas* 4:1–31.

———. 1939. Geology and oil resources of Logan, Gove and Trego Counties, Kansas. *Min. Res. Circ. Geol. Surv. Kansas* 11:1–45.

Lawrence, G. H. M. 1951. *Taxonomy of Vascular Plants.* New York: Macmillan.

Leriche, M. 1909. Sur les fossils de la craie phosphate de la Picardie, à *Actinocamax quadratus. C.R. Ass. Franc. Avanc. Sci. Congr. Clermont-Ferrand* 37:494–503.

Lesquereux, L. 1874. Contribution to the fossil flora of the western territories. Part I. The Cretaceous flora. *Rept. U.S. Geol. Surv.* 6:136.

———. 1883. The fossil floras of the western territories. Part III. The Cretaceous and Tertiary floras. *Rept. U.S. Geol. Surv.* 8:283.

———. 1892. The flora of the Dakota Group. *Monogr. U.S. Geol. Surv.* 17:256.

Luyendyk, B. P., D. Forsyth, and J. D. Phillips. 1972. Experimental approach to the paleocirculation of the oceanic surface waters. *Bull. Geol. Soc. Am.* 83:2649–2664.

Mack, L. E. 1962. Geology and ground-water resources of Ottawa County, Kansas. *Bull. Geol. Surv. Kansas* 154:1–145.

Mackowsky, M.-TH. 1975. Comparative petrography of Gondwana and northern hemisphere coals related to their origin. In *Gondwana Geology* (K. S. W. Campbell, ed.). Canberra: Australian National University Press, pp. 195–220.

MacNeal, D. L. 1958. The flora of the Upper Cretaceous Woodbine Sand in Denton County, Texas. *Monogr. Acad. Nat. Sci. Philadelphia* 10:152.

Martin, H. A. 1977. The history of *Ilex* (Aquifoliaceae) with special reference to Australia: evidence from pollen. *Aust. J. Bot.* 25:655–673.

May, F. E., and A. Traverse. 1973. Palynology of the Dakota Sandstone (Middle Cretaceous), near Bryce Canyon National Park, southern Utah. *Geosci. and Man* 7:57–64.

Medwell, L. M. 1954. Fossil plants from Killara, near Casterton, Victoria. *Proc. R. Soc. Vict.* 66:16–23.

Mellon, G. B. 1967. Stratigraphy and petrology of the Lower Cretaceous Blairmore and Mannville Groups, Alberta foothills and plains. *Bull. Res. Coun. Alberta* 21:270.

Merriam, D. F., W. R. Atkinson, P. C. Franks, N. Plummer, and F. W. Preston. 1959. Description of a Dakota (Cretaceous) core from Cheyenne County, Kansas. *Bull. Geol. Surv. Kansas* 134:5–104.

Middlemiss, F. A. 1976. Studies in the sedimentation of the Lower Greensand of the Weald, 1875–1975: a review and commentary. *Proc. Geol. Assoc.* 86:457–473.

Miller, C. N. 1977. Mesozoic conifers. *Bot. Rev.* 43:217–280.

Moore, D. M. 1972. *Trees of Arkansas.* Little Rock: Arkansas Forestry Commission.

Morgan, J. P. 1970. Depositional processes and products in the deltaic environment. In *Deltaic Sedimentation, Modern and Ancient* (J. P. Morgan, ed.). *Spec. Publ. Soc. Econ. Miner. Paleont. Tulsa* 15:31–47.

Muller, J. 1970. Palynological evidence on early differentiation of angiosperms. *Biol. Rev.* 45:417–450.

Muller, J., and C. Caratini. 1977. Pollen of *Rhizophora* (Rhizophoraceae) as a guide fossil. *Pollen et Spores* 19:361–389.

Murphy, M. A. 1956. Lower Cretaceous stratigraphic units of northern California. *Bull. Am. Ass. Petrol. Geol.* 40:2098–2119.

Newberry, J. S. 1895. The flora of the Amboy Clays. *Monogr. U.S. Geol. Surv.* 26:137.

Obradovich, J. D., and W. A. Cobban. 1975. A time scale for the Late Cretaceous of the western interior of North America. In *The Cretaceous System in the Western Interior of North America* (W. G. E. Caldwell, ed.). *Spec. Pap. Geol. Assoc. Can.* 13:31–54.

Oishi, S. 1940. The Mesozoic flora of Japan. *J. Fac. Sci. Hokkaido Imp. Univ. Series* 4,5:437–502.

Oldham, T. C. B. 1976. Flora of the Wealden plant debris beds of England. *Palaeontology* 19:437–502.

Pacltova, B. 1978. Evolutionary trends of platanoid pollen in Europe during the Cenophytic. *Cour. Forsch.-Inst. Senckenberg* 30:70–76.

Perkins, B. F., W. Langston, and J. F. Stone. 1979. *Field Trip Guide: Lower Cretaceous Shallow Marine Environments in the Glen Rose Formation: Dinosaur Tracks and Plants*. Dallas: American Association of Stratigraphic Palynologists 12th Annual Meeting.

Phleger, F. B. 1969. Some general features of coastal lagoons. In *Lagunas Costeras* (A. A. Castañares and F. B. Phleger, eds.) Mexico City: Universidad Nacional Autónoma de Mexico, pp. 5–26.

Pierce, R. L. 1961. Lower Upper Cretaceous plant microfossils from Minnesota. *Bull. Geol. Surv. Minnesota* 42:82.

Plumstead, E. P. 1969. Three thousand years of plant life in Africa. *Trans. Geol. Soc. S. Afr.* annex to 72:72.

Pons, D., and J. Broutin, 1978. Les organes reproducteurs de *Frenelopsis oligostomata* (Crétacé, Portugal). *103ᵉ Congr. Nat. Soc. Sav. Nancy, Sci.* 2:139–159.

Popenoe, W. P., R. W. Imlay, and M. A. Murphy. 1960. Correlation of the Cretaceous formations of the Pacific coast (United States and northwestern Mexico). *Bull. Geol. Soc. Am.* 71:1491–1540.

Proctor, M. C. F. 1978. Insect pollination syndromes in an evolutionary and systematic context. In *The Pollination of Flowers by Insects* (A. J. Richards, ed.). *Symp. Series Linnaean Soc. (Lond.)* 6:105–116.

Raven, P. H. 1977. A suggestion concerning the Cretaceous rise to dominance of the angiosperms. *Evolution* 31:451–452.

Raven, P. H., and D. I. Axelrod. 1974. Angiosperm biogeography and past continental movements. *Ann. Miss. Bot. Gard.* 61:539–673.

Read, C. B., and S. R. Ash. 1961. Stratigraphic significance of the genus *Tempskya* in the western United States. *Prof. Pap. U.S. Geol. Surv.* 424D:250–254.

Regal, P. J. 1977. Ecology and evolution of flowering plant dominance. *Science* 196:622–629.

Retallack, G. J. 1975. The life and times of a Triassic lycopod. *Alcheringa* 1:3–29.

——. 1977a. Triassic paleosols in the upper Narrabeen Group of New South Wales. Part II. Classification and reconstruction. *J. Geol. Soc. Aust.* 24:19–24.

——. 1977b. Reconstructing Triassic vegetation of eastern Australasia: a new approach for the biostratigraphy of Gondwanaland. *Alcheringa* 1:247–277.

Reymanówna, M., and J. Watson. 1976. The genus *Frenelopsis* and the type species *Frenelopsis hoheneggeri* (Ettingshausen) Schenk. *Acta Palaeobot. Cracov.* 17:17–26.

Reyment, R. A., and E. A. Tait. 1972. Biostratigraphical dating of the early history of the South Atlantic Ocean. *Phil. Trans. R. Soc. (Lond.)* B264:55–95.

Reyre, D. 1966. Histoire geologique du bassin du Douala (Cameroun). In *Sedimentary Basins of the African Coasts. Part 1, Atlantic Coast* (D. Reyre, ed.). Paris: Association of African Geological Surveys, pp. 143–161.

Reyre, D., Y. Belmonte, F. Derumaux, and R. Wenger. 1966. Evolution geologique du Bassin Gabonais. In *Sedimentary Basins of the African Coasts. Part 1, Atlantic Coast* (D. Reyre, ed.). Paris: Association of African Geological Surveys, pp. 171–191.

Romans, R. C. 1975. Palynology of some Upper Cretaceous coals of Black Mesa, Arizona, *Pollen et Spores* 17:273–329.

Rorison, I. H. 1973. Seed ecology-present and future. In *Seed Ecology* (W. Heydecker, ed.). University Park: Pennsylvania State University Press, pp. 497–519.

Rubey, W. W., and N. W. Bass. 1925. The geology of Russell County, Kansas. *Bull. Geol. Surv. Kansas* 10:1–86.

Rushforth, S. R. 1970. Notes on the fern family Matoniaceae from the western United States. *Geol. Stud. Brigham Young Univ.* 16:3–34.

———. 1971. A flora from the Dakota Sandstone Formation (Cenomanian), near Westwater, Grand County, Utah. *Sci. Bull. Brigham Young Univ. Biol. Ser.* 14:44.

Sahni, B. 1936. The occurrence of *Matonidium* and *Weichselia* in India. *Rec. Geol. Surv. India* 71:152–165.

Samylina, V. A. 1968. Early Cretaceous angiosperms of the Soviet Union, based on leaf and fruit remains. *J. Linn. Soc. London Bot.* 61:207–218.

Schlee, J., J. C. Behrendt, J. A. Grow, J. M. Robb, R. E. Mattick, P. T. Taylor, and B. J. Lawson. 1976. Regional geologic framework of northeastern United States. *Bull. Am. Ass. Petrol. Geol.* 60:926–951.

Schlee, J., R. G. Martin, R. E. Mattick, W. P. Dillon, and M. M. Ball. 1977. Petroleum geology on the United States, Atlantic-Gulf of Mexico margins. *Proc. S. W. Legal Foundation Explor. Econ. Petrol. Indust.* 15:47–93.

Schoewe, H. W. 1952. Coal resources of the Cretaceous System (Dakota Formation) in central Kansas. *Bull. Geol. Surv. Kansas* 96:69–156.

Scholle, P. A. 1977. Geological studies on the COST no. B-2 well U. S. mid-Atlantic outer continental shelf area. *Circ. U.S. Geol. Surv.* 750:71.

Scott, R. A., and C. J. Smiley. 1979. Some Cretaceous plant megafossils and microfossils from the Nanushuk Group, northern Alaska. *Circ. U.S. Geol. Surv.* 794:89–117.

Scott, R. W., and A. M. Taylor. 1977. Early Cretaceous environments and paleocommunities in the southern western interior. *Mountain Geol.* 14:155–173.

Senior, B. R., A. Mond, and H. Harrison. 1978. Geology of the Eromanga Basin. *Bull. Bur. Min. Resour. Geol. Geophys. Canberra* 167:102.

Seward, A. C. 1926. The Cretaceous plant-bearing rocks of western Greenland. *Phil. Trans. R. Soc.* (Lond.) B215:57–174.

———. 1933. *Plant Life Through the Ages.* Cambridge: Cambridge University Press.

Shepard, F. P. 1952. Revised nomenclature for depositional coastal features. *Bull. Am. Ass. Petrol. Geol.* 36:1902–1912.

———. 1956. Marginal sediments of the Mississippi Delta. *Bull. Am. Ass. Petrol. Geol.* 40:2537–2623.

Siemers, C. T. 1971. Stratigraphy, Paleoecology, and Environmental Analysis of the Upper Part of the Dakota Formation. Ph.D. dissertation, Indiana University, Bloomington.

———. 1976. Sedimentology of the Rocktown channel sandstone, upper part of the Dakota Formation (Cretaceous), central Kansas. *J. Sedim. Petrol.* 46:97–123.

Skvarla, J. J., and D. A. Larson. 1965. An electron microscope study of pollen morphology in the Compositae, with special reference to the Ambrosiinae. *Grana Palynol.* 6:210–269.

Smiley, C. J. 1969. Cretaceous floras of Chandler-Colville region, Alaska: stratigraphy and preliminary floristics. *Bull. Am. Ass. Petrol. Geol.* 53:482–502.

Sohl, N. F. 1971. North American Cretaceous biotic provinces delineated by gastropods. *Proc. Nth. Am. Paleont. Conv.* 1L:1610–1637.

Spencer, A. M. 1974. Mesozoic-Cenozoic orogenic belts. *Spec. Pub. Geol. Soc. (Lond.)* 4:809.

Srivastava, S. K. 1976. The fossil pollen genus *Classopollis. Lethaia* 9:437–457.

Stebbins, G. L. 1974. *Flowering Plants: Evolution Above the Species Level.* Cambridge: Belknap Press of Harvard University Press.

——— . 1977. *Processes of Organic Evolution.* Englewood Cliffs, New Jersey: Prentice-Hall.

Stephenson, L. W. 1948. Cretaceous mollusca from depths of 1894 to 1896 feet in the Bethards Well. *Bull. Board Nat. Resour. Maryland* 2:125–126.

Stevens, G. R. 1971. Relationships of isotopic temperatures and faunal realms to Jurassic-Cretaceous palaeogeography, particularly of the south-west Pacific. *J. Roy. Soc. New Zealand* 1:145–158.

——— . 1973. Cretaceous belemnites. In *Atlas of Palaeobiogeography*, (A. Hallam, ed.). Amsterdam: Elsevier, pp. 385–401.

Stott, D. F. 1974. Lower Cretaceous coal measures of the foothills of west-central Alberta and northeastern British Columbia. *Canad. Min. Metall. Bull.* 67:87–100.

Swineford, A., and H. L. Williams. 1945. The Cheyenne Sandstone and adjacent formations of a part of Russell County, Kansas. *Bull. Geol. Surv. Kansas* 60:101–168.

Takhtajan, A. 1969. *Flowering Plants: Origin and Dispersal.* Edinburgh: Oliver and Boyd.

Teslenko, Y. V., A. V. Golbert, and I. D. Poliakova. 1966. Puti rasseleniya drevneischikh pokr'itosemaynikh v zapadnoi sibiri (The routes of dispersal of the most ancient angiosperms in western Siberia). *Bot. Zh.* 56:801–803.

Thom, B. G. 1967. Mangrove ecology and deltaic geomorphology: Tabasco, Mexico. *J. Ecol.* 55:301–343.

Tidwell, W. D., N. E. Hebbert, J. D. Shane, and S. R. Ash. 1977. Petrified angiosperms within *Tempskya* false trunks from the Cedar Mountain Formation, Utah. *Abstr. Prog. Geol. Soc. Am.* 9:515.

Tidwell, W. D., G. F. Thayn, and J. L. Roth. 1976. Cretaceous and early Tertiary floras of the Intermountain area. *Geol. Stud. Brigham Young Univ.* 22:77–98.

Tomlinson, P. B., R. B. Primack, and J. S. Bunt. 1979. Preliminary observations on floral biology in mangrove Rhizophoraceae. *Biotropica* 11:256–277.

Townrow, J. A. 1960. The Peltaspermaceae, a pteridosperm family of Permian and Triassic age. *Palaeontology* 3:333–361.

Tralau, H. 1964. The genus *Nypa* van Wurmb. *K. Svenska VetenskAkad. Handl.* 10:29.

Vachrameev, V. A. 1952. *Regionalinay Stratigrafii S.S.S.R. Tom 1. Stratigrafii i Iskopaemay Flora Meloviih Othozhenii Zapadnogo Kazakhstana (Regional stratigraphy of the U.S.S.R. Vol. 1. Stratigraphy and Fossil Flora of the Cretaceous Deposits of Western Kazkhstan).* Moscow: Isdatelistvo Akademia Nauk S.S.S.R.

——— . 1978. The climates of the northern hemisphere in the Cretaceous in the light of paleobotanical data. *Paleont. J.* 12:143–154.

Vachrameev, V. A., and V. A. Krassilov. 1979. Reproduktinie organi tsvetkoviikh iz aliba kazakhstana (Reproductive organs of a shoot of a flowering plant from the Albian of Kazakhstan). *Paleont. Zh. for 1979*, pp. 121–128.

van der Pijl, L. 1972. *Principles of Dispersal in Higher Plants.* Berlin: Springer Verlag.

van Steenis, C. G. G. J. 1962. The distribution of mangrove plant genera and its significance for palaeogeography. *Proc. K. Ned. Akad. Wet.* C65:164–169.

Velenovsky, J. 1889. Kvĕtena ĕeskeńo cenomanu. *Roz. Král. Česke Společnost. Nauk Mat.-Přír. Čís.* 7:1–75.

Vokes, H. E. 1948. Cretaceous mollusca from depths of 4875 to 4885 feet in the Maryland Esso well. *Bull. Board Nat. Resour. Maryland* 2:126–151.

Walker, J. W. 1976. Evolutionary significance of the exine in pollen of primitive angiosperms. In *The Evolutionary Significance of the Exine* (I. K. Ferguson and J. Muller, eds.). London: Academic Press, pp. 251–308.

Walkom, A. B. 1919. Mesozoic floras of Queensland. Parts 3 and 4. The floras of the Burrum and Styx River Series. *Publs. Geol. Survey Queensland* 263:77.

Ward, L. F. 1899. The Cretaceous formation of the Black Hills as indicated by the fossil plants. *Ann. Rept. U.S. Geol. Surv.* 19:527–712.

———. 1905. Status of the Mesozoic floras of the United States. *Monogr. U.S. Geol. Surv.* 48:616.

Watson, J. 1977. Some Lower Cretaceous conifers of the Cheirolepidiaceae from the U.S.A. and England. *Palaeontology* 20:715–749.

West, R. C. 1977. Tidal salt-marsh and mangal formations of middle and South America. In *Wet coastal Ecosystems* (V. J. Chapman, ed.). Amsterdam: Elsevier, pp. 193–213.

Whitehead, D. R. 1969. Wind pollination in the angiosperms: evolutionary and environmental considerations. *Evolution* 23:28–35.

Wieland, G. R. 1906. American fossil cycads. Vol. I. *Publ. Carnegie Inst. Wash.* 34:295.

———. 1916. American fossil cycads. Vol. II. *Publ. Carnegie Inst. Wash.* 34:277.

———. 1921. Two new North American cycadeoids. *Bull. Dept. Mines Geol. Surv. Canada* 33:79–85.

———. 1928a. Paleobotany. *Yrbk Carnegie Inst. Wash.* 27:390–391.

———. 1928b. Certain fossil plants erroneously referred to cycadeoids. *Bot. Gaz.* 86:32–49.

———. 1934. Fossil cycads, with special reference to *Raumeria Reichenbachiana* Goeppert sp. of the Zwinger of Dresden. *Palaeontographica* B79:85–130.

———. 1942. Cycadeoid types of the Kansas Cretaceous. *Am. J. Sci.* 240:192–203.

Wodehouse, R. P. 1935. *Pollen Grains.* New York: McGraw-Hill.

Wolfe, J. A., J. A. Doyle, and V. M. Page. 1975. The bases of angiosperm phylogeny: paleobotany. *Ann. Miss. Bot. Gard.* 62:801–824.

Wolfe, J. A., and H. M. Pakiser. 1971. Stratigraphic interpretation of some Cretaceous microfossil floras of the middle Atlantic states. *Prof. Pap. U.S. Geol. Surv.* 750B:35–47.

Zeiller, R. 1882. Observations sur quelques cuticules fossiles. *Ann. Sci. Nat. Bot.* 13:217–238.

3

PALEOCLIMATIC SIGNIFICANCE OF THE OLIGOCENE AND NEOGENE FLORAS OF THE NORTHWESTERN UNITED STATES

Jack A. Wolfe

INTRODUCTION

The Oligocene and Neogene floras of Oregon, Washington, and parts of adjacent Idaho have been the subject of description and discussion for over a century. Most of the work has been based on leaf impressions and associated fructifications, although during the last 20 years some palynological information has been accumulated. Early in these studies it became evident that many families and genera no longer extant in the Pacific Northwest were represented in these floras, which would perhaps indicate that the vegetation represented was markedly different than now occurs in the region. After some abortive attempts to interpret many of these floras in terms of the modern redwood forest of coastal southern Oregon and northern California, most workers came to regard the modern analogs (or even homologs) of the Oligocene and Neogene floras to be in eastern Asia or eastern North America. Most of the paleoclimatic inferences based on these floras have been derived from the present distributions of their contained taxa; this chapter will attempt to analyze these floras, primarily on the basis of their inferred physiognomy.

METHODS

A traditional method of reconstructing paleoclimates for fossil floras is to attempt to find a supposed homologous association. That is, a worker attempts to find an extant association that has numerous lineages in common with the fossil assemblage; then, the climatic parameters of this extant association are assigned to the fossil assemblage. In some instances, certain taxa are given more weight than others in relating fossil and extant associations, because no Tertiary association can be exactly matched by any living association (MacGinitie 1969). This point is well emphasized by the joint occurrence of several taxodiaceous genera in the early Tertiary of Spitsbergen (Schweitzer 1974); the present geographical distributions of these genera are, of course, generally mutually exclusive. More significantly, the temperature parameters of these genera are also generally mutually exclusive (Wolfe 1980). To assign temperature parameters to a fossil assemblage that contains these genera using the present climatic distribution of these genera would be most questionable.

Even Neogene assemblages may contain apparently dissonant genera, that is, genera that today have mutually exclusive climatic parameters. Other than the fact that many of these same taxodiaceous genera occur regularly in the same Neogene associations, assemblages that on palynological evidence must represent coniferous forest contain genera such as *Liquidambar*, *Platanus*, and *Liriodendron*—genera that today are confined to broadleaved vegetation.

The work of Van Devender and Spaulding (1979) on Quaternary assemblages in the southwestern United States has demonstrated clearly that the various plant associations have been extremely transitory. The Tertiary associations have also been subject to much change, and thus the paleo-temperature interpretations based on the associational methodology can be little more than gross approximations at best (MacGinitie 1969) and, at worst, may be in serious error if some genera are given unwarranted weight.

That lineages must have changed their tolerances has been suggested on theoretical grounds (Mason 1947). Paleobotanical evidence other than dissonant associations also indicates shifts in physiology. For example, the leaves of *Acer oregonianum*—a Miocene maple showing a high degree of morphological similarity (and probably ancestral) to the extant *A. macrophyllum*—have a chemistry most similar to the extant east Asian *A. carpinifolium* (Niklas and Gianassi 1978). This chemical similarity could well be interpreted to indicate that the ancestor of *A. macrophyllum* was physiologically closer to a maple that lives under a markedly different climatic regime than the descendant species. That is, the studies of paleo-biochemistry will assist greatly in interpreting the climates in which various lineages have evolved. This suggestion is reinforced by the physiognomical data presented later that indicate that Niklas and Gianassi's specimen of *A. oregonianum* represents part of an assemblage that lived under climatic conditions similar to those in the habitat of *A. carpinifolium*. Paleobiochemical analyses may thus be as much—possibly more—indicative of adaptive similarities as of phylogenetical relationships.

Another basic problem in many paleotemperature estimates from paleobotanical evidence is that many comparisons have been based on inadequate knowledge of temperatures that apply directly to modern vegetation. For example, Axelrod and Bailey (1969) have given precise temperature estimates for the Tertiary in the Pacific Northwest, but such temperatures would never produce the broadleaved deciduous vegetation that is so well documented in the Oligocene and Neogene of this region. One problem with Axelrod and Bailey's paleotemperature estimates is that, in part, they applied data for certain stations to vegetation 1,000 m altitudinally above those stations (Wolfe 1971).

Still another problem has been using the temperature-vegetation relations that prevail currently in much of the eastern United States. I have shown (Wolfe 1979) that much of the broadleaved deciduous forest of the eastern United States occurs in major temperature parameters that elsewhere in the world produce broadleaved evergreen vegetation. Until the initiation of intensive Arctic cold fronts in the late Neogene, the vegetation of much of the eastern United States was, in fact, broadleaved evergreen. The now anomalous temperature-vegetation relations in the eastern United States cannot validly form the basis for the reconstruction of early Neogene or older climates.

The best methodology of determining paleoclimate from fossil plant assemblages is that suggested by Bailey and Sinnott (1915): attempt to find physiognomical characteristics of foliage that can be correlated with climatic parameters. The physiognomy—including foliar—of plants largely represents adaptive responses of plants to their environment, and thus physiognomical analysis of fossil leaf assemblages offers, at the present time, the most direct way of gaining estimates of paleoclimates.

Although physiognomical analyses of fossil leaf assemblages are more readily made than floristic analyses, physiognomical analyses are dependent on an accurate correlation between physiognomy and climate.

The classification of modern physiognomical units used as a framework for the following discussion is based on the distribution of humid to mesic vegetation in eastern Asia (Wolfe 1979). Vegetation in regions where the mean of the cold month is 1°C (or higher) and the mean of the warm month is between 20 and 30°C is dominantly broadleaved evergreen. In regions of cooler summers, the vegetation is dominantly coniferous. In regions of colder winters—where the mean of the cold month is between 1 and -2°C— the vegetation is dominantly broadleaved deciduous, but with a significant broadleaved evergreen element. In regions of still colder winters, but of some summer warmth, the vegetation is almost exclusively broadleaved deciduous. In all these major vegetational types, the ratios of the three major elements—coniferous (or needle-leaved), broadleaved evergreen, and broadleaved deciduous—are typically characteristic of each type. In some types, the foliage has a high percentage of palmately lobed species, and in other types pinnately lobed species are common. Other physiognomical characters that can be used include leaf size, type of leaf margin, and diversity of woody climbers.

The vegetational units relevant to the discussion that follows are:

1. Notophyllous Broadleaved Evergreen forest. The so-called *oak-laurel* forest of eastern Asia is dominantly broadleaved evergreen with an admixture of broadleaved deciduous trees and shrubs. Except in secondary vegetation, conifers are not common. The leaf size is dominantly notophyllous (Webb 1959), that is, the smaller half of Raunkiaer's (1934) mesophyllous class. Woody climbers are profuse. Palmately lobed leaves are rare—typically less than 4 percent of the species. Under conditions of summer drought, vegetation in the same temperature parameters forms a woodland, as in both California and the Mediterranean region.

2. Mixed Broadleaved Evergreen and Coniferous forest. This forest in eastern Asia has been typically included in either Notophyllous Broadleaved Evergreen forest or in Mixed Mesophytic forest. The proportion of broadleaved evergreens, which dominate the crown, is, however, far too high for Mixed Mesophytic forest, and the proportion of conifers is higher than in primary Notophyllous Broadleaved Evergreen forest. In other aspects of structure, however, the vegetation is most similar to the notophyllous forest. In areas of summer drought, vegetation in these temperature parameters forms a more open forest in which conifers, broadleaved deciduous, and broadleaved evergreen trees occur about equally; such vegetation is present in the interior valleys of southern Oregon and at moderate altitudes of the coast ranges of California.

3. Mixed Mesophytic forest. This vegetation is dominantly broadleaved deciduous. Some notophyllous broadleaved evergreens are present, but typically as small trees and shrubs; about 15 to 30 percent of the species are broadleaved evergreen. Conifers are typically a minor element. In leaf size, the size classes smaller than notophyll are well represented. Palmately lobed leaves are present in about 8 to 10 percent of the species.

4. Mixed Northern Hardwood forest and Mixed Broadleaved Deciduous forest. Both these forests are dominantly broadleaved deciduous with some conifers intermixed. Notophyllous broadleaved evergreens are absent. Palmately lobed leaves are present on about 15 to 20 percent of the species in the Mixed Northern Hardwood forest and occur in several families (e.g., Rosaceae, Aceraceae, and Araliaceae). Pinnately lobed leaves are common in the Mixed Broadleaved Deciduous forest. Woody climbers may be present, but are not diverse.

5. Mixed Coniferous forest. Dominantly coniferous, this forest may contain an admixture of small broadleaved trees and shrubs. In areas in which the mean of the cold month is about −2°C, notophyllous broadleaved evergreens occur. In areas in which the mean of the warm month is below 15°C, the diversity and abundance of broadleaved woody plants markedly decreases. In secondary vegetation, broadleaved plants may be dominant. Except in secondary vegetation, woody climbers are represented poorly.

Modern vegetation that is dominantly coniferous in terms of basal area may floristically appear to be broadleaved. In detailed quadrat studies in northern California, conifers occupy between 79 and almost 100 percent of the basal area; in terms of woody species, however, conifers represent from 17 to 38 percent (Wolfe 1979). Similarly, in the Mixed Broadleaved Evergreen and Coniferous forest in Japan, broadleaved deciduous species represent only 3 percent in basal area, but are 28 percent of the total flora (Yoshino 1968). Another factor that must be considered is that most fossil assemblages contain an overrepresentation of streamside vegetation (Mac-Ginitie 1953), which is dominantly broadleaved—even in regions occupied by coniferous forest. In areas of Broadleaved Evergreen Forest, the stream-side vegetation can be composed of deciduous species, such as *Populus*, *Salix*, *Pterocarya*, and *Alnus*. Thus, the broadleaved evergreen element will tend to be underrepresented, particularly in small fossil assemblages in which streamside vegetation predominates. Even in vegetation such as the Paratropical Rain forest of Viet Nam, banks of major rivers are occupied by the broadleaved deciduous *Pterocarya*.

The determination of which species in a fossil assemblage represent broadleaved evergreens is not without some difficulties. Generally, broad-leaved evergreens have a thick texture, which can be observed even in impressions. Some broadleaved deciduous plants (e.g., *Platanus*) can, however, produce thick-textured leaves, and in the analyses, species that are members of genera that today have an exclusively broadleaved deciduous habit were relegated to the broadleaved deciduous category.

The general relation between leaf margin type (whether entire or nonentire) of woody dicotyledonous and climate was pointed out by Bailey and Sinnott (1915). From an overwhelming preponderance of entire-margined species in Tropical Rain forest, the percentage of species that have nonentire (toothed or lobed) margins increases with increasing altitude or latitude; the decrease in entire-margined species is thus generally correlative with lowered levels of heat. Bailey and Sinnott (1916), however, pointed out that the percentages obtained for dry vegetation could be anomalous, that is, that some deciduous plants in dry vegetation can have leaves that are predominantly nonentire. Thus, the leaf margin analysis has been generally applied to fossil assemblages that, on other physiognomical criteria, are thought to represent humid to mesic vegetation.

A recent compilation of leaf margin analyses for humid to mesic vegetation in eastern Asia indicated a pronounced correlation between the leaf margin percentages and mean annual temperature (Wolfe 1978, 1979); this compilation was based on quadrat studies (e.g., Sato 1946) or local floras (e.g., Hongkong). In contrast, Dolph (1979) and Dolph and Dilcher (1979) have suggested that the correlation is not precise, particularly, if the vegetation of small areas is concerned. In regard to Dolph's (1979) studies, it should be pointed out that of the 38 samples he considered, 84 percent had less than 30 species; I (1971) have emphasized that in applications to the fossil record, leaf margin percentages obtained on assemblages of less than 30 species can be subject to considerable alteration if more species are collected and added to the data base. Of the six samples that Dolph (1979) analyzed and that contained 30 or more (37 was the maximum) species, two were in dry (i.e., moist to dry as opposed to wet to rain forests) vegetation, and the percentages for this vegetation were in fact anomalously low. It would be difficult to accept the validity of generalizations based on the four remaining quadrats.

The study by Dolph and Dilcher (1979) of leaf margins in North and South Carolina was on a county basis. These workers (1979, p. 170) concluded that ". . . there is no mathematical correlation between leaf form and climate which would allow for a precise value to be given for . . . mean annual temperature . . . based on a specific expression of leaf form." I suggest that, to the contrary, the data presented by Dolph and Dilcher (1979) support a strong correlation between leaf margin and mean annual temperature. First, these workers did not apply mean annual temperature, but rather mean annual biotemperature. Second, leaf margin percentages were based on increments of 5 percent. Third, if leaf margin data are averaged for a county, then so also should the temperature data. If the approximate correlation of 3 percent/1°C mean annual temperature indicated in the vegetation of eastern Asia (Wolfe 1979) is applied to the situation in North and South Carolina, the 35 percent entire-margined datum line should be approximately coincidental with the 12°C isotherm and the 50 percent entire-margined datum line should be approximately coincidental with the 17°C isotherm. Considering the gross nature of a county-by-county compilation, and the fact that the temperatures are not county averages, the approximate coincidences evidenced in Figure 3.1 are remarkable. The problems of drawing accurate leaf margin datum lines are illustrated by the data for Dare County, North Carolina, which includes land stretching from coastal North Carolina at 35°N latitude north to the Virginia border; including leaf margin data for southern Dare County in the county total yields a high percentage and skews Dolph and Dilcher's (1979, fig. 3) data lines in the northeastern sector. In any case, if the leaf margin data compiled by Dolph and Dilcher (1979) had been based on fossil assemblages and paleoisotherms reconstructed from those data, a remarkably accurate approximation of paleotemperatures would have been provided. A method that would, apparently, place paleoisotherms to within 100 km (or even 200 km) of their actual positions should be of great value in paleoclimatic reconstructions.

If physiognomical data directly applicable with high accuracy to the fossil record is desired, then what is needed are studies of vegetation in depositional basins. If a fossil assemblage represents a sampling of the vegetation in a basin, then several quadrats in a given modern basin should

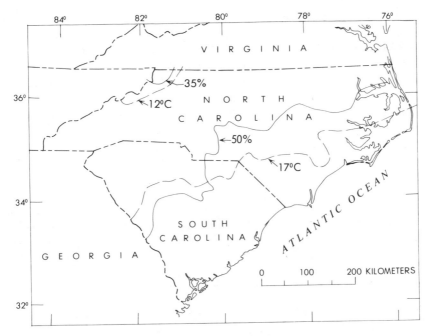

FIGURE 3.1. Comparison of leaf margin type and temperature data in North and South Carolina. Solid lines are percent of species that have entire-margined leaves (from Dolph and Dilcher 1979, fig. 3); dashed lines are mean annual temperature isotherms (compiled from publications of the National Oceanographic and Atmospheric Administration).

be sampled in order to arrive at a concept of the physiognomical characters of the vegetation of that basin. In this connection, Dolph's (1979) study would have been more relevant had he sampled several quadrats in the vicinity of a given meteorological station. Also needed are taphonomical studies to determine the relation of the source vegetation in a basin to the plant remains actually being deposited. I emphasize that until such studies are carried out, an analysis of one element of foliar physiognomy for a fossil assemblage can be expected to yield only an approximate correlation to climate.

Leaf sizes are a valuable criterion. The distribution of the size classes can, however, be altered greatly between the source vegetation and the sample actually deposited and collected due to mechanical fragmentation.

THE RECORD

The dominantly broadleaved evergreen vegetation of the Eocene in the Pacific Northwest has been discussed previously (Wolfe 1971). This vegetation, at times, represented Paratropical or even Tropical Rain forest; during at least the Middle Eocene, Paratropical Rain forest is known to have extended north to beyond latitude 60° in Alaska (Wolfe 1977). The low mean annual range of temperature that can be documented during the Eocene precludes the existence of any mesic temperate broadleaved deciduous forest (Wolfe 1978). At the end of the Eocene, a major decline in mean annual temperature occurred in the Pacific Northwest concomitant with a major increase in mean annual range of temperature. It is the vegetation following this terminal Eocene event that will be discussed in more detail, the vegetation of about the last 32 m.y. (million years).

The Early Oligocene Rujada (Lakhanpal 1958), Willamette, and Lyons (Meyer 1973) floras occur in volcanic rocks that were laid down along the western margin of the Cascade Range following the withdrawal of a marine embayment from the major part of what is now the Willamette Valley (Fig. 3.2). The Willamette and slightly more interior Rujada in particular have a

FIGURE 3.2. Location of some Oligocene and Neogene floras in the Pacific Northwest.

high proportion of conifers and broadleaved evergreens that can be best compared to the admixture in the Mixed Broadleaved Evergreen and Coniferous forest (Table 3.1). The leaf margin percentage indicates a mean annual temperature of 12 to 13°C. At almost the same latitude, but more interior by about 200 km, the correlative assemblage from near Post (Chaney 1927) represents dominantly Broadleaved Deciduous forest that had a reduced coniferous element and a moderate broadleaved evergreen element. The physiognomy of the fossil assemblage supports Chaney's (1952) conclusion that this flora represents Mixed Mesophytic forest. In order to account for the transition from a Mixed Broadleaved Evergreen and Coniferous forest to a somewhat more interior Mixed Mesophytic forest, a mean annual range of temperature of about 23°C can be inferred.

In the Willamette Valley of Oregon, today mean annual range of temperature is 14 to 16°C and mean annual temperature is about 11 to 12°C. Although a slight decrease in mean annual temperature can be inferred since about 30 m.y., clearly the major change has been in a significant decrease in mean annual range of temperature of at least 7°C.

The Lyons flora, although correlative with the Willamette and Rujada, has a significantly lesser representation of broadleaved evergreens than these two other floras (Table 3.2). Such a difference in a distance of only 65 km, concomitant with the fact that the Lyons flora would have been geographically closer to the remnant of the marine embayment than either the Willamette or Rujada, is anomalous. However, in central Oregon, the contemporaneous floras from Post north to Fossil, which is about the same latitude as the Lyons, display a similar reduction in the broadleaved evergreen element (Table 3.3). That the Willamette and Bridge Creek assemblages are correlative has been established clearly by radiometric age determinations (Evernden and James 1964). It is notable that these more northern assemblages, which are part of the Bridge Creek flora, *s.l.*, do contain some notophyllous broadleaved evergreens. This is significant in

TABLE 3.1: Physiognomical Elements in Some Early Oligocene Assemblages from Coastal to Central Oregon

Assemblage	Species	Margin Entire (%)	Conifer (%)	Broadleaved Evergreen (%)	Broadleaved Deciduous (%)
Willamette	36	41	18	41	41
Rujada	37	33	24	35	41
Post	37	34	11	22	67

Source: Compiled by the author (Chaney 1927).

TABLE 3.2: Broadleaved Evergreen Representation in Some Early Oligocene Assemblages from Western Oregon

Assemblage	Latitude	Species	Broadleaved Evergreen	
			Specimens (%)	Species (%)
Lyons	44°45′	37	<1	19
Willamette	44°00′	36	[abundant]	41
Rujada	43°45′	37	23	35

Source: Compiled by the author.

indicating that the mean of the cold month was above −2°C, that is, that these assemblages must represent Mixed Mesophytic forest. Even in areas of China occupied by the Mixed Mesophytic forest where the cold month is about −2°C (e.g., T'ien-mu-shan), broadleaved evergreens still represent about 15 percent of the woody flora. Thus, the reduced broadleaved evergreen element in these fossil assemblages is anomalous relative to extant vegetation. That this reduced broadleaved evergreen element is the result of depositional factors is improbable; the Rujada, Willamette, and Lyons floras are all based on collections from similar lithologies.

The Late Oligocene floras of the Pacific Northwest are few and poorly known. McClammer (1978) has studied the Yaquina flora from coastal Oregon that represents Notophyllous Broadleaved Evergreen forest. Coeval assemblages are known as far north as latitude 45°30′; these assemblages also contain a diversity of notophyllous broadleaved evergreens. One of

TABLE 3.3: Broadleaved Evergreen Representation in Some Early Oligocene Assemblages from Central Oregon

Assemblage	Latitude	Species	Broadleaved Evergreen	
			Specimens (%)	Species (%)
Fossil	45°00′	28	1	7
Knox Ranch	44°57′	30	1	7
Twickenham	44°45′	34	1	9
Bridge Creek	44°37′	34	14	12
Post	44°10′	37	49	22

Source: Compiled by the author (Chaney 1927).

these floras, the Sandstone Creek, has a physiognomy indicative of Broad-leaved Evergreen and Coniferous forest or marginal Notophyllous Broad-leaved Evergreen forest (Table 3.4). About 350 m stratigraphically above the Sandstone Creek locality is a definite Miocene flora (the Collawash), suggesting that the Sandstone Creek flora is very late in the Oligocene and, thus, probably, about 23 or 24 m.y.

The vegetation of the Early to Early Middle Miocene (Seldovian Age, about 15 to 24 m.y.) is well-known. Floras of this age include the Collawash (Wolfe unpublished data), Eagle Creek (Chaney 1920), Grand Coulee (Berry 1931), Latah (Knowlton 1926), Whitebird (Berry 1934), Clarkia (Smiley et al. 1975), and Rockville (part of Graham 1965, Sucker Creek flora), as well as an assemblage from the coastal area on Wishkaw River (Wolfe unpublished data).

To group under one time interval floras such as the Collawash, Latah, Rockville, and Mascall is not valid from the standpoint of determining paleoisotherms. The age spread of these assemblages is known to be several million years and some assemblages represent a warm interval and others a relatively cool interval. Thus, during the early part of the Early Miocene (Collawash and Eagle Creek floras) mean annual temperatures were probably somewhat lower overall than during the Late Early and Early Middle Miocene (Latah and equivalents, Wishkaw, Mascall). The so-called mid-Miocene warming has been discussed elsewhere (Wolfe 1971).

The Collawash is the most diverse Neogene flora in the Pacific North-west, with at least 140 megafossil species represented. From the distribution and thickness of the basalt flows that covered the post-Collawash and Eagle Creek landscape, it can be inferred that the Collawash beds were deposited at an altitude of about 500 m and the Eagle Creek beds at about 100 to 200 m. The Collawash assemblage (Table 3.4) represents Mixed Mesophytic forest as indicated by all physiognomical criteria available. Further, from the leaf margin percentage, a mean annual temperature of about 10°C can be inferred, which, in turn, indicates a mean annual range of temperature of about 20 to 22°C. The leaf margin percentage for the Eagle Creek (Table 3.5) indicates a mean annual temperature of about 11°C. As in the instances of the Early Oligocene floras, the Collawash and Eagle Creek floras indicate a continued decrease in mean annual range of temperature, but little, if any, change in mean annual temperature. (See Figure 3.3)

Floras about 500 km more interior than the Collawash and slightly more north are the Whitebird and Clarkia. These assemblages are dominantly broadleaved deciduous, but they have some notophyllous broad-leaved evergreens and a minor coniferous element. They thus represent Mixed Mesophytic forest and would indicate very little increase in mean annual range of temperature going into the interior. This situation is paralleled by that today in China at about 35°N latitude.

TABLE 3.4: Physiognomical Analysis of Late Oligocene and Early Miocene Assemblages from Northwestern Oregon

Assemblage	Species	Margin Entire (%)	Conifer (%)	Broadleaved Evergreen (%)	Broadleaved Deciduous (%)	Broadleaved, Leaf Size Mesophyll (%)	Broadleaved, Leaf Size Notophyll (%)	Broadleaved, Leaf Size Microphyll (%)
Collawash	142	28	8	20	72	4	37	57
Sandstone Creek	35	38	9	48	43	17	60	23

Source: Compiled by the author.

TABLE 3.5: **Physiognomical Elements in Some Early and Middle Miocene Assemblages from the Pacific Northwest**

Assemblage	Latitude	Species	Margin Entire (%)	Broadleaved Evergreen (%)
Grand Coulee	47°55'	18	30	0
Latah	47°30'	70	30	10
Wishkaw	47°00'	30	40	17
Whitebird	45°40'	28	36	16
Eagle Creek	45°38'	30	31	26
Fish Creek	45°03'	28	32	30
Collawash	45°02'	142	28	20
Mascall	44°25'	56	28	18
Rockville	44°15'	43	30	23

Source: Compiled by the author.

To the north, the Latah flora represents the second largest assemblage known in the Pacific Northwest in the Early to Middle Miocene interval. The fact that some notophyllous broadleaved evergreens occur in the Latah indicates that the mean of the cold month was above −2°C, and, combined with the diverse and dominant broadleaved deciduous flora, indicates that a Mixed Mesophytic forest was represented. The broadleaved evergreens in the Latah flora, however, are few: a laurel, *Exbucklandia*, and *Arbutus* are the only broadleaved evergreens that occur at more than one of the several Latah localities. *Quercus*, *Ilex*, and *Lyonothamnus* comprise the remainder of the broadleaved evergreen element. Other evergreen laurels and oaks, *Castanopsis* and *Mahonia*, which are common in more southerly floras are lacking in the Latah. Mixed Mesophytic forest should, in fact, have more representation of broadleaved evergreens than occurs in the Latah.

The Grand Coulee flora is also problematic. No broadleaved evergreens are represented. Although this might be indicative of Mixed Northern Hardwood or Mixed Deciduous forest, the Grand Coulee has too low of a representation of lobed leaves—either palmately or pinnately lobed—to represent either of these vegetational types. Indeed, the fact that the Grand Coulee area was considerably closer to the coast than the Latah would argue against a lower cold month mean for the Grand Coulee necessary to produce Mixed Northern Hardwood forest.

Just as anomalous is the coastal Wishkaw assemblage at a latitude intermediate between the Collawash and Latah. A coastal location should have produced a Mixed Broadleaved Evergreen and Coniferous forest. The broadleaved evergreen element in the Wishkaw River flora is, however, much less than would be expected. Again, there appears to be an unexpected

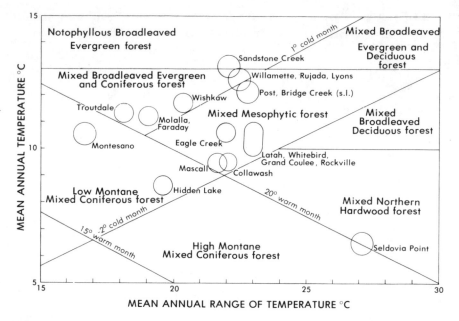

FIGURE 3.3. Inferred temperature parameters for some Oligocene and Neogene floras from the Pacific Northwest. (Vegetational classification from Wolfe [1979].)

reduction going northward in the Broadleaved Evergreen element that is apparently not related to temperature factors.

In southern Alaska at latitude 59°N, the rich Seldovia Point assemblage (Wolfe and Tanai 1980) is known from beds that are about the same age as the Latah and coeval floras in the Pacific Northwest. This correlation is, in part, based on radiometric ages obtained from Alaskan sections that have pollen assemblages locally correlative with those from the Seldovia Point beds (Turner et al. 1980). Analysis of the Seldovia Point flora indicates a mean annual temperature of about 6 to 7°C and a mean annual range of temperature of about 26 to 27°C. The overall temperature change since deposition of the Seldovia Point beds has thus been a lowering of mean annual temperature of about 4°C and a decrease in mean annual range of temperature of about 9 to 10°C. A change of these parameters further indicates that winter temperatures have been unchanged overall, but the mean of the warm month has declined from about 20 to 11°C.

One important aspect of the temperature changes since the Middle Miocene is that the difference in mean annual temperature between latitudes 44 and 59°N today is about 9°C, whereas the difference in the Middle Miocene was about 5°C. That is, the latitudinal temperature gradient along the coast has markedly increased; an increase in this gradient would, as I have discussed previously (Wolfe 1978), lead to an intensified subtropical high pressure system and concomitantly increasing summer drought.

Most of the late Middle to early Late Miocene (Homerian age) assemblages from the Cascade Range and Columbia Plateau represent Mixed Coniferous forest (Wolfe 1969). The diverse broadleaved accessories, in combination with the presence of notophyllous broadleaved evergreens is indicative of Low Montane Mixed Coniferous forest (Wolfe 1979). In the area occupied by this vegetation, the mean of the cold month is above $-2°C$ and the mean of the warm month is between 15 and 20°C.

In the lowlands west of the Cascades, however, the Homerian vegetation apparently represents broadleaved forest, at least in the interior valleys. An assemblage such as the Faraday, from near Portland, is difficult to interpret unless it is put in the context of the preceding and succeeding vegetation. The slightly older Molalla flora, which was also lowland, has a higher representation of broadleaved evergreens, particularly those such as *Magnolia* and *Meliosma* that now live in areas of high summer precipitation. Instead, the Faraday flora has broadleaved evergreens that, in part, now occur in summer-dry areas, for example, *Arbutus*, *Rhododendron*, *Gaultheria*, and *Leucothoe*. Whereas the probable vine element in the Molalla flora represents 14 percent of the woody broadleaved flora, including genera such as *Vitis* and *Toxicodendron*, the vine element in the Faraday flora represents only 6 percent of the woody broadleaved flora and comprises genera such as *Clematis* and *Smilax*. A reduction in the vine element is in itself an indication of greater dryness, as well as the elimination of genera now typical of areas of summer precipitation. The available evidence from the preceding vegetation indicates that the Faraday vegetation resulted from decreased summer precipitation.

The succeeding latest Miocene Troutdale assemblage (Chaney 1944) is even more indicative of summer dryness. In the pollen spectra, the genera now restricted to areas of summer precipitation represent about 5 percent as opposed to 20 percent for the Faraday. The vine element is virtually lacking in the Troutdale. Further, in the Troutdale samples, pollen of probable herbaceous groups is diverse (Table 3.6), indicating the possibility of an open forest.

One of the major reasons for relying more heavily on floristic evidence in interpreting the Late Miocene vegetation and climate is that, if, indeed, summer drought had started, physiognomical comparisons of the fossil assemblages to the summer-wet vegetation of eastern Asia are not valid. Perhaps, the closest physiognomical analog to the Faraday and Troutdale floras is in the interior valleys of southern Oregon, where the vegetation is a mixture of conifers in a basically Broadleaved forest that contains some sclerophylls, that is, the Interior Valley zone of Franklin and Dyrness (1969). This comparison would indicate temperature parameters about intermediate

TABLE 3.6: Composition of Some Pollen Samples from Late Neogene Formations in Western Oregon and Washington

| | Latest Miocene | | Pliocene |
	Interior	Coastal	Coastal
Abies	0.5	–	–
aff. *Cedrus*	3.0	11.0	1.0
Keteleeria	+	–	–
Picea	1.0	8.5	16.0
Pinus	11.5	37.5	31.5
Pseudotsuga	–	+	2.5
Tsuga	0.5	6.5	3.5
Taxodiaceae/Cupressaceae	2.5	3.0	6.5
Conifers	18.0	66.5	71.0
Liquidambar	0.5	0.5	–
Platanus	10.0	–	–
Ulmus/Zelkova	1.0	–	+
Fagus	0.5	0.5	0.5
Quercus	40.0	4.5	0.5
Alnus	2.0	9.0	9.5
Ostrya/Carpinus	+	0.5	1.0
Carya	–	1.0	+
Juglans	14.5	–	–
Pterocarya	3.5	5.0	0.5
Tilia	+	–	–
Pachysandra	–	0.5	–
Acer	0.5	–	–
Ilex	+	0.5	1.0
Sambucus	–	0.5	+
Centrospermae	+	1.0	–
Onagraceae	+	–	–
Umbelliferae	+	–	–
Polemoniaceae	+	–	–
Compositae	2.0	0.5	0.5
Graminae	1.0	–	–

Source: Compiled by the author.

between those suggested for the Seldovian vegetation and those that occur in the region today.

Along the coast, vegetation approximately coeval with the Troutdale is represented by pollen from the upper part of the Montesano Formation (Table 3.6). This vegetation was dominantly coniferous, although a diverse broadleaved element persisted. By the Pliocene, the broadleaved element was reduced, particularly, in abundance and somewhat in diversity, as indicated by analysis of a sample from the Quinault Formation. Noteworthy is the persistence of an extinct genus allied to *Cedrus*, which was one of the dominants in the Mixed Coniferous forest of the Cascades and Columbia Plateau during the Middle to Late Miocene (Wolfe 1969).

To the east of the Cascades, vegetational and climatic change in the Late Miocene and Pliocene has the added factor of the piling up of the volcanics of the High Cascades (Smiley 1963). The Ellensburg assemblages record the replacement of a more mesic by a subhumid broadleaved vegetation at low altitudes. By the Pliocene, the Deschutes flora represents a depauperate broadleaved vegetation of *Populus*, deciduous *Quercus*, and *Ulmus* (Chaney 1938; Wolfe unpublished data).

DISCUSSION

In the previous section, temperature parameters have been estimated for many assemblages in the Oligocene and Neogene of the northwestern United States. These assemblages represent vegetation that occupied sites that varied latitudinally, longitudinally (i.e., distance from the coast), and altitudinally. Taking into account these various factors, I have constructed a model of temperature change from the Early Oligocene to the Pliocene for sites about 100 km from the coast at about latitude 45°N and of less than 100 m altitude; the present city of Salem, Oregon, is at a site that satisfies these criteria (Fig. 3.4). As emphasized previously, the overall trend is characterized by primarily a decrease in mean annual range of temperature; mean annual temperature appears to have declined very slightly. Winter temperatures have increased obviously since the Early Oligocene by a value somewhat greater than the decline in summer temperatures. Superimposed on these overall trends are some fluctuations in mean annual temperature.

The plant assemblages following the Middle Miocene record increasing summer drought. Waring and Franklin (1979) have pointed out a number of physiological characters of conifers that apparently favor these trees over broadleaved trees under conditions of summer drought. One of the basic premises is the supposedly unique climatic regime and supposedly unique coniferous forests of the Pacific Northwest west of the Cascade crest. In temperature, however, large areas of Asia—for example, upland Taiwan, the Southwestern Plateau of China, and much of the Himalayas—have virtually identical temperature regimes to this area of the summer-dry Pacific Northwest (Fig. 3.5). Although these Asian areas are summer-wet, they are occupied by coniferous forests. Indeed, these coniferous forests may be fully as great in stature as those of the Pacific Northwest: Wang (1961) cites trees of *Tsuga* that have a diameter at breast height (DBH) of 1.8 m and trees of *Picea* that have a height of 60 m, which certainly compares favorably with

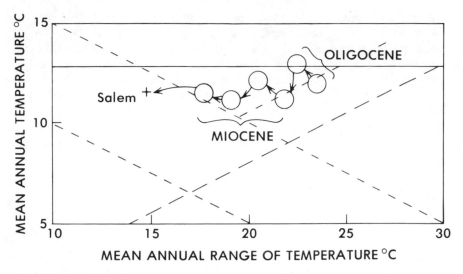

FIGURE 3.4. Inferred temperature changes in the Pacific Northwest since the Early Oligocene standardized to localities at about latitude 45° N that were less than 100 km inland and were less than 100 m altitude.

the size data presented by Waring and Franklin. In diversity, the Asian forests are clearly richer than those of the Pacific Northwest: the Southwestern Plateau, for example, has 46 species of conifers.

The greater diversity in genera and species of conifers in these Asian coniferous forests could be indicative of the reverse of what Waring and Franklin (1979) suggest; that is, these data could indicate that summer-wet climate is more favorable to coniferous forest than is summer-dry climate. Moisture regimes are, however, irrelevant. If Waring and Franklin's (1979) basic thesis about the better adaptability of conifers versus broadleaves were valid, then how have broadleaved trees come to dominate in the pronouncedly summer-dry climates at low altitudes in the mountains of California?

In fact, the increasing summer drought during the Neogene led to a decline in diversity of both conifers and broadleaved trees. A late Middle to early Late Miocene assemblage, such as the Hidden Lake flora, represents Mixed Coniferous forest. The rich broadleaved element that was accessory to the conifers has suffered great extinction; genera such as *Ulmus*, *Zelkova*, *Pterocarya*, *Liquidambar*, and *Liriodendron* are not extant in the Pacific Northwest. But, the rich coniferous element, a major part of which was a genus allied to *Cedrus*, has also suffered extinction. At least 35 percent of the coniferous lineages of the Neogene of the Pacific Northwest have suffered extinction. In terms of genera, 42 percent of the broadleaved genera have

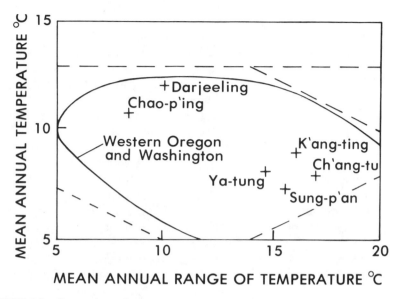

FIGURE 3.5. Comparison of temperature parameters of summer-wet coniferous forests in southeastern Asia to temperature parameters of summer-dry coniferous forests in northwestern United States. (Data based on Wolfe [1979].)

become extinct in western North America during the Neogene in comparison to 36 percent of the coniferous genera.

From the preceding, I think that it is clear that the change in precipitation during the Neogene has not caused the change from a broadleaved to a coniferous forest in the Pacific Northwest and summer-dry climate does not maintain the present dominance of conifers. The sole climatic factor in the assumption of dominance by conifers is the decline in summer temperature. Except for parts of western Europe, where historical factors related to Quaternary glaciation have resulted in an anomalous situation, areas of only moderate summer temperature are typically occupied by coniferous forest (Wolfe 1979).

This analysis suggests that the angiosperms have come to dominate forests in warm summer areas because of factors related to heat. That is, the angiospermous trees have at least three adaptations that allow them to dominate over conifers: (1) deep root systems to tap moisture sources, (2) vessels to conduct water more effectively, and (3) large leaves that, combined with the two preceding adaptations, assist in the rapid movement of water upward. The higher the temperature, the greater advantage broadleaved have over coniferous trees, and, in fact, the diversity and dominance of broadleaved versus coniferous trees appear correlative with summer temperature.

I will now return to the anomalous situations regarding broadleaved evergreen distribution in the Oligocene and Miocene. In the Early Oligocene, assemblages at about latitude 44°N have a diversity of broadleaved evergreens, whether these assemblages are near-coastal or interior. In central Oregon a series of localities, the material from which has comprised the Bridge Creek flora *sensu lato*, extends from latitudes 44 to 45°N. The southernmost localities have both a diversity and abundance of broadleaved evergreens (Chaney 1927). In the localities along Bridge Creek (about lat. 44°35′N), the broadleaved evergreens are neither abundant nor diverse. The localities near 45°N produce only rare broadleaved evergreens.

In the early Late to early Middle Miocene, the floras at latitudes 45 to 46°N have a diversity of broadleaved evergreens, but the floras at latitudes 47° to 48°N do not. This phenomenon is apparently unrelated to temperature, as I suggested previously based on the paleoclimatic analyses. Today, despite the dominance of coniferous forest, members of ten species and six genera of large-leaved (i.e., notophyllous) broadleaved evergreens are found at latitude 49°N in the Pacific Northwest (Wolfe 1980). At latitude 50°N, however, only one species of notophyllous broadleaved evergreen persists. As I have elsewhere discussed, areas of southeastern Alaska and southern British Columbia have temperatures suitable for certain broadleaved evergreens that, however, have their northern distributional limits far to the south. The elimination of most broadleaved evergreens at about latitude 50° is possibly a function of the low intensity and duration of winter light poleward of that latitude.

The distribution of broadleaved evergreens in the Neogene floras elsewhere in the Northern Hemisphere neither supports nor contradicts the hypothesis suggested here in regard to the limitation of broadleaved evergreens. The northernmost Neogene floras now known in Europe are at about latitude 56°N in Jutland; several of the genera and families represented in this fruit and seed assemblage are today exclusively broadleaved evergreen. Europe, however, has been subject to considerable northward movement during the Tertiary, and since the Early Miocene, Jutland has moved over 10° northward (Smith and Briden 1977). In eastern North America, Neogene floras are unknown north of latitude 42°N. In eastern Asia, the northern temperature limit (cold month <−2°C) of broadleaved evergreens has, during the Oligocene and Neogene, been south of paleolatitudes of 40 to 45°N.

Paleoclimatic analyses of Neogene floras in the Northern Hemisphere indicate that, in general, mean annual temperatures at lower latitudes have increased (largely winter temperatures), mean annual temperatures at middle latitudes have changed little if at all, and mean annual temperatures at high latitudes have decreased (largely summer temperatures). In all areas, mean annual range of temperature has decreased, the decrease being directly proportional to latitude. Such temperature changes could be the result of a decrease in the inclination of the earth's axis of rotation. Milankovitch (1938) calculated that with decreasing inclination, the resulting insolational and temperature changes at various latitudes would be in the directions that the paleobotanical evidence indicates. If this suggestion of a decreasing inclination is valid, then the progressive northward decrease in broadleaved evergreens in the Pacific Northwest Oligocene and Miocene floras can be explained. If the inclination were as great as 28 to 29° in the Early Oligocene, then broadleaved evergreens would be limited to areas equatorward of about latitude 45°. If the inclination has since then changed at a uniform rate, then by the Middle Miocene, broadleaved evergreens would be limited to areas equatorward of about latitudes 47 to 48°N. I suggest that the climatic trends during the last 32 m.y. and the changing distribution of broadleaved evergreens during that time represent independent lines of evidence that indicate that the inclination of the rotational axis has decreased.

REFERENCES

Axelrod, D. I., and H. P. Bailey. 1969. Paleotemperature analysis of Tertiary floras. *Palaeogeog., Palaeoclimatol., Palaeoecol.* 6(3):163–195.

Bailey, I. W., and E. W. Sinnott. 1915. A botanical index of Cretaceous and Tertiary climates. *Science* 41:831–834.

———. 1916. The climatic distribution of certain types of angiosperm leaves. *Amer. J. Bot.* 3:24–39.

Berry, E. W. 1931. A Miocene flora from Grand Coulee, Washington. *Prof. Pap. U.S. Geol. Sur.* 170C:31–42.

———. 1934. Miocene plants from Idaho. *Prof. Pap. U.S. Geol. Sur. Prof.* 185E:97–125.

Chaney, R. W. 1920. The flora of the Eagle Creek formation. *Walker Mus. Contrib.* 2:115–181.

———. 1927. Geology and palaeontology of the Crooked River basin with special reference to the Bridge Creek flora. *Publ. Carnegie Inst. Wash.* 346:45–138.

———. 1938. The Deschutes flora of eastern Oregon. *Publ. Carnegie Inst. Wash.* 476:187–216.

———. 1944. The Troutdale flora. *Publ. Carnegie Inst. Wash.* 553:323–351.

———. 1952. Conifer dominants in the middle Tertiary of the John Day basin, Oregon. *Palaeobotanist* 1:105–113.

Dolph, G. E. 1979. Variation in leaf margin with respect to climate in Costa Rica. *Torrey Bot. Club Bull.* 106:104–109.

Dolph, G. E., and D. L. Dilcher. 1979. Foliar physiognomy as an aid in determining paleoclimate. *Palaeontographica* 170(B):151–172.

Evernden, J. F., and G. T. James. 1964. Potassium-argon dates and the Tertiary floras of North America. *Amer. J. Sci.* 262:945–974.

Franklin, J. S., and C. T. Dyrness. 1969. Vegetation of Oregon and Washington. *U.S. For. Serv. Res. Pap.* PNW 80:216.

Graham, A. 1965. The Sucker Creek and Trout Creek Miocene floras of southeastern Oregon. *Kent State Univ. Bull.* 53(10):103.

Knowlton, F. H. 1926. Flora of the Latah formation of Spokane, Washington, and Coeur d'Alene, Idaho. *Prof. Pap. U.S. Geol. Sur.* 140A:17–81.

Lakhanpal, R. N. 1958. The Rujada flora of west central Oregon. *Univ. Cal. Publ. Geol. Sci.* 35:1–66.

McClammer, J. U., Jr. 1978. Paleobotany and Stratigraphy of the Yaquina Flora (Latest Oligocene-Earliest Miocene) of Western Oregon. University of Maryland: Unpub. M.A. thesis.

MacGinitie, H. D. 1969. The Eocene Green river flora of northwestern Colorado and northeastern Utah. *Univ. Cali. Publ. Geol. Sci.* 83:140.

Mason, H. L. 1947. Evolution of certain floristic associations in western North America. *Ecol. Mono.* 17:201–210.

Meyer, H. 1973. The Oligocene Lyons flora of northwestern Oregon. *Ore Bin* 35:37–51.

Milankovitch, M. 1938. Astronomische Mittel zur Erforschung der erdgeschichtlichen Klimate. *Hand. der Geophy.* 9(3):593–698.

Niklas, K. J., and D. E. Gianassi. 1978. Angiosperm paleobiochemistry of the Succor Creek flora (Miocene), Oregon. *Amer. J. Bot.* 65:943–952.

Raunkiaer, C. 1934. *The Life Forms of Plants and Statistical Plant Geography.* Oxford: Clarendon Press.

Sato, W. 1946. Studies on the plant-climate in the south-western half of Japan. *Kanazawa Nor. Coll., J. Sci.* 1:110.

Schweitzer, H. J. 1974. Die 'tertiären' Koniferen Spitzbergens. *Palaeontographica* 149(B):1–89.

Smiley, C. J. 1963. The Ellensburg flora of Washington. *Univ. Cal. Publ. Geol. Sci.* 35:159–276.

Smiley, C. J., J. Gray, and L. M. Huggins. 1975. Preservation of Miocene fossils in unoxidized lake deposits, Clarkia, Idaho. *J. Paleontol.* 49:833–844.

Smith, A. G., and J. C. Briden. 1977. *Mesozoic and Cenozoic Paleocontinental Maps.* Cambridge Earth Science Series.

Turner, D. L., D. M. Triplehorn, C. W. Naeser, and J. A. Wolfe. 1980. Radiometric dating of ash partings in Alaskan coal beds and upper Tertiary paleobotanical stages. *Geology* 8:92–96.

Van Devender, T. R., and W. G. Spaulding. 1979. Development of vegetation and climate in the southwestern United States. *Science* 204:701–710.

Wang, Chi-Wu. 1961. The forests of China. *Publ. Harvard Univ., Maria Moors Cabot Found.* 5:313.

Waring, R. H., and J. F. Franklin. 1979. Evergreen coniferous forests of the Pacific Northwest. *Science* 204:1380–1386.

Webb, L. J. 1959. Physiognomic classification of Australian rain forests. *J. of Ecol.* 47:551–570.

Wolfe, J. A. 1969. Neogene floristic and vegetational history of the Pacific Northwest. *Madrono* 20(3):83–110.

———. 1971. Tertiary climatic fluctuations and methods of analysis of Tertiary floras. *Palaeogeog., Palaeoclimatol., Palaeoecol.* 9:27–57.

———. 1977. Paleogene floras from the Gulf of Alaska region. *Prof. Pap. U.S. Geol. Sur.* 997:108.

———. 1978. A paleobotanical interpretation of Tertiary climates in the Northern Hemisphere. *Amer. Sci.* 66:694–703.

———. 1979. Temperature parameters of humid to mesic forests of eastern Asia and relation to forests of other regions of the Northern Hemisphere and Australasia. *Prof. Pap. U.S. Geol. Sur.* 1106:37.

———. 1980. Tertiary climates and floristic relationships at high latitudes in the Northern Hemisphere. *Palaeogeog., Palaeoclimatol., Palaeoecol.* 30:313–323.

Wolfe, J. A., and T. Tanai. 1980. The Miocene Seldovia Point flora from the Kenai Group, Alaska. *Prof. Pap. U.S. Geol. Sur.* 1105:52.

Yoshino, M. 1968. Distribution of evergreen broad-leaved forests in Kanto District, Japan. *Geograph. Rev. Jap.* 41:674–694.

4

THE STATUS OF CERTAIN FAMILIES OF THE AMENTIFERAE DURING THE MIDDLE EOCENE AND SOME HYPOTHESES REGARDING THE EVOLUTION OF WIND POLLINATION IN DICOTYLEDONOUS ANGIOSPERMS*

William L. Crepet

INTRODUCTION

The Amentiferae share simple florets, predominant wind pollination (except in certain Fagaceae), mostly an arborescent habit, small, somewhat smooth pollen, and a predominantly Northern Hemisphere distribution, except the Celtidoideae (Ulmaceae) and Myricaceae (Hjelmquist 1948; Soepadmo 1972; Abbe 1974). Although the Amentiferae have been considered a natural, hamamelidalean-derived group with the removal of certain strategic families, for example, Salicaceae (Cronquist 1968; Takhtajan 1969), other phylogenists have considered the Didymelaceae, Leitneriaceae, urticalean families, and the Juglandaceae to be allied with other taxa (Airy Shaw 1966; Thorne 1973; Wolfe 1973). It is generally agreed that the Amentiferae are united by adaptations thought to be associated with the

---

*This contribution is an exercise in extrapolation from the concept that there may be competition for pollinators among species of flowering plants. Since this possibility is controversial and since many aspects of early angiosperm evolution/environment are still open to discussion, I have been careful to include the term *hypotheses* in the title. As the paleoecological evidence is interpreted for this paper, it appears as though wind pollination evolved at a time and at latitudes that make the possibility of a cold climate/season being involved unlikely. Although this is not consistent with suggestions by Heinrich and Raven (1972) based on energetics, it seems to be the most plausible interpretation of available fossil evidence. Perhaps suggestions offered in this paper explain the evolution of wind pollination without the intervention of a cold season or, indeed, future paleontological evidence may indicate that a cold season actually was a significant factor. Various alternative possibilities will be explored by the author in the future.

phylogenetic change from entomophily to anemophily. My intention is to consider certain aspects of the history of the Amentiferae and even of wind-pollinated dicots in general as follows:

1. Updating the fossil record of the Amentiferae, last summarized by Wolfe (1973), by including contributions to their fossil history based on the angiosperm flower record.
2. Attempting to determine the point in time where approximately modern levels of adaptation for wind dispersal of pollen were reached in the Amentiferae.
3. Considering the possible time of origin of wind pollination in the Amentiferae, their ancestral stock, and in dicots in general.
4. Suggesting hypotheses related to a series of factors that may have been significant in the development and radiation of wind pollination in the Amentiferae, and synthesizing these into a general hypothesis for the origin of wind pollination in the dicotyledonous angiosperms.

Certain basic assumptions are accepted in formulating these hypotheses. These are given in the following list. Justifications for certain of these assumptions will be elaborated upon later in this chapter.

1. Wind-pollinated Amentiferae are derived from insect-pollinated ancestors. Their flowers, inflorescences, and pollen demonstrate various modifications, depending on the family. This is the contemporary view based on the comparative morphology of extant taxa (Cronquist 1968; Takhtajan 1969). In addition, there is now a fossil intermediate consistent with these assumptions in one family of the Amentiferae—the Ulmaceae (Zavada and Crepet 1981).
2. As suggested by Whitehead (1969), the deciduous habit evolved early in angiosperm history in response to a dry or seasonally dry tropical environment. This type of environment, whether widely distributed or regional, is considered often to have been important in the radiation of the angiosperms, as well as their origin (Stebbins 1965, 1974; Axelrod 1966). This assumption is consistent with the paleoclimatic literature based on oxygen ratios (Lowenstam 1964), and with most logical interpretations of the paleobotanical, paleopalynological, and paleophytogeographical evidence (Vakhrameyev 1952, 1978; Brenner 1976; Doyle and Hickey 1976; Doyle 1977; Doyle et al. 1977; Hickey and Doyle 1977).
3. Pollen of the proper size, shape, and exine ornamentation is considered suggestive of wind dispersal (Whitehead 1969).
4. The fossil record of insects is accepted more or less at face value and inferences as to the state of advancement of insects based on the fossil leaf record are not accepted for the purpose of formulating the impending hypotheses. To suggest that certain insect types (and by inference certain

modes of pollination) existed at a particular time in the past by extrapolation from the fossil record of leaves is misleading. Implicit in this reasoning is the idea that relative rates of evolution of different plant organs are the same and that a particular leaf type is suggestive of a particular flower type, implying that a specific type of pollinator may exist. This is *a priori*, an unreasonable approach, since it is obvious that different selective forces operate on leaves and flowers and they are, therefore, likely to have the capacity to evolve at different rates. Leaves are subject to selection related to physical factors that may not have changed substantially in quality, although variation in these factors would occur obviously through time (relative humidity and light conditions, maximum use of available light, etc.). Flowers, however, are subject to selection involving coevolution with pollinators and selective forces are apt to change considerably through time as pollinators continue to evolve. It would not be surprising, then, to have a plant with leaves at a relatively advanced level of adaptation, while flowers were considerably more primitive, since the flower could only reflect adaptations to pollinators available at a given time. Thus, it would be risky to consider dispersed angiosperm leaves reflective of the general state of advancement of the species of family until more is known about relative rates of evolution among various plant organs.

The above problem exists in assessing the history of any taxon from the fossil record. When considering the Amentiferae specifically, there are additional problems related to the nature of amentiferous pollen and to the state of understanding the fossil record of leaves of the Amentiferae. Pollen of the Amentiferae includes one type that may be confused with the pollen of other families. The tricolporate pollen characteristic of the Fagaceae could be confused with the tricolporate pollen of other dicotyledonous families in the fossil record, especially since many reports of fagoid pollen are included in floras and the individual pollen grains have not been given the attention necessary to instill confidence in the suggested affinities. Another palynological problem in the Amentiferae is the difficulty in distinguishing among the pollen of certain families (e.g., Betulaceae and Myricaceae). This is difficult with modern material and could easily be a source of error in paleopalynological analyses. The fossil record of leaves has not contributed as much as it could to our understanding the history of the Amentiferae because most reports of amentiferous Cretaceous and Paleogene leaves lack the morphological features or details of fine venation necessary to verify the suggested affinities (Wolfe 1973, 1977). Due to the level of resolution available presently in the fossil record of the Amentiferae, only the most reliable fossil evidence is used in the following summary and hypotheses. This includes floral, fruit, pollen, and, in the Tertiary, well-documented leaf evidence. The net result of this conservative approach might be, if anything, to understate the degree of advancement of the Amentiferae at a given time.

STATUS OF THE AMENTIFERAE BY THE MIDDLE EOCENE

Juglandaceae

The fossil record of the Juglandaceae is excellent and the family is well represented by inflorescences and fruit remains, as well as by dispersed pollen (Wolfe 1973, Crepet, Dilcher, and Potter 1975; Dilcher, Potter, and Crepet 1976). Although many fossil leaves have been ascribed to the Juglandaceae, careful fine venation analyses or other morphological studies necessary to justify the suggested affinities are lacking in the majority of these reports.

The megafossil record of the *Alfaroa-Oreomunnea-Engelhardia* complex of the Juglandaceae is particularly well-known from the Middle Eocene and two types of staminate inflorescence have been reported from the Claiborne Formation in the southeastern United States. *Eokachyra* (Crepet, Dilcher, and Potter 1975), has helically arranged florets, 3.3 × 2.7 mm, with three floral envelope parts and up to 15 stamins. Florets are subtended by shallowly three-lobed bracts (Figs. 4.1–4.3). Well-preserved cuticle reveals large peltate trichomes and anomocytic stomata. Pollen is preserved within the anthers and is triporate, 19.6 μm in diameter, and similar to the dispersed *Momipites coryloides* complex (Fig. 4.4; Nichols 1973). These catkins are very similar to the staminate catkins of genera within the modern *Alfaroa-Oreomunnea-Engelhardia* complex, but they are different enough from the catkins of individual modern genera to be considered representative of an extinct genus (Crepet, Dilcher, and Potter 1975).

The other type of staminate catkin known from the Claiborne Formation, *Eoengelhardia puryearensis*, is distinct from *Eokachyra* in having smaller florets (Fig. 4.5; Crepet, Daghlian, and Zavada 1980), more acutely lobed subtending bracts, and distinctly more triangular in amb and much smaller pollen (14.6 μm versus 19.6 μm, Fig. 4.6). These are very similar to the catkins of the modern Old World genus *Engelhardia* (Crepet, Daghlian, and Zavada 1980).

Although *Eokachyra* cannot be compared with a specific modern genus, both of these fossil inflorescences are comparable to those of the modern complex in their level of adaptation to wind dispersal of pollen. The pollen is of optimal shape and size for wind dispersal (Whitehead 1969); the floral envelopes are small and offer no impediment to potentially dispersing air currents; and pollen is produced in copious amounts (Faegri and Van der Pijl 1971).

Further insights into the status of this complex are available from the Middle Eocene record of winged fruits. These have diagnostic value within the *Alfaroa-Oreomunnea-Engelhardia* complex (Stone 1973), are well known from a variety of Middle Eocene localities of Europe and North

America (Reid and Chandler 1926; MacGinitie 1969; Akhmetyev and Bratzeva 1973), and are especially well-known from the Claiborne Formation (Dilcher, Potter, and Crepet 1976). Winged fruits related to modern New and Old World genera and representing certain extinct genera (Reid and Chandler 1926; Dilcher, Potter, and Crepet 1976) have revealed that the complex was more diverse during the Middle Eocene and that certain of the genera were more widespread then than at present (Dilcher, Potter, and Crepet 1976).

The record of the *Alfaroa-Oreomunnea-Engelhardia* complex suggests that the entire family may have diversified during the Upper Cretaceous. Reports of well-preserved inflorescences of *Platycarya strobilacea* and fruits of *Pterocarya* from the Paleocene Golden Valley Formation (Hickey 1977) support this possibility. Further support is provided by the dispersed pollen flora. Pollen of *Carya* and *Juglans* has been reported from the Paleocene (Wolfe 1973); pollen of the *Alfaroa-Oreomunnea-Engelhardia* complex radiated early (Nichols 1973; Wolfe 1976), and is widely reported from the Paleocene (Wolfe 1973; Hickey 1977). It seems reasonable to assume, then, that the Juglandaceae diversified during the Upper Cretaceous and reached their peak during the Middle Eocene after an uppermost Cretaceous-Paleogene radiation.

Fagaceae

Recent discovery of castaneoid catkins in conjunction with the record of fagoid fruits indicates the level of advancement of the family by the Middle Eocene. These catkins from the Claiborne Formation are up to 9 cm in length and have staminate florets with connate, lobed, floral envelopes (Figs. 4.7, 4.8). Florets have at least nine stamens (each less than 0.25 mm in length) and are borne in helically arranged, bract subtended, dichasia of three florets each (Figs. 4.9, 4.10). Pollen is a good diagnostic feature to subfamily in the Fagaceae (Crepet and Daghlian 1980). The pollen, which is smooth, small, and tricolporate (Fig. 4.11), in conjunction with the floral features and the relatively robust inflorescence axes, indicate castaneoid affinities. The robust inflorescence axis is similar to those of modern castaneoids, which suggests a similar upright habit and concomitantly the possibility of insect pollination.

Since insect pollination is considered a derived condition in the Fagaceae, the occurrence of inflorescences during the Middle Eocene as well adapted for insect pollination as those of modern insect-pollinated Fagaceae suggests a considerable history of the family prior to the Middle Eocene. Fruits of *Quercus* are known from the Clarno Formation (Manchester, personal communication) and fruits of *Fagus* have been reported from the Eocene (Chandler 1964). It is not surprising that fruits of *Quercus* should be

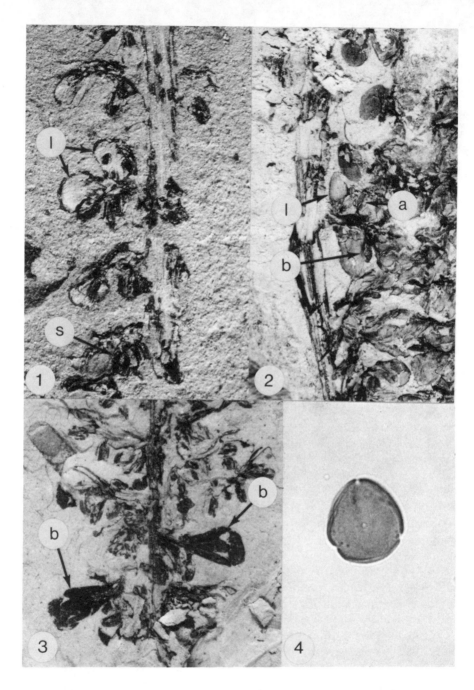

FIGURES 4.1–4.4. Staminate inflorescences (*Eokachyra*). **4.1.** Part of an inflorescence showing several florets. Note the stamens (s) and the floral envelope lobes (l). × 7. **4.2.** Another inflorescence of *Eokachyra* having a floret with most of the floral envelope parts in one plane. Note the floral envelope lobes (l), anthers (a), and part of a subtending bract (b). × 4. **4.3.** Several florets illustrating the three-lobed subtending bracts (b). × 4. **4.4.** Pollen isolated from an anther. × 1000.

found in the Middle Eocene, since staminate catkins similar to those of modern *Quercus* are found in the Oligocene of Texas (Figs. 4.12, 4.13). Although the diversification of the three subfamilies of the Fagaceae by the Middle Eocene is suggestive of an earlier radiation, the fossil record of the family prior to the Middle Eocene has not been useful. Although fagoid leaves are definitely known from the Paleocene (Wolfe 1973), the Paleogene pollen and leaf records have not been investigated carefully enough to contribute to our understanding of evolution within the family prior to the Middle Eocene.

Ulmaceae

The Neogene fossil record of the Ulmaceae is excellent (Weyland et al. 1958; Wolfe 1973, 1977) and includes some of the most spectacularly preserved fossil leaves (Niklas and Giannasi 1977), but the Paleogene record of the family is more problematical. Ulmoid leaves are known from the Paleocene of Alaska (Wolfe 1966), the Rocky Mountains (Brown 1962), and the Golden Valley Formation of North Dakota (Hickey 1977), but these leaves cannot always be reliably assigned to modern genera (Wolfe 1977). The earliest report of the Ulmaceae is from the Turonian of the U.S.S.R. (Grudzinskaya 1967). Pollen has been reported as early as the Maestrichtian in the Rocky Mountain region (Wolfe 1973) and is also known from the Golden Valley Formation (Hickey 1977). By the Middle Eocene, the pollen record of the Ulmaceae indicates that the family was diverse (work in progress, Claiborne Formation), but the megafossil record of the family is still not well-known. The Ulmaceae, then, appear to have evolved by the Upper Cretaceous and to have radiated during the Paleogene, but at present, few details of their radiation are available from the fossil record.

Floral megafossil evidence of the Ulmaceae from the Middle Eocene Claiborne Formation provides welcome insight into the history of the family. These flowers occur in pairs, are up to 9 mm in diameter (Fig. 4.14), have four tepaloid perianth parts (Fig. 4.15), and are densely pubescent (Fig. 4.16). Trichomes are elongate hairs. Flowers are unisexual and have up to ten stamens with elongate anthers (Fig. 4.17). Pollen is triporate with circular to longulate pores, 40 μm in diameter, is circular in polar view, and has a finely scabrate, minutely perforate exine (Fig. 4.18). In section, the exine is quite interesting and has granules interspersed among short stout columellae with which they appear to anastomose (Fig. 4.19). Exine structure appears to be intermediate between columellate and granular.

To determine the possible affinities of these flowers, the combined characteristics of floral and pollen morphology were compared with those of the flowers of extant families having triporate pollen. Only the subfamily Celtidoideae (Ulmaceae) shared all the features with the fossil flowers and they are considered to be allied most closely with the modern Ulmaceae

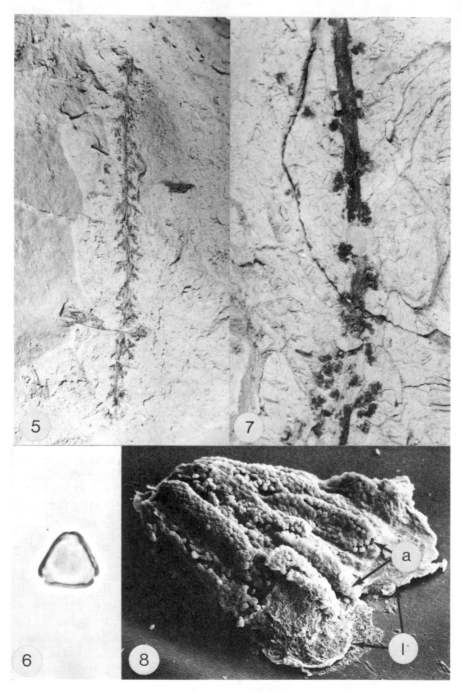

FIGURES 4.5–4.8. **4.5.** *Eoengelhardia*—staminate inflorescence. × 1.5. **4.6.** *Eoengelhardia*-pollen grain isolated from an anther. × 1000. **4.7.** *Castaneoidea*-overall view illustrating the robust catkin axis and the small florets. × 3. **4.8.** SEM of a floret fragment of *Castaneoidea*. Note perigon lobes (l) and anthers (a). × 136.

(Zavada and Crepet, in press). These flowers are of further interest because, although they are clearly ulmoid, they differ from the flowers of the modern Ulmaceae in several ways: the flowers are much larger than those of the modern Ulmaceae (9 versus 2 mm maximum diameter), the pollen is larger than pollen of the modern subfamily (40 versus 23 μm), and exine structure is intermediate between columellar and granular. These features are fascinating because they are the type one might expect in a flower that represented an intermediate in a reduction sequence going from species with large insect-pollinated types of flowers having typically larger columellate pollen to species with small, reduced flowers having smaller pollen with granular exine structure. These fossil flowers may represent the first fossil evidence supporting the widely held notion that the Amentiferae are derived from insect-pollinated ancestors through reduction of flowers and pollen. Because the prior fossil record of the Ulmaceae suggests that the family originated during the Upper Cretaceous, these flowers should not be interpreted as being reflective of the level of adaptation of the subfamily during the Middle Eocene. Instead, it seems probable that these flowers represent a genus that originated much earlier (when they truly represented the level of adaptation of the subfamily) and persisted until the Middle Eocene. Full appreciation of the significance of these flowers will, however, require a better understanding of the Upper Cretaceous and Paleogene records of the subfamily than is now available.

In addition to these flowers and inflorescences, there have been reports of the inflorescences of two additional families of the Amentiferae from Paleogene sediments. Although these inflorescences have not been subjects of individual morphological investigations (both appear in floras), their state of preservation and gross morphology seem supportive of the suggested affinities. One of these families is the Casuarinaceae, known from staminate and pistillate inflorescences from the Middle Eocene of Australia (Lange 1970); the other family, the Betulaceae, is known from inflorescences of *Betula* and *Corylus* in the Paleocene Golden Valley Formation (Hickey 1977). Both of these reports are consistent with the dispersed pollen floras of their respective times. Two other reports of amentiferous families are based on apparently reliable evidence, although the inflorescences of neither have been reported from the fossil record. These are the Moraceae from Eocene fruit remains (Chandler 1964) and the Rhoipteleaceae from Maestrichtian pollen (Wolfe 1973).

The fossil record of inflorescences provides corroboration for inference from other fossil evidence that a substantial number of families of the Amentiferae are as well adapted for wind dispersal of pollen by the Middle Eocene as their modern counterparts, and perhaps even as early as the Paleocene in certain instances (Fig. 4.20). Certainly, such early adaptation invites scrutiny of the pre-Paleogene fossil record of the Amentiferae and stimulates speculation as to the origin of wind pollination in the aments. The

FIGURES 4.9–4.12. **4.9–4.11.** *Castaneoidea.* **4.9.** Bract (b) subtended dichasium. × 19. **4.10.** View of a single dichasium showing three florets (f), two lateral and one central one that has been radially compressed on the axis. × 17. **4.11.** SEM of in situ pollen. × 4100. **4.12.** *Quercus*-like staminate catkin from the Oligocene of Texas. × 6.

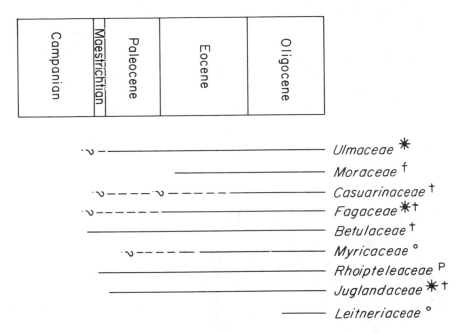

FIGURE 4.20. Geologic ranges of extant families of the Amentiferae (modified from Wolfe 1973).

Note: *—flowers or inflorescences known from Middle Eocene Claiborne Formation.
†—flowers or inflorescences known from other Paleogene or Neogene Localities.
°—Paleogene Fossil record questionable.
P—based on pollen only.
f—earliest record based on fruits.

earlier megafossil record of the Amentiferae, however, is spotty and relatively unreliable (Wolfe 1973, 1977).

The record of dispersed pollen has more potential in considering the history of the Amentiferae. The pollen record of the Upper Cretaceous suggests that certain modern families of the Amentiferae had evolved by that time, but since corroborative evidence from other fossil plant organs is lacking and most of the reports of pollen are not based on definitive studies, it seems best to consider such reports as possibly problematical. The report of one modern family identified on the basis of pollen from the Upper Cretaceous, the Rhoipteleaceae, seems convincing because of the rather distinctive features of this grain (Wolfe 1973). Wolfe suggests that pollen of the Rhoipteleaceae represents a link between the Juglandaceae, already diverse by the Paleocene, and the taxa represented by the *Normapolles* complex of dispersed pollen that first appears in the Cenomanian (although this complex doubtlessly contains antecedents of amentiferous families other than the Juglandaceae). When the *Normapolles* complex first appears in the

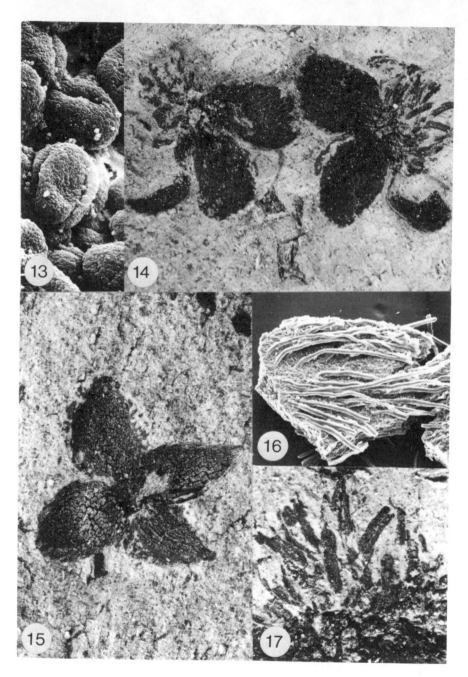

FIGURES 4.13–4.17. 4.13. SEM of pollen isolated from the anthers of the specimen illustrated in Figure 4.12. × 1200. **4.14–4.17.** Celtoid flowers from the Middle Eocene. **4.14.** Overall view of a pair of laterally compressed flowers. × 6. **4.15.** One radially compressed flower illustrating the four tepaloid perianth parts. × 7.5. **4.16.** Fragment of pedicel from the specimen illustrated in Figure **4.14.** Note the elongate trichomes. × 40. **4.17.** Closeup of a celtoid flower illustrating the elongate anthers. × 14.

Cenomanian, it is quintessentially adapted for dispersal by wind (Tschudy 1975) in size, shape, and in exine ornamentation (Whitehead 1969).*

At about the same time, an unrelated complex of triporates, the *Triorites* complex, appears in equatorial Africa (Jardiné and Magloire 1965; Doyle et al. 1977). These grains also confrom well to the features considered optimal in wind-dispersed pollen. This implies a history of selection favoring improved wind dispersability prior to the appearance of the *Normapolles-Triorites* complex in the Cenomanian; especially since a well-documented evolutionary trend (Doyle 1969) that leads from Albian tricolpates, which are themselves sometimes small, smooth, and possibly adapted to wind dispersal (Doyle 1977), to the *Normapolles* complex. The modifications of the tricolpate condition in this sequence suggest increasing adaptation to and selection for wind dispersal. Doyle (1977) has suggested that the appearance of *Normapolles* triporates in the fossil record is the first evidence of wind pollination. I suggest that, since the appearance of triporates is obviously the culmination of a trend involving continuing selective pressure, the inception of this trend marks the appearance of, at least, facultative wind pollination. If one rigidly assumes that wind pollination requires adaptation of pollen to the advanced level noted in the *Normapolles* complex, then from pollen alone, one would be forced to the misleading conclusion that the temperate Fagaceae are insect pollinated today.

One could go even further in suggesting early adaptations to wind dispersal of pollen. Some of the earliest tricolpates are small and smooth enough to suggest wind dispersal (Brenner 1976), and it is conceivable that some of these could represent the end of a line selected for wind dispersal. We can conceive of three possible "waves" of radiation in response to early selection for wind pollination: (1) culminating with the appearance of small, smooth, tricolpate pollen; (2) culminating with the appearance of the *Normapolles* and *Triorites* complexes; and (3) beginning with the radiation of the triporate complexes.

CONDITIONS OF CLIMATE IN THE AREA OF SELECTION FOR WIND DISPERSAL OF POLLEN

During this early selection for wind dispersal of pollen, the climate was warm (Lowenstam 1964) and much more equable than today (Axelrod and Bailey 1968) so latitudes even as high as the Atlantic Coastal Plain (about 30°N latitude at that time) might have been comparable to modern subtropics or tropics in temperature. Even the upper Albian-Cenomanian climatic

---

*Known Cenomanian reproductive organs are consistent with the possibility that wind pollination had evolved by the Cenomanian (Dilcher, 1979). Dilcher, however, interprets such remains as reflecting an original, rather than derived condition.

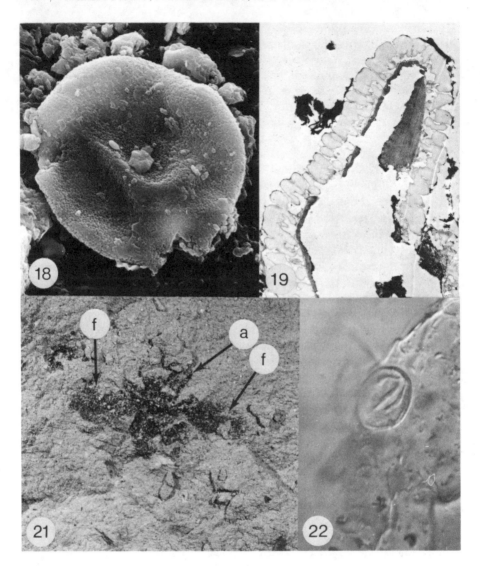

FIGURES 4.18, 4.19, 4.21, 4.22. Pollen isolated from the anthers of the flowers illustrated in Figure **4.14. 4.18.** SEM view of pollen illustrating the fine scabrate exine and three pores. Note that one of the pores has ruptured due to compression. × 1300. **4.19.** TEM of pollen illustrated in Figure **4.18** showing exine stratification. ×7100. **4.21–4.22.** Flower from the Cenomanian of the Atlantic Coastal Plain. **4.21.** Overall view. Note floral envelope (f), anthers (a) × 3. **4.22.** Pollen isolated from an anther by transfer technique. × 1000.

decay took place at levels well above today's temperatures (Lowenstam 1964).

Whitehead (1969) suggests that adaptation for wind pollination took place while the radiating angiosperms encountered seasonally or otherwise dry tropical environments and that the origin of wind pollination followed the origin of the deciduous habit in such regions (Axelrod 1966, 1967). Considering the paleopalynological record and paleophytogeography, this possibility seems quite certain. Trends favoring better dispersal of pollen by wind culminate with extremely well-adapted grains in low latitude areas during a period with high temperature and equability. The deciduous habit probably evolved in these same areas or wind dispersal of pollen would not have been physically feasible. In warm equable environments, there is only one possible selective agent to precipitate the deciduous condition: seasonal dry periods or other aridity.

The transition to triporates of the *Triorites* complex in equatorial Africa is consistent with the above analysis. The climate was certainly quite arid and undoubtedly tropical in nature (Vakhrameyev 1952, 1978; Brenner 1976; Doyle et al. 1977). Interpretation of the evidence for the Atlantic Coastal Plain where transition to the *Normapolles* complex took place is more complicated. It has been suggested that this region (and all of southern Laurasia) had a rather uniformly moist environment (Brenner 1976; Hickey and Doyle 1977). This is based primarily on interpretations of the dispersed pollen and spore flora (Brenner 1976) and the presence in that flora of a diverse pteridophyte assemblage (Vakhrameyev 1978). However, it is difficult to rationalize the trend toward wind-dispersed pollen in a uniformly moist environment. Selection for wind dispersal probably followed the origin of the deciduous habit and as noted previously, the warm climate precludes the possibility that the deciduous habit evolved in response to winter-like conditions. This indicates that seasonal aridity was the precipitating factor in selection for the deciduous condition.

The suggestion that southern Laurasia was uniformly moist has also caused difficulties in accounting for the dispersed leaf flora. Hickey and Doyle (1977) note that compound *Sapindopsis*-type leaves (occuring as early as the mid-Albian in the sediments of the Atlantic Coastal Plain), are highly suggestive of the deciduous habit, especially in accordance with interpretations of functional morphology by Givnesh (1978). Hickey and Doyle (1977) then rationalize the occurrence of the deciduous habit in a uniformly moist environment by suggesting that this character is an artifact in species that migrated into the Atlantic Coastal Plain from drier tropical regions where the deciduous habit would be selectively advantageous. This does not seem consistent with evolutionary theory. If a taxon with the deciduous habit migrated to the Atlantic Coastal Plain region entering a uniformly moist environment, it seems likely that the deciduous habit would be strongly selected against. Yet these leaves persist in the post-Albian Cretaceous record. It is more logical to suppose that there was at least regional aridity in that area. This possibility is consistent with the trends observed in pollen evolution and with the occurrence of deciduous leaves. It is also consistent with the palynological assemblages. High frequencies of pteridophyte pollen in no way preclude seasonally dry climates. Even though pteridophyte spores have been suggested as indicators of uniformly moist climate, these plants are common in seasonally dry tropical forests. Barro Colorado Island (Panama), for example, has a severe dry season and deciduous dicot forest vegetation, yet a wide variety of ferns abound, including Schizeaceae, Gleicheniaceae, and tree ferns (Croat 1978). Another factor complicating the observation of a relatively short dry season would be local variation in habitat (i.e., some plants will always be in moister areas) and riparian species that are less affected by dry seasons and that are well represented in the pollen and spore flora. It is obvious that should a modern seasonally dry tropical forest be characterized by its pollen rain, it would be considered a lush uniformly moist environment, and this would be accurate for most of the season. It seems as though it would be extremely difficult to identify a seasonally dry tropical forest from palynological data. The logical necessity for the existence of a dry season, at least regionally, in southern Laurasia, in combination with the most internally consistent interpretation of the paleobotanical and palynological data from the Atlantic Coastal Plain, mandates a hypothesis including some periodicity in moisture regime during the Lower-mid-Cretaceous.

Discussion

Hypotheses proposed above are in conformity with contemporary views on the origin of wind pollination in the dicotyledonous angiosperms (Whitehead 1969). In view of this, it is interesting to consider the pollination spectrum in modern areas that approximate the environmental conditions hypothesized for the regions where wind pollination originated during the Lower-mid-Cretaceous—the seasonally dry tropics. These are presently environments physically ideal for the dispersal of pollen by wind (Baker 1959; Whitehead 1969; Daubenmire 1972; Bawa 1974; Frankie, Baker, and Opler 1974a). With the exception of the riparian species, most of the trees are deciduous (Frankie, Baker, and Opler 1974a), removing the impediment of leaves from successful wind pollination (Whitehead 1969), and there are strong seasonal winds that would be more than adequate for effective dispersal of pollen. Even certain fruits are wind dispersed in these communities (Janzen 1967). It is, therefore, remarkable that there is no wind pollination among the dicot trees in the seasonally dry deciduous tropical forest today. This has been noted frequently, and in view of the suitability of such environments for wind pollination, has been the subject of extensive commentary (Baker 1959; Whitehead 1969; Daubenmire 1972). One factor mitigating against wind-dispersed pollen in seasonally dry tropical forests, and the possibility that wind pollination might have evolved under such conditions, is the spatial arrangement of species. It is certainly the conventional wisdom of botany that there is great heterogeneity in the tropics with little of the clumping of species that makes wind pollination feasible. However, Hubbell (1979) has demonstrated recently that there is actually more clumping of species in tropical forests than has been previously supposed. This is plausible, since even most insect-pollinated species require some clumping for reasonable efficiency. If there was little clumping in the tropics today, it would have little bearing on the situation during the mid-Cretaceous, because the heterogeneity found presently in the tropics is generally considered to be a relatively recent phenomenon (Flenley 1979).

Wind pollination does not occur in seasonally dry tropical forests, even though the physical requirements for wind pollination are available. The dominance of insect pollination in such ecosystems, as suggested by modern observations, may result from the superior energetics of insect pollination (Pohl 1937; Cruden 1977) and, in the long run, from the ability of insect pollination to promote speciation (Grant 1950; Baker 1959).

We are, therefore, presented with a dilemma: it has been hypothesized, that wind pollination evolved in the dicots in environments similar to modern seasonally dry tropics, yet wind pollination does not exist in such environments today where insect pollination is apparently superior. How, then, can we reasonably entertain the possibility that wind pollination evolved from insect-pollinated ancestors in such environments?

The resolution of this apparent difficulty lies in a careful examination of dicot pollination biology in modern seasonally dry tropics. The phenology of such plants has been the object of extensive careful studies (Janzen 1967; Frankie, Baker, and Opler 1974a, 1974b). These studies have revealed that flowering occurs during the dry season. This has two major advantages: (a) flowers stand out better and are observed more easily by pollinators when there are no obscuring leaves; and (b) pollination during the dry season allows fruit development and dispersal to occur before the wet season and the dispersed fruits can then take advantage of the wet season to germinate (Janzen 1967). One of the most interesting aspects of dry season-flowering is the nonoverlap of flowering times in large numbers of species. Frankie, Baker, and Opler (1974a, 1974b; Frankie 1975) noted that the flowering periods of 59 species (including congeneric species) were staggered so as to minimize overlap. They interpret this as a result of strong selection to reduce competition for pollinators. Nonrandomness in the array is suggested by the nonoverlap of congeneric species and the observation that weak nectar producers tend to flower at suboptimal times near the end of the dry season, presumably to minimize competition with producers of copious nectar that flower nearer to the favorable peak of the season. The conclusions of

Frankie, Baker, and Opler (1974a, 1974b) seem quite reasonable in view of the steadily mounting evidence pointing to competition for pollinators as a selective agent (Hocking 1968; Levin and Anderson 1970; Mosquin 1971; Heinrich and Raven 1972, Straw 1972; Waser 1978) and to the power of pollinators as selective agents as demonstrated by the importance of pollinator faunal quality in determining the array of floral structures and floral colors in a given community (Ostler and Harper 1978).

We may now imagine a situation where wind pollination could be selectively advantageous in seasonally dry tropics. Suppose competition for pollinators, already potent enough to affect the timing of flowering among species, began increasing over the present levels. Eventually, even the strictest staggering of flowering times would not alleviate the competitive pressure and could produce a situation where facultative wind pollination would be selectively advantageous. This seems reasonable, but can this scenario be applied to the Lower-mid-Cretaceous? Such an extrapolation requires evidence that competition for pollinators in environments equivalent to modern seasonally dry tropics was more severe at that time.

The nature of insect pollination in modern seasonally dry tropics provides the means of applying this scenario to the Cretaceous. Bees are the overwhelmingly important insect pollinators in modern tropical deciduous forests (Janzen 1967; Frankie 1975; Frankie, Opler, and Bawa 1976). Furthermore, bees are more effective pollinators during dry climatic conditions than other anthophilous insects (Frankie, Opler, and Bawa 1976) and are generally better adapted to dry conditions than other anthophilous insects (Janzen and Schoener 1968; Michener 1974). Bees, however, are missing from the pollinator fauna of the Lower-mid-Cretaceous. The first bees appear in the Lower Oligocene (Burnham 1978) and even allowing for the imperfections of the fossil record, it would be unreasonable to assume that bees existed as early as the Lower Cretaceous without leaving a trace of their existence. Thus, the most crucial pollinators in seasonally dry tropical forests today were missing from the pollinator fauna during the period when wind pollination evolved. Insect pollination during the Lower-mid-Cretaceous would have to have been effected by less constant, less intelligent, and less perceptive members of other anthophilous orders—probably Coleoptera and Diptera—and their modern counterparts are known not to function particularly well as pollinators under dry conditions.

Presumably, the net result of this qualitative (and probably quantitative) difference in the pollinator fauna would have been intense competition for insect pollinators. This would, presumably, have placed a selective value on facultative wind pollination and provided the impetus for the origin of wind pollination in the dicots. We then might expect to find flowers in the Cretaceous fossil record with combinations of features suggestive of possible facultative wind dispersal of pollen. One type might be flowers with relatively well-developed floral envelopes and small smooth pollen that could be dispersed easily by wind. In fact, such flowers have been recovered recently from the Cenomanian of New Jersey. These flowers have floral envelopes 19 mm in diameter, are radially symmetrical, and have small, smooth, tricopate pollen averaging 14 μm in diameter (Figs. 4.21, 4.22). Another regular type of flower with a rather large perianth and small pollen has been reported from the Cenomanian of Nebraska (Basinger and Dilcher, 1980). In this instance, the pollen is so small that it seems unlikely that it could be effectively trapped by a stigma and is likely to follow the air stream around an object instead of colliding with it (Whitehead 1969), but it illustrates that there were general combinations of features that are unusual by today's standards. These sets of characters may have been related to selection for facultative wind dispersal in several taxa.

Refined adaptations to wind pollination can be envisioned as having attended the incursion of the angiosperms into progressively more hostile environments. The radiation of the *Normapolles* complex, and perhaps even the latter phases of the evolutionary line leading to this complex, could be considered examples of this phenomenon. This possibility, as well as other aspects of the above hypotheses, will be evaluated more easily when more reliable information is available on Cretaceous climates. In any case, the concept of competition for insect pollinators as a selective agent important in precipitating wind dispersal of pollen should remain viable. If Cretaceous climates prove to have been more severe than has been hypothesized previously, the only effect would be to intensify competition for pollinators and to place wind pollination at an even greater selective advantage.

It is interesting to consider why there is no wind pollination in the seasonally dry tropics today. I have suggested that there was severe competition for pollinators during the Lower-mid-Cretaceous in similar environments. Obviously, this situation would be changed by the evolution of bees (as well as changes in other anthophilous insect groups) and their increasing availability in the tropics (the bees are not considered a basically tropical group, Michener 1979).

We are beginning with relatively low diversity of pollinators at a given time during the Cretaceous when wind pollination would have been at a selective optimum (Fig. 4.23). As the number of available pollinators increased, wind pollination could be conceived of as having been less and less selectively advantageous and could be imagined as having decreased in frequency with increasing insect diversity (Fig. 4.23). However, wind would have maintained its selective advantage in areas where pollinator availability remained relatively low. By the time wind pollination had died out in the tropics, wind-pollinated angiosperms would have migrated into more temperate areas where the cool or winter season would have kept anthophilous insect diversity relatively lower and where the preadapted deciduous

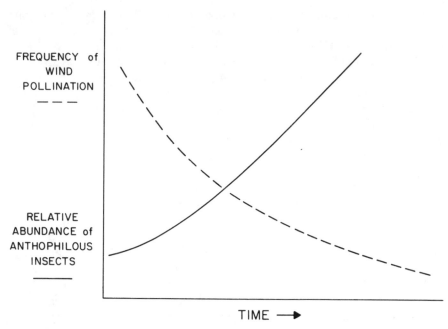

FIGURE 4.23. Theoretical relationship through time between frequency of wind pollination and anthophilous insect abundance in seasonally dry tropics.

habit would have an obvious selective value. This, of course, suggests that the present distribution of the Amentiferae may be related to the availability of insect pollinators. The evidence is consistent with the possibility that the modern aments are dominant where competition for insect pollinators places them at a competitive advantage. The frequency of anthophilous insects declines with increasing latitude as the frequency of the aments increases with latitude (Fig. 4.24).

While these data are suggestive, they are in no way proof of the hypothesis. More data are needed, especially related to the frequency changes in anthophilous insects with latitude. This type of information is very difficult to obtain because it requires extensive effort (Janzen 1973a, 1973b; Janzen and Pond 1975) and ecological and seasonal variables make it somewhat difficult to interpret (although these factors have been taken into consideration [Janzen 1973a, 1973b; Janzen and Pond 1975]).

There are definite exceptions to the generally temperate distribution of the Amentiferae. First, there are subtropical aments that are wind polli- nated, e.g., the *Alfaroa-Oreomunnea-Engelhardia* complex of the Juglanda- ceae. Does this suggest that wind-pollinated aments can be successful in areas where there is great pollinator availability? Not at all, most of these

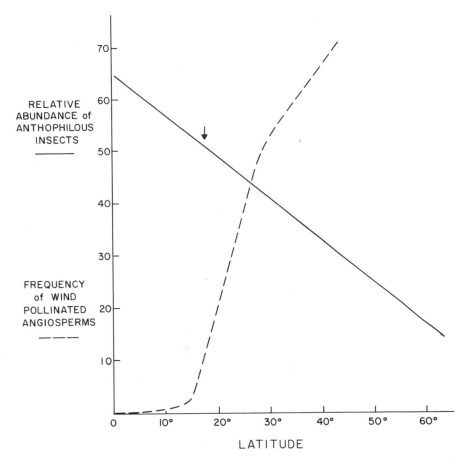

FIGURE 4.24. Relationship between the availability of insect pollinators and the frequency of wind pollination with changing latitude. Arrow indicates the point at which wind pollination becomes selectively advantageous in certain taxa.

species live in montane habitats and the abundance of pollinators decreases with altitude, as well as decreasing with latitude (Janetschek et al. 1976). Second, the Fagaceae are quite diverse in tropical-subtropical southeast Asia. While certain species inhabit montane areas (Soepadmo 1972), others definitely occur in lowland habitats where pollinator diversity is high. However, these are the species of the Fagaceae that have reverted to insect pollination and, having maintained themselves in or having successfully reinvaded areas with abundant pollinators, has required reversion to insect pollination. This is quite consistent with the hypotheses presented previously.

CONCLUSIONS

Many families of the Amentiferae were well adapted by the Middle Eocene and perhaps even as early as the Paleocene for the dispersal of pollen by wind.

Paleopalynological, paleobotanical, paleoclimatological, and paleophytogeographical data suggest that much of the selection for wind pollination in the dicots took place in environments similar to the modern seasonally dry tropics.

Fossil evidence coupled with modern observations suggests that the existence of the appropriate physical conditions alone was not enough to precipitate the origin of wind pollination. Competition for pollinators and the appearance of proper physical conditions are suggested as having been equally important in the evolution of wind pollination.

ACKNOWLEDGMENTS

Research for this publication was supported by NSF Grant DEB-7811120. The author gratefully acknowledges the thoughtful criticisms of Dr. Peter H. Raven, Director of the Missouri Botanical Garden; Dr. David L. Dilcher, Department of Biology, Indiana University; Dr. T. Delevoryas, Department of Botany, The University of Texas; and the generous assistance of Dr. C. W. Rettenmeyer, Biological Sciences, The University of Connecticut.

REFERENCES

Abbe, E. C. 1974. Flowers and inflorescences of the "Amentiferae." *Bot. Rev.* 40:159–261.

Airy Shaw, H. K. 1966. *A Dictionary of Flowering Plants and Ferns.* Cambridge: Cambridge University Press.

Akhmetyev, M. A., and G. M. Bratzeva. 1973. Fossil remains of the genus *Engelhardtia* from Cenozoic deposits of Sikhote-Alin and southern Primorye. *Rev. Palaeobot. and Palynol.* 16:123–132.

Axelrod, D. I. 1966. Origin of deciduous and evergreen habits in temperate forests. *Evolution* 20:1–15.

Axelrod, D. I. 1967. Drought, diastrophism, and quantum evolution. *Evol.* 21:201–209.

Axelrod, D. I., and H. P. Bailey. 1968. Cretaceous dinosaur extinction. *Evolution* 22:595–611.

Baker, H. G. 1959. Reproductive methods as factors in speciation in flowering plants. *Quant. Biol.* 24:177–191.

Basinger, J. F. and D. L. Dilcher. 1980. Bisexual flowers from the mid-Cretaceous of Kansas. Abstracts of the papers to be presented at the University of British Columbia, Vancouver. July 12–16.

Bawa, K. S. 1974. Breeding systems of tree species of a lowland tropical community. *Evolution* 28:85–92.

Brenner, G. J. 1976. Middle Cretaceous floral provinces and early migrations of angiosperms. In *Origin and Early Evolution of Angiosperms* (C. B. Beck, ed.). New York: Columbia University Press, pp. 23–47.

Brown, R. W. 1962. Paleocene flora of the Rocky Mountains and Great Plains. *Prof. Pap. U.S. Geol. Sur.* 375:1–119.

Burnham, L. 1978. Survey of social insects in the fossil record. *Psyche* 85:85–133.

Chandler, M. E. J. 1964. *The lower Tertiary floras of southern England, IV.* London: British Museum of Natural History, pp. 1–151.

Crepet, W. L. and C. P. Daghlian. 1980. Castenoid inflorescences from the middle Eocene of Tennessee and the diagnostic value of pollen (at the subfamily level) within the Fagaceae. *Amer. J. Bot.* 67:739–757.

Crepet, W. L., C. P. Daghlian, and M. Zavada. 1980. Investigations of angiosperms from the Eocene of North America: a new juglandaceous catkin. *Rev. Palaeobot. and Palynol.* 30:361–370.

Crepet, W. L., D. L. Dilcher, and F. W. Potter. 1975. Investigations of angiosperms from the Eocene of North America: a catkin with juglandaceous affinities. *Amer. J. Bot.* 62:813–823.

Croat, T. B. 1978. *Flora of Barro Colorado Island.* Stanford: Stanford University Press.

Cronquist, A. 1968. *The evolution and classification of flowering plants.* Boston: Houghton Mifflin.

Cruden, W. C. 1977. Pollen-ovule ratios: a conservative indicator of breeding systems in flowering plants. *Evolution* 31:32–42.

Daubenmire, R. 1972. Phenology and other characteristics of tropical semi-deciduous forest in northwestern Costa Rica. *J. Ecol.* 60:147–170.

Dilcher, D. L. 1979. Early angiosperm reproduction: an introductory report. *Review of Palaeobotany and Palynology.* 27:291–328.

Dilcher, D. L., F. W. Potter, and W. L. Crepet. 1976. Investigations of angiosperms from the Eocene of North America: juglandaceous winged fruits. *Amer. J. Bot.* 63:532–544.

Doyle, J. A. 1969. Cretaceous angiosperm pollen of the Atlantic Coastal Plain and its evolutionary significance. *J. Arnold Arbor.* 50:1–35.

———. 1977. Patterns of evolution in early angiosperms. In *Patterns of Evolution* (A. Hallem ed.). Amsterdam: Elsevier.

Doyle, J. A., P. Biens, A. Doerenkamp, and S. Jardiné. 1977. Angiosperm pollen from the pre-Albian Lower Cretaceous of Equatorial Africa. *Bull. Cent. Rech. Explor.-Prod. Elf-Aquitane* 1:451–473.

Doyle, J. A. and L. J. Hickey. 1976. "Pollen and leaves from the mid-Cretaceous Potomac Group and their bearing on early angiosperm evolution." In C. B. Beck (Editor), *Origin and Early evolution of the angiosperms.* New York: Columbia University Press, pp. 139–206.

Faegri, K., and L. Van der Pijl. 1971. *The Principles of Pollination Ecology.* Oxford: Pergamon.

Flenley, J. R. 1979. *The Equatorial Rain Forest. A Geological History.* Boston: Butterworth.

Frankie, G. W. 1975. Tropical forest phenology and pollinator plant co-evolution. In *Coevolution of Animals and Plants* (L. E. Gilbert and P. H. Raven eds.). Austin: University of Texas Press.

Frankie, G. W. H. G. Baker, and P. A. Opler. 1974a. Comparative phenological studies of trees in tropical wet and dry forests in the lowlands of Costa Rica. *J. Ecol.* 62:881–919.

———. 1974b. Tropical plant phenology: applications for studies in community ecology. *Phenology and Seasonality Modeling* (H. Lieth ed.). New York: Springer-Verlag.

Frankie, G. W., P. A. Opler, and K. S. Bawa. 1976. Foraging behavior of solitary bees: implications for outcrossing of a neotropical forest tree species. *J. Ecol.* 64:1049–1057.

Givnish, T. J. 1978. The adaptive significance of compound leaves, with particular reference to tropical trees. In *Tropical Trees as Living Systems.* (P. B. Tomlinson and M. H. Zimmermann eds.). Cambridge: Cambridge University Press.

Grant, V. 1950. The flower constancy of bees. *Bot. Rev.* 16:379–398.

Grudzinskaya, I. A. 1967. Ulmaceae and reasons for distinguishing Celtidoideae as a separate family, Celtidaceae. *Bot. Zhur.* 52(12):1723–1749.

Heinrich, B., and P. H. Raven. 1972. Energetics and pollination ecology. *Science* 176:597–602.

Hickey, L. J. 1977. Stratigraphy and paleobotany of the Golden Valley Formation (early Tertiary) of western North Dakota. *Geol. Soc. Amer. Memoir* 150:1–156.

Hickey, L. J., and J. A. Doyle. 1977. Early Cretaceous fossil evidence for angiosperm evolution. *Bot. Rev.* 43:3–104.

Hjelmquist, H. 1948. Studies on the floral morphology and phenology of the Amentiferae. *Bot. Not.* (Suppl.) 2(1):1–171.

Hocking, B. 1968. Insect-flower associations in the high Arctic with special reference to nectar. *Oikos* 19:359–388.

Hubbell, S. P. 1979. Spatial structure in the tropical dry forest. Paper read at the 30th annual AIBS meeting August 15, 1979. Oklahoma State University.

Janetschek, H., I. De Zoro, E. Meyer, H. Troger, and H. Schatz. 1976. Altitude and time related changes in arthropod faunation (Central High Alps: Obergurgl-area, Tyrol). *Proc. of 15th Int. Cong. Entomol.* 185–207.

Janzen, D. H. 1967. Synchronization of sexual reproduction of trees within the dry season in Central America. *Evolution* 21:620–637.

———. 1973a. Sweep samples of tropical foliage insects: description of study sites, with data on species abundances and size distributions. *Ecology* 54:659–686.

———. 1973b. Sweep samples of tropical foliage insects: effects of seasons, vegetation types, elevation, time of day, and insularity. *Ecology* 54:687–708.

Janzen, D. H., and C. M. Pond. 1975. A comparison, by sweep sampling, of the arthropod fauna of secondary vegetation in Michigan, England, and Costa Rica. *Trans. R. Ent. Soc.* (Lond.) 127:33–50.

Janzen, D. H., and T. W. Schoener. 1968. Differences in insect abundance and diversity between wetter and drier sites during a tropical dry season. *Ecology* 49:96–110.

Jardiné, S., and L. Magloire. 1965. Palynologie et stratigraphie du Crétacé des bassins du Sénégal et de Côte d'Ivoire. *Mem. Bur. Rech. Geol. Minières* 32:187–245.

Lange, R. T. 1970. 1. The Maslin Bay flora, South Australia. 2. The assemblage of fossils. *Neues Jahrb. Geol. Palaontol. Monatsh.* 8:486–490.

Levin, D. A., and W. W. Anderson. 1970. Competition for pollinators between simultaneously flowering species. *Amer. Nat.* 104:455–467.

Lowenstam, H. A. 1964. Palaeotemperatures of the Permian and Cretaceous periods. In *Problems in Palaeoclimatology* (A. E. M. Nairn ed.). London: Interscience, pp. 227–252.

MacGinitie, H. D. 1969. *The Eocene Green River Flora of Northwestern Colorado and Northeastern Utah.* Berkeley: University of California Press.

Michener, C. D. 1974. *The Social Behavior of the Bees.* Cambridge: Belknap Press.

Michener, C. D. 1979. Biogeography of the bees. *Annals Missouri. Bot Gard.* 66: 277–347.

Mosquin, T. 1971. Competition for pollinators as a stimulus for the evolution of flowering time. *Oikos* 22:398–402.

Nichols, D. J. 1973. North American and European species of *Momipites* ("Engelhardia") and related genera. *Geosci. and Man* 7:103–117.

Niklas, K. J., and D. E. Giannasi. 1977. Flavonoids and other chemical constituents of fossil Miocene Zelkova (Ulmaceae). *Science* 196:877–878.

Ostler, W. K., and K. T. Harper. 1978. Floral ecology in relation to plant species diversity in the Wasatch Mountains of Utah and Idaho. *Ecology* 59:848–861.

Pohl, F. 1937. Die Pollenerzeugung der Windblüter. Eine vergleichende Untersuchung mit Ausbliken auf den Bestäubungshaushalt tierblütiger Gewächse und die pollenanalytische Waldgeschichtsborschung. *Beih. Bot. Zbl.* 56:365–470.

Reid, E. M., and M. E. J. Chandler. 1926. Catalogue of Cainozoic plants in the Department of Geology. Vol. I, *The Bembridge Flora.* London: British Museum of Natural History.

Soepadmo, E. 1972. Fagaceae. In *Flora Malesiana.* Ser. 1, Vol. 7, Pt. 2:265–403.

Stebbins, G. L. 1965. The probable growth habit of the earliest flowering plants. *Ann. Mo. Bot. Garden.* 52:457–468.

———. 1974. *Flowering Plants: Evolution above the Species Level.* Cambridge: Harvard University Press.

Stone, D. E. 1973. Patterns in the evolution of amentiferous fruits. *Brittonia* 25:371–384.

Straw, M. 1972. A markov model for pollinator constancy and competition. *Amer. Nat.* 106:597–620.

Takhtajan, A. 1969. *Flowering Plants—Origin and Dispersal.* Edinburgh: Oliver and Boyd.

Thorne, R. F. 1973. The "Amentiferae" or Hamamelidae as an artificial group: a summary statement. *Brittonia* 25:395–405.

Tschudy, R. H. 1975. *Normapolles* pollen from the Mississippi Embayment. *Prof. Pap. U.S. Geol. Sur.* 865:1–42.

Vakhrameyev, V. A. 1952. Stratigrafiya: iskopayemaya flora melorykh otlozheniy Zapadnogo Kazakhstana. *Reg. Stratigraf. S.S.S.R.* 1:1–340.

———. 1978. The climates of the Northern Hemisphere in the Cretaceous in light of

Paleobotanical data. *Paleont. J.* 2:143–154.

Waser, N. M. 1978. Competition for pollinators (hummingbird) cause selective force sufficient to maintain divergent flowering times in nature. *Ecology* 59:934–944.

Weyland, W., H. D. Pflug, and H. Jahnichen. 1958. *Celtoidanthus pseudorobustus* n. gen., N. sp. eine Ulmaceen-Blute aus Braunkolhe der Niederlausitz. *Palaeontographica* 105B:67–74.

Whitehead, D. R. 1969. Wind pollination in the angiosperms: evolutionary and environmental considerations. *Evolution* 23:28–35.

Wolfe, J. A. 1966. Tertiary plants from the Cook inlet region, Alaska. *Prof. Pap. U.S. Geol. Sur.* 398B:1–32.

———. 1973. Fossil forms of Amentiferae. *Brittonia* 25:334–355.

———. 1976. Stratigraphic distribution of some pollen types from the Campanian and lower Maestrichtian rocks (Upper Cretaceous) of the Middle Atlantic States. *Prof. Pap. U.S. Geol. Sur.* 977:1–18.

———. 1977. Revisions of *Ulmus* and *Zelkova* in the middle and late Tertiary of western North America. *Prof. Pap. U.S. Geol. Sur.* 1026:1–11.

Zavada, M., and W. L. Crepet. 1981. Ulmoid flowers from the middle Eocene of Tennessee. *Amer. J. Bot.* In press.

5

PALEOECOLOGY, ORIGIN, DISTRIBUTION THROUGH TIME, AND EVOLUTION OF HEPATICAE AND ANTHOCEROTAE

Rudolf M. Schuster

INTRODUCTION

Darwin referred to the origin of the angiosperms as a "dark chapter" in botanical history; some light has been shed in recent decades to illuminate at least its perimeters. An even darker chapter has been that dealing with the origin of land floras. Here, again, we find some aspects becoming clarified, in large part because of the contributions of Harlan Banks and his students to the knowledge of Early Devonian vascular plants. Much conceptual progress was made only *after* Tracheophyte botanists rejected the once-attractive idea that an intermediate step in evolution between early vascular plants and the ancestral Chlorophyta was formed by the Hepaticae. In this chapter, I shall try to deeply bury this idea; without such interment, real progress in understanding evolution of early land plants is impossible. Yet, even today, the idea is not wholly abandoned by the authors of some textbooks. In the graphic words of Arnell (1922, p. 16), impossible ideas, like scientific names that have been long buried, "werden, wie oft man sie auch totzuschlagen versucht, wie der Vogel Phoenix wieder aufleben und neue Verwirrung anrichten."

Modern evidence suggests that plants once circumscribed as the class Hepaticae fall into two unrelated groups, the Hepaticae, s. str., and a group, Anthocerotales, that probably represent the remnant of an unsuccessful attempt at land invasion, by a group in which gametophytic stomata had evolved. The Hepaticae, by contrast, show probable evolution from ancestral types in which both generations were radially symmetric, but both

129

lacked leaf-like appendages. Evidence is presented that leads to a number of major conclusions: (1) Migration from water to land by the members of the Jungermanniidae, probably, was from an estuarine-riverine *amphibious* habitat. By contrast, the basically continental range of the Marchantiidae, and the aquatic nature (with pond and lake habitats prevalent) of certain taxa (*Naidita, Riella, Riccia,* and *Ricciocarpus*), the rarity in areas with running fresh waters suggest that this subclass *may* have had a drastically different evolutionary history: its ancestors may have undergone transmigration from areas of still waters. (2) Geographical and ecological characteristics of the two subclasses and their different life-styles suggest they represent an early (Devonian-Carboniferous, or earlier) dichotomy in evolution—with divergent ecological parameters becoming ever more pronounced. (3) An undue concentration of erect, isophyllous taxa with malleable and generalized morphology, lacking means of asexual propagation, and with unisexual gametophytes, occurs peripheral to the shores of the ancient Pacific, or Panthallassa. This may reflect mere persistence there of suitable sites (oceanic or hyperoceanic climates), but is so marked a feature as to suggest that these taxa are the remnants, or the rear guard, of the initial invaders—and that invasion of the land surface occurred from estuarine-riverine loci along margins of a Pangaea-like land mass. (4) Subsequent to early adaptation to land, the major division of Hepaticae into orders and families occurred, apparently, from Late Devonian to Late Permian time. (5) With evolution of *relatively* dense coniferous forests by Late Paleozoic to Early Mesozoic time, the initial diversification of modern, "successful" taxa of Jungermanniidae (Radulaceae, Porellaceae, Jubulaceae, Lejeuneaceae, also *Plagiochila* of the Plagiochilaceae) occurred; the early generalized ancestors of these families, all inhabiting vertical rock walls, invaded the bark of conifers. This seems to have occurred initially chiefly in Gondwanalandic areas. (6) With advent of angiosperms in the Early Cretaceous, a huge number of suitable niches, not existing previous to Cretaceous times, were formed. This stimulated the Cretaceous and Tertiary explosion of modern genera of these families. Since over 70 percent of all taxa belong to these five families (the other 30 percent belong to some 65 to 68 other families), it is obvious that the Hepaticae are not "on the downward slide to oblivion," and that we may dry our figurative tears: so long as angiosperms exist and flourish, so will the liverworts.

The history of the Hepaticae, thus, is complex: we find the same mix of stenotypic, primitive types and modern groups in an explosive state of evolution as in Angiosperms.

No detailed recent attempt at analyzing the origin of the Hepaticae (and Anthocerotae) exists; that in Schuster (1966a), although detailed, is now beginning to be dated. I have attempted to develop the following broad topics in this chapter: (1) When and where did the Hepaticae originate?

(2) What were early types like? (3) What were the initial steps in their evolution? (4) What were their intrinsic limitations, as defined by the inferior life-cycle "opted" for? (5) What role did the early vascular flora play in ameliorating the initial steps in hepatic evolution? (6) What was the effect of the Cretaceous angiosperm explosion on hepatic evolution and diversification? (7) What role did the spectacular—and now relatively clearly documented—reorientation of sialic plates, their migration—and the coincident change in climate—have on distribution and evolution of Hepaticae, especially of primitive types?

To shed light on some of these topics, it is first necessary to destroy some commonly accepted ideas that have almost acquired the status of dogma. Dismissed must be such pseudofacts as: (1) Hepaticae evolved from organisms with strong antithetic alternation—thalloid genera like *Riccia*, with its strikingly simple sporophytes, being regarded as close to ancestral types; (2) Hepaticae did not evolve from such thalloid types, but from erect isophyllous types similar to mosses such as *Mnium*; and (3) Hepaticae show gradual elaboration of a structurally more sophisticated sporophyte. Corollaries of assumptions 1 and 3 are that: (a) *Riccia* is "primitive," the leafy Jungermanniales derivative; (b) the massive, stomate-producing Anthocerotae are even more derivative, and from them vascular plants may have originated.[1]

In recent decades, hepaticologists have arranged themselves into two mutually exclusive groups: those who adopt assumption no. 1, and hence reject assumption no. 2, and those who adopt 2, but reject 1. Such neat and orderly ideas are basically atomistic and simplistic. Evolution is, ultimately, the adaptation of existing genotypes to the environmental stresses of the moment; it does not plan for the future. Furthermore, a single group, rarely, is exposed to merely one set of stresses. Most existing schemata that try to reconstruct past evolution suffer from one form of myopia or another. The commonest mistake is that of trying to inject a certain order and logic into a process that, by its very nature, is random and opportunistic. Furthermore, selection pressures that, *at a particular time*, may result in evolution of types like Bryophyta, that, *with the stresses of the moment*, are clearly *adaptive* in the evolutionist's lexicon, may lead them into what may, ultimately, prove to be evolutionary blind alleys. And, once in such a blind alley, no matter how the group may grope for an escape, it is ultimately trapped. The Hepaticae are such a trapped group, with all that this implies.

The great Palaeozoic drama is not simply that of plants invading land; the real drama concerns itself with the varied attempts to cope with the initially very difficult conditions—and how some groups succeeded, whereas others soon perished. Initial steps in evolution gave us *at least* three major units of vascular plants (the Psilophytes, Trimerophytes, and Zosterophyllophytes), but essentially simultaneously many other experiments in land plant evolution were tried and soon failed; several such experiments involved plants which opted for progressively longer epiphytism of sporophyte on gametophyte. We must conceptualize extant mosses, hepatics, and anthocerotes as the survivors of a series of such organisms. In many ways, haploid-dominant groups such as these are inferior to Psilophytes, Trimerophytes, and Zosterophyllophytes—all of which were extinct by the end of the Devonian. It seems likely, indeed, that these early vascular plants or their descendants played a crucial role in ameliorating the initially harsh environments that plants with a gametophyte-dependent sporophyte inhabited; in fact, the former probably created the environments frequented by the latter. In this chapter, this theme is examined in detail. Indeed, probably two major stadia occurred in diversification of the Hepaticae—a Devonian-Carboniferous phase, and a Cretaceous-Tertiary phase—which were conditioned, if not caused, by major explosions in numbers and types among vascular plants. It is, thus, futile and simplistic to pursue the entire topic of the evolution of the Hepaticae (and the only very loosely allied groups, Musci and Anthocerotae) in a vacuum. Not only can such ideas not be pursued in a vacuum, but before they are pursued, much of the evolutionary framework erected in the past must be torn down.

Some Fictions versus Some Facts

The topic of the evolution of the Hepaticae is bedeviled with a series of unique problems: (1) No consensus exists as to what is primitive in Hepaticae versus what is advanced. Thus, there is no complete agreement as to direction(s) of evolution. The mutable subject of evolutionary progressions and past adaptations, intimately linked with paleoecology, can be approached only cautiously and indirectly. Two examples document this problem: As recently as 1969, R. Grolle insisted that evolution in Hepaticae progressed from thalloid-aplanate to leafy types; H. Bold still uses an evolutionary progression (in his 1973 plant morphology text) starting with *Riccia*-like thalloid organisms and ending with leafy types. If such ideas are correct, everything developed here is arrant nonsense! It is obvious that there is no common ground between the classical theories which have evolution start with aplanate *Riccia*-like types and the theory defended here, that evolution began with erect, radial types. (2) There has been a recent revolution with regard to both the accretion and understanding of vascular plants fossils, but fossil data for the Hepaticae are only slightly less inadequate today than two or three decades ago. For two orders regarded as uniquely primitive, the Calobryales and Monocleales, no shred of fossil evidence exists. For the Anthocerotes, which should go into a separate division, Anthocerotophyta (Schuster 1977), there is also no fossil evidence, that is of any significance, aside from the recent demonstration of *Anthoceros*-like spores in the Upper Cretaceous (Jarzen 1979).

Under these conditions, the topics treated here are dangerously subject to massive errors in judgment. However, after over three decades of study of modern types, and of speculation derived therefrom, I find I have had to only slightly shift my ideas. (This may merely mean that I stopped learning at an early age!) During this time, I have had opportunities to study, in the field, a nearly complete ensemble of primitive taxa and thereby learned not only their individual ecologies, but, equally important, their reproductive strategies and shortcomings. This synthesis derives from a unique direct knowledge of what I regard as primitive—and what I regard as primitive ecologies—linked with an analysis of the fragmentary fossil evidence.

Perhaps even more important is the fact that *if* one starts from the viewpoint here presented, then one is almost compelled by what evidence we have to adopt all or almost all the conclusions that follow. In essence, one has to "buy the whole package." The reader should thus consider himself forewarned. Because of the controversial nature of much that follows—due to differing perceptions of what is reality here—I have arranged my ideas in two columns, entitled "Facts versus Fictions" (see Table 5.1). It goes without saying that my facts may be, to others, fictions. All my paleoecological speculations are predicated clearly on the "facts" as I see them.

The organisms described under my "fictions" list correspond very closely to the Marchantiales, and, specifically, the Ricciaceae. The organisms in my "facts" list correspond closely to the Calobryales. Such a radial organism is not incompatible with that derived by Niklas (1979). It is striking, that on the basis of their low chromosome number ($n = 4$ or 5), some Calobryales, specifically *Takakia*, are exceedingly primitive. Another striking fact is that organisms approaching those of my "facts" list belong to stenotypic groups with a relict range; organisms falling into my "fictions" list show strong speciation (*Riccia* is a taxonomically complex genus with intricate speciation, with a nearly cosmopolitan range in areas that are not particularly cool). From study of an ensemble of what I regard as primitive types, I would argue that early types arose in areas with cool, moist climates; extant members of the Ricciineae and Corsiniineae—which nearly correspond to my "fictions" list, inhabit areas with arid, or intermittently arid, climates in warm to hot regions.

Sources of Data

Among my sources of data are: (1) The explosion in knowledge about early vascular land plants (Banks 1968, 1969). Extrapolation from the history and paleoecology of early vascular plants permits speculation about the environments early hepatics inhabited and, especially, to explore the hypothesis that vascular plants *appear to have created* by the Early Devonian the very niches that permitted initial diversification of the Hepaticae!

TABLE 5.1:   Evolutionary and Ecological Facts versus Fictions

Facts	Fictions
1. Early types of Hepaticae were erect, radial (but probably leafless) and grew by a tetrahedral apical cell.	Early types were flat thalli, growing by an apical cell with two cutting faces.
2. Early types protected otherwise naked gametangia by keeping them lubricated by means of overarching slime papillae.	Early types protected gametangia by "submerging" them within the thallus, or by having thallus tissue grow up and around them.
3. Early types were drought-intolerant, mesophytic types that once even partly desiccated, perished.	Early types were thallose and able to undergo repeated and prolonged desiccation.
4. Early types had chlorophyllose, complex, massive sporophytes with foot and seta massive, and a thick-walled (8–12-stratose) capsule wall; elaters were already present.	Early types had simple, chlorophyll-free, immersed sporophytes lacking foot and seta; elaters were lacking.
5. Early types had histologically simple gametophytes, but the gametophyte axis probably had at least vestigial vascularization.	Early types were histologically complex, had air chambers and pores and lacked all trace of vascularization.
6. Primitive types showed an identity in early ontogeny of the gametangia; gametangia were massive, solitary.	Primitive types already showed striking differences in initial stages of gametangial ontogeny; gametangia were small.
7. Primitive types were uniformly unisexual.	Primitive types were bisexual (as is true of most Marchantiales, including most taxa of *Riccia*).
8. Early types were restricted to near the oceanic, moist portions of land masses.	Early types probably evolved in areas with strong seasonal alternations in precipitation, and with long dry periods.

Source: Compiled by the author.

(2) Plate tectonics theory has given us a picture, still lacunose, with time sequences often a bit fuzzy, of the physical world during relevant times—chiefly, the early Paleozoic (Bambach, Scotese, and Ziegler, 1980; Ziegler et al. 1977). (3) Recent developments in our ideas as to when and where the angiosperms evolved and what they looked like (Doyle and Hickey 1976). Subsequent to when early vascular plants created environments allowing

early Bryophytes to diversify initially, a second proliferation occurred in Cretaceous-Tertiary times, conditioned by the simultaneous explosion of the angiosperms. Accurately fixing early patterns of angiosperm evolution, in time and place, facilitates understanding of both timing and dimensions of modern diversification of the Hepaticae. (4) Finally, recent major advances in our understanding of evolution in the Hepaticae (Schuster 1966a, 1972c, 1979a), and, indeed, of what is primitive and what is advanced, and of where primitive versus advanced taxa are concentrated, allow us to speculate about loci of origin and the ecological sites inhabited by the early Hepaticae.[2] A modern understanding of how the basic bryophyte cycle evolved— especially, recognition that the Hepaticae are a *sporophyte reduction series* probably evolved from ancestors with nearly isomorphic alternation (Schuster 1966a), conditions our entire approach to this topic. The following proceeds directly from the basic thesis that the paleoecology of the Hepaticae was directly influenced by the evolution of the bryophyte life cycle. Evolution, adjusting the organisms to the environmental stresses of the moment and unable to "plan" for the future, dictated that certain early adaptations of the Bryophyta (and, especially, of the Hepaticae) forever after strikingly limited that group: early sporophyte internalization by the Hepaticae carries with it numerous profound implications, relevant to their entire future history. In one sense, the Hepaticae, early in their history, made one of the great evolutionary blunders of all time, and thus are doomed to be limited throughout subsequent history by that simple fact.

The basic bryophyte life cycle, especially that of the Hepaticae, is so self-limiting that from this fact alone several operative generalizations derive, among them the following: (1) With permanent attachment of diploid to haploid generation, the intrinsic limitations of an amphibious gametophyte on land are, in effect, transferred to the sporophyte—a generation which, due to its water-free reproduction is, in effect, preadapted to a land environment. (2) Concomitant with gradual adaptation to land environments, the gametophyte was under continuous and severe selection pressure to become progressively more perfectly bilateral and dorsiventral and/or foreshortened, in order that the essentially terminally produced gametangia would be posited close to ground level. The reduced and simplified gametophytes of some Bryophyta are similar basically to those of several other groups of early land plants—and the similarity hinges on the exigencies of the land environment. (3) Exigencies of early adaptations to land *probably* were such that the transmigration, in many cases, was facilitated by limited symbiosis with fungi. (Some Hepaticae, including ancient and isolated types like *Haplomitrium*, *Takakia*, *Verdoornia*, and *Treubia*, possess mucilage *papillae*, sometimes of more than one type, which may secrete massive quantities of mucilaginous matter in which certain fungi may occur.) (4) In Hepaticae, efficiency and physiological economy dictated that, early in their evolution, they opted for progressive sporophyte reduction to the point where it (in all extant groups and, so far as knowledge of known fossils permits extrapolation, in fossil groups back to the Devonian) is internalized to the time when spores are mature. Opportunities inherent in the archegonium induced early epiphytism of sporophyte on gametophyte; simple efficiency dictated that the former be reduced in order not to put undue demands on the now-dominant gametophyte. Hence, bryophyte-like plants could have evolved rapidly from prebryophytic types with free sporophytes.

The following speculations derive from extrapolation from these and correlated facts and hypotheses. In effect, they derive from environmental pressures during relevant geological time (estimated to be from the Devonian, if not the Silurian, to Pennsylvanian time). Response of the early bryophyte types and, specifically, of the Hepaticae, conditioned the evolution of the group in all subsequent geological periods. Bryophyta, and especially Hepaticae, at a very early date committed a profound biological error that subsequently limited them to niches where they exist in Tracheophyte-dominated communities and by their sufferance. Exceptions, today, occur chiefly in modern types in various extreme environments, such as the Arctic Tundra—an environment of basically modern evolution. The error to which I refer, of course, is to put their faith into the future of the wrong generation—the gametophyte. Limitations of a terrestrial, but aquatically reproducing, gametophyte and the restrictions placed on such gametophytes on land are obvious; they need no further emphasis. An important implication is that Hepaticae must have undergone the greater part of their diversification *subsequent to* the origin of forests, and their major evolutionary explosion occurred in the period from the start of the Cretaceous to the end of the Tertiary—an explosion not only correlated in time with that of the angiosperms, but stimulated and conditioned by them. Much taxonomic proliferation (of genera and species) in the Hepaticae was induced largely by angiosperms creating numerous ecological niches which did not exist prior to mid-Cretaceous times! This thesis does not contradict the conclusion that *most* early *macroevolution*—into subclasses, orders, and families—goes back a very long time and, indeed, seems to have occurred principally concurrent with Devonian-Mississippian proliferation of early vascular plants.

All internal evidence derived from study of evolutionary patterns in the Hepaticae thus suggests that these patterns can be broken down into two types: (1) Early diversification, into the larger taxa, with limited speciation (and evolution of relatively few generic types), during early Paleozoic time; this was probably conditioned by creation of suitable microenvironments by the emerging tracheophyte flora. (2) Modern diversification, largely limited to a small series of families (Porellaceae, Plagiochilaceae, Radulaceae, Jubulaceae—and, above all, Lejeuneaceae—altogether accounting for over

70 percent of all extant taxa), which exploit angiosperm-created habitats (bark and living leaves of woody plants); this, equally, was conditioned by the creation of a large series of new and highly suitable microenvironments by evolving angiosperms. A knowledge of the paleoecology of tracheophytes, thus, serves, to a large extent, to explain both timing and patterns of evolution of the Hepaticae.

The Paradox of the Anthocerotes

It was once fashionable to claim that the lower vascular plants derived from forms similar to the Anthocerotae. Campbell (1925) emphasizes the "possible origin" of vascular plants "from forms related to the Anthocerotae." He interprets the relatively "independent" sporophyte of the Anthocerotales as "a strong argument in favor of . . . the correctness of the antithetic theory," and claims the "extraordinarily close resemblances which . . . large *Anthoceros* sporophytes bear to . . . the Rhyniaceae [is] remarkable." Smith (1938), influenced by Campbell, indeed conceptualized this viewpoint in a diagram (Fig. 5.1). In Schuster (1966a), these ideas are rejected: *Anthoceros* is regarded as "a relatively unspecialized organism" in which we find retention of "alternation [that] is more nearly homologous than in any other hepatic, *s. lat.*: the similarity in stomatal apparati, the almost equally developed chlorenchyma of the two generations, the indeterminate growth of both generations (cessation of growth being largely externally imposed)." However, Anthocerotae in *all* cases have evolved a basically bryophytan life cycle: the sporophyte remains at least marginally dependent on the gametophyte throughout its life. (There is no basis for statements [Mielinski 1926] that one can find a gametophyte which was "restlos abgestorben" while the sporophyte became "selbstständig.") Furthermore, *"general decrease in bulk and complexity of both sporophyte and gametophyte"* of *Anthoceros* is emphasized, and the group is regarded as "isolated"—being placed (pp. 386, 388) into an autonomous class, the Anthocerotae, subsequently (Schuster, 1977) elevated to the rank of a separate division, Anthocerotophyta. In the nonsynchronously developing spores, the nature of the chloroplast and pyrenoid, the nature of gametangial development (especially, total lack of a free archegonial wall; indeed, in one sense Anthocerotae lack archegonia, as usually defined), and in the lack of mucilage papillae, the group is wholly isolated; it lacks *any* real evolutionary contact point with Bryophyta, and is probably not derived from Psilophytes (see legend to Fig. 5.1), nor gave rise to Psilophytes. They may represent relatively modern derivatives of an algal group distinct from those Chlorophyta that gave rise to all other land plants.

We must, thus, reject the probability of *any* evolutionary contact point with other land plant groups. Rather, Anthocerotophyta represent a parallel experiment with land invasion—one that never really became viable (the five

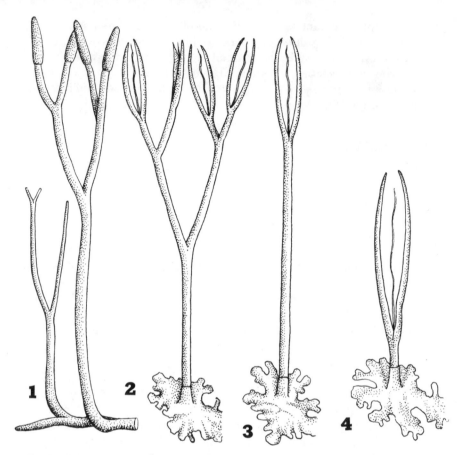

FIGURE 5.1. One possible interpretation of the evolution of Anthocerotae from a Psilophytalean ancestral type—showing progressive reduction of the sporophyte with the advent of its permanent epiphytism on the gametophyte. Smith (1938, 1955), from whom the figure is adapted, believed in the opposite evolutionary sequence (thus, evolution proceeded from 4 to 1). It seems unlikely that the Psilophytales can be interpreted as the direct ancestors of the Anthocerotae. (From Schuster [1966a].)

to six extant genera probably include fewer than 100 species). Thus, they are to be regarded as simply another unsuccessful attempt at land invasion. The paradox is that even though *most* such early attempts, which resulted in the Psilophytes, Trimerophytes, Zosterophyllophytes, and the probably loosely bryophytic Sporogonitales, among early Devonian assemblages, died out by the end of the Devonian, the Anthocerotophyta persisted. They persisted in spite of their singular lack of drought resistance.[3] The presence of stomata of both the *n* and *2n* generation suggests modern types derived relatively recently by reduction from anatomically more sophisticated and presumably considerably larger ancestral types. The persistent meristem of the sporophyte—Anthocerotae are unique among bryophyte-like plants in this respect—is remarkable and points in the same direction.

Perhaps even more remarkable is that there is *no* credible evidence that the gametophyte evolved from a radial ancestral type—the type here visualized as being the common ancestor of all Bryophyta and probably of all other land plants. The gametophyte lacks any evidence of the usual tetrahedral apical cell and, indeed, it has been assumed that growth was by several apical cells of equal rank, situated side-by-side (Leitgeb 1879; Mehra and Handoo 1953) although Renzaglia (1978) claims there is always a single apical cell whose segmentation patterns differ from those of Hepaticae.

On balance, therefore, we must reject the oft-expressed assumption that *Anthoceros* and its allies played any role in evolution of other types of early land plants. Lack of pre-Cretaceous fossils credibly assignable to Anthocerotophyta complicates things measurably; the group, as a consequence, remains a mystery. The bilaterally symmetrical spermatozoid (Carothers and Duckett 1979) is different fundamentally from that of the Hepaticae and Musci, suggesting there is more basis for a phylogeny in which the two latter groups are segregated as a division, Bryophyta, as contrasted to an isolated division, Anthocerotophyta (Schuster 1977), than for one in which three divisions (Hepatophyta, Bryophyta, and Anthocerotophyta) are recognized (Stotler and Crandall-Stotler 1977).

ECOLOGICAL PERIMETERS OF MODERN AND FOSSIL HEPATICAE

Space constraints dictate my giving generalizations, tempered by isolated examples. One subclass, the Jungermanniidae, includes virtually all primitive types; in this, drought resistance is usually limited (of the four orders, Metzgeriales, Calobryales, and Treubiales almost uniformly lack drought resistance; a minority of Jungermanniales possess considerable resistance to desiccation). The second subclass, the Marchantiidae, demonstrates no gametophytic drought resistance in two orders (Sphaerocarpales and Monocleales), while the highly advanced taxa comprising the third order (Marchantiales) have evolved sophisticated structural devices for desiccation control coupled with physiological drought resistance of a high order. Fossil evidence shows clearly that diversification of Jungermanniidae occurred in Paleozoic times with some diversification as early as Devonian-Carboniferous times; there is no reliable evidence for existence of *any* Marchantiidae prior to the Early Mesozoic. Since Marchantiales preserve much better than other groups of Hepaticae and numerous Cretaceous-Tertiary fossils are known, their absence in collections from early epochs is surely *not* due to oversight. These facts have phylogenetic implications.

In Jungermanniidae, water, during both reproductive and vegetative regimes, is *the* limiting factor in nearly all modern groups. There is no evidence that, in the past, the situation was ever different. (In dealing with the paleoecology of the Hepaticae, we can almost neglect temperature since many taxa today tolerate a wide range of temperatures. *Aneura pinguis*, for example, is found from at least 82°32'N to 52°33'S, and I have collected plants that are inseparable morphologically at 82°32'N on Ellesmere Island, and at less than 30' from the Equator, along the Rio Negro in Brazil—as well as at 52°33'S on Campbell Island. Admittedly, this is an extreme example, but compared to the importance of availability of water, temperature is of minor significance.)[4]

Reasons for the crucial role played by water in the life of Hepaticae lie in several facts. (1) True cutinization hardly occurs in modern groups, aside from the modern Marchantiales; it is not known to have occurred in fossil groups, hence there are, primitively at least, no devices for controlling water loss. Most taxa are *ectohydric*—in the sense that water is absorbed equally efficiently from all parts of the surface and, by the same token, lost as rapidly. (2) Physiological drought resistance—the ability of protoplasm to rapidly and repeatedly go from a sol to a gel state—is absent or limited in primitive groups, unlike in mosses. (3) Structural devices to promote water uptake from within the ground are usually lacking. (4) Fertilization is aquatic.

These factors together characterize all but certain advanced taxa today; they equally (or even more so) characterized prior taxa; the assumption that adaptation has involved loss of drought resistance cannot be supported. The basic gametophytic architecture is—and was—equally limiting: a tetrahedral or cuneiform apical cell, with a delicate, exposed free face, cutting off segments in spiral or pendular sequence. These derivatives, when still very immature—thus, very close to the exposed apical cell—give rise to gametangial initials. Such an architecture is singularly unsuitable for a true land organism. Indeed, primary problems of early hepatics undoubtedly were how to protect the apical region and developing gametangia from desiccation. *Even if* the mature gametophyte evolved limited resistance to desiccation, these juvenile structures always lack such resistance. In modern types, a number of devices have developed to ease this situation. However, at the time (Devonian-Carboniferous) when the primary stages of hepatic diversification occurred, no sophisticated, efficient devices to protect shoot apices and gametangia appear to have been evolved.

These factors almost preclude the Hepaticae from having been successful—as measured in numbers of individuals and phylogenetic types—prior to the creation of environmental niches by vascular plants which afforded some protection. Suitable microenvironments could not have occurred prior to the mid- and late Devonian explosion of early tracheophytes. The first stages in hepatic evolution seem to have been conditioned by such a phenomenon. (See the section entitled "Paleozoic Evolution: the Initial Creation of Suitable Sites".) The group thus appears to have undergone true diversification only in Cretaceous to Tertiary times, coincident with angiosperm evolution. Again, the creation of a host of microenvironments by the evolving angiosperms conditioned and stimulated this "modern" proliferation of genera and species, especially in modern families that are often epiphytic. (See the section entitled "The Mesozoic Explosion of the Angiosperms and the Coevaluation of Modern Jungermanniales and Metzgeriales".)

Ecology of "Modern" Taxa and Its Relationship to Paleoecology

The ecology of the forerunners of today's Hepaticae can be fairly safely deduced from that of modern taxa: (1) Structural or physiological drought resistance occurs in a number of modern types (water-sacs in Jubulaceae, Frullaniaceae, and Lepidolaenaceae—all modern families; closely shingled leaves; filiform julaceous shoots, as in *Gymnomitrion*, a modern alpine and arctic genus); such devices are lacking or only incipiently developed in primitive taxa (e.g., Vetaformaceae, Blepharostomataceae, Lepicoleaceae, Takakiaceae, and Haplomitriaceae) and in all known pre-Jurassic fossils. Devices and strategies for limiting water loss were, if anything, less effective in fossil groups than in modern ones. (2) Fossil groups, as well as most generalized or primitive groups, occur on rock faces or mineral soil; epiphytism (in many advanced Jungermanniales; in some taxa of the advanced families Metzgeriaceae and Aneuraceae of the Metzgeriales; in *no* other taxa) is a modern phenomenon, which appears to have evolved largely in and since the Cretaceous. (3) Toleration of temperature fluctuations is high in many taxa today (Schuster 1957, pp. 217–20)[4]; there is no evidence that the temperature factor was ever more relevant a limiting factor in the paleoecology of the Hepaticae.

A number of ecological progressions can be deduced: (1) Primitive taxa were apparently all terricolous or, less often, saxicolous; thus, modern allies of the Paleozoic *Pallaviciniites*, *Blasiites*, and *Treubiites* all occur on purely mineral substrates (aside from rare invasion of peat or decayed wood by *Pallavicinia*). (2) Competition and probably rapid accumulation of organic debris on forest floors early stimulated evolution of devices that allowed survival on rock faces; many primitive groups already inhabit such sites, and it seems a safe assumption that saxicolous niches, such as boulders and cliffs, were invaded very early in hepatic evolution. (3) Various taxa growing appressed to rock walls (such as Radulaceae, Metzgeriaceae, Porellaceae, Jubulaceae, and Lejeuneaceae) were, because of the closely adnate, flat growth on vertical surfaces and their growth confined to two planes, *preadapted* to invade the bark of trees—as soon as that niche evolved. We know nothing about the adequacy of early gymnosperm bark as a suitable habitat for epiphytes; however, of Laurasian gymnosperms very few genera *today* provide good habitats for epiphytes (chiefly *Cryptomeria, Thuja,* and *Chamaecyparis*; Pinaceae *s. str.* are almost uniformly devoid of epiphytes), and in Gondwanalandic areas, chiefly the Podocarpaceae and *Agathis* possess suitable bark characteristics and are often invaded. Few tree ferns have a suitable bark. I assume that with only limited early experiments involving gymnosperm hosts (the Cycads almost never seem to host epiphytes), epiphytism became widespread only in the Cretaceous, coincident to angiosperm diversification. Angiosperms provided not only good and diverse bark habitats, but epiphytism on living leaves evolved chiefly because angiosperm leaves provided greater levels of shade and higher humidity (in part because, in general, angiosperm communities are more nearly closed than most gymnosperm communities). With angiosperm expansion in the Cretaceous, thus, innumerable niches were created for Hepaticae. The explosion of genera like *Frullania* (about 400 taxa, almost all on bark), *Lejeunea* (about 11 subgenera with over 300 species) and *Cololejeunea* (with about 250 species, almost all epiphyllous) must be understood in terms of coevolution with their normal angiosperm hosts. Thus, the drastic environmental changes created by the angiosperm explosion stimulated a massive change in the ecology of, especially, the Jungermanniales. Orders which, for one reason or another, could not exploit such angiosperm habitats are today stenotypic (Monocleales has one genus, two species; Calobryales has two genera, about 10 to 11 species; Metzgeriales has about 26 genera and— *Metzgeria* and *Riccardia* aside—only about 50 to 60 species); only *Riccardia* and *Metzgeria*, in part able to exploit angiosperm habitats, are species-rich. This explosion of corticolous hepatics is well documented in early studies of Gottsche (1886), who found numerous modern epiphytes enclosed in amber from Eocene-Oligocene times; included were about 15 *Frullania* species, two of *Radula,* and five of *Lejeunea s. lat.* (= Lejeuneaceae). Caspary (1906) found six *Frullania* species, one of *Radula,* one of *Porella,* and three Lejeuneaceae Holostipae ("Phragmicoma") and four Lejeuneaceae Schizostipae ("Lejeunea") enclosed in amber.

TIME OF ORIGIN AND
PREVALENT ENVIRONMENTAL PERIMETERS

Early Fossils

Recognizable Hepaticae are known in small numbers from Carboniferous times (*Hepaticites*, a form genus including, in part, plants superficially like the extant genus *Riccardia*, *Blasiites* Schust., *Metzgeriites* Steere, *Treubiites* Schust., and *Metzgeriothallus* Schust.). A single rather well-documented case is known from the Devonian (*Pallaviciniites devonicus* [Hueb.] Schust.). In addition, the poorly known sporophyte, *Sporogonites* Halle (1916, 1936) *may* have had a liverwort-like cycle (Andrews 1960); there is nothing in the size and structure of the sporophyte to exclude it from Hepaticae.[5]

All of these organisms appear allied to, or belong, insofar as can be determined, to the modern order Metzgeriales. There is, thus, no good evidence that members of the two other large groups, the Jungermanniales and Marchantiales, existed prior to the Mesozoic. The main groups *seem* to be ancient, but fossil history supports such claims of antiquity only for Metzgeriales. Implications of this are examined later.

Problems in fixing the time of origin are formidable: (1) Early thallose gametophytes may be so similar to fern gametophytes that certain identification of fragments is difficult. (2) Most hepatic gametophytes are delicate and the middle lamella seems to break down readily—much more so than in Musci (Schuster 1966a, p. 355)—so that fossilization is infrequent. (3) Convergence between early vascular plants and early bryophyte gametophytes, at the Silurian–Early Devonian level, may be so great that only sporophytes (and their permanent versus transitory attachment to gametophytes) may allow positive attribution. Thus, according to Andrews (1960), *Sporogonites*, with apparently permanently epiphytic unbranched sporophytes, would have to be considered an Early Devonian liverwort (see Schuster 1966a, p. 353, where it is assigned to the order Sporogonitales). These fragmentary data suggest that Hepaticae occurred, if in small numbers, during the period of the Devonian explosion of early vascular plant types.

These conclusions are admittedly flimsily based. Early fossil types (*Pallaviciniites*, *Treubiites*, *Blasiites*, and *Metzgeriites*) all belong to an order (Metzgeriales) in which—in modern types— physiological drought resistance is almost unknown, in which structural devices to retard water loss hardly exist, and in which there is no effective cutinization. It is difficult to conceive that modern types would show *less* drought resistance and would have *lost* a cuticle and structural devices for preventing water loss. The early fossil types must have been even less capable of coping with water loss and were presumably even more restricted than modern genera to niches in which desiccation was not a factor. With a relative humidity of 90 to 95 percent, cells of a wide range of Jungermanniales and Metzgeriales suffer irreversible damage, even after as little as 24 hours; thus, even after some 350 to 400 million years of evolution, drought resistance has not evolved effectively in the bulk of taxa (Clausen 1952). Ecological restrictions of early types of gametophyte-dominant *prebryophyta* must have been equally, if not more, acute. Early evolution and diversification of the Hepaticae were of necessity conditioned by environmental amelioration by early Tracheophyta.

The early origin of Hepaticae, and their occurrence in limited diversity in Carboniferous strata, linked with a mesophytic morphology, is striking. The, typically, more drought tolerant Musci are hardly known from paleozoic fossils. This is especially notable in view of the fact that the relatively high resistance to decay of the latter, as contrasted to the former, would lead one to expect a higher level of fossilization. The fossil evidence thus suggests the Musci either differentiated later, or at least, became numerically significant later than the Hepaticae. (Conceivably, with early development of drought-resistance, early moss types may already have inhabited loci, such as rock faces, where opportunities for fossilization hardly existed.)

Evidence from Phylogeny and Fossil History as to the Relative Antiquity of the Various Orders of the Hepaticae

Six to eight major groups of Hepaticae are accepted generally, usually called orders, which fit into two subclasses: (1) Subclass Jungermanniidae: Orders Calobryales (the Takakiales are often segregated as a separate order), Metzgeriales (the Treubiales are, perhaps on sound basis, to be segregated as a distinct order), and Jungermanniales. (2) Subclass Marchantiidae: Orders Monocleales (the sole phylogenetic contact point with Jungermanniidae among living taxa), Sphaerocarpales, and Marchantiales. Fossil evidence— such as it is—is available from several taxa of Metzgeriales (including the only Paleozoic fossils)[6] and from Jungermanniales and Marchantiales (both known from Mesozoic fossils, but not surely from Paleozoic ones); the Rhaetic *Naiadita* Buckland is probably a primitive type allied to Sphaerocarpales (Harris 1938, 1939; Schuster 1966a). Lack of fossil data for the most primitive members of both subclasses, the Calobryales and Monocleales, is severely limiting. With these limitations, the following is about all that can be safely—or unsafely—ventured.

The Antiquity of the Metzgeriales

All Paleozoic hepatic fossils known appear referable to the Metzgeriales (although *Sporogonites* may represent an autonomous order, the Sporogonitales; Schuster 1966a, p. 353), suggesting, indeed, that this group may be more primitive than the Jungermanniales. Known Carboniferous Metzgeriales appear to represent several discrete suborders: *Treubiites* almost surely is allied to Blasiineae (or Blasiales); *Blasiites* probably is allied to Pelliineae and may be allied to *Noteroclada*; *Pallaviciniites* may be allied to the Pallaviciniineae; *Hepaticites metzgerioides* Walton (1925) seems "superficially very similar to the modern genus *Metzgeria* [=suborder Metzgeriineae]" (Schuster 1966a).[7] From this, one could conclude that the Metzgeri- ales go back to the beginnings of hepatic evolution; one could also conclude that, therefore, *thalloid* gametophytes preceded leafy types. However, the clearly leafy shoot of *Treubiites* and the leaf-like lateral scallops of *Blasiites* suggest that bilaterally leafy types existed by Carboniferous times. Hence, currently available fossil evidence allows no room for major phylogenetic conclusions.

It is suggestive that the Metzgeriales (together with the stenotypic Calobryales and Monocleales, for which no fossils are known) are the most mesophytic of all Hepaticae; except for a few modern taxa of *Riccardia* and *Metzgeria*—specialized, derivative genera forming the very apices of evolu- tion in the order—Metzgeriales are notoriously unable to tolerate desicca- tion, even for very short time intervals. They are bound, in general, to permanently moist substrates. Equally suggestive is that the 25 to 26 extant genera are very sharply separated from each other and are, *Metzgeria* and *Riccardia* aside, very stenotypic (16 genera each contain one or two species only). A large number (especially of stenotypic groups) are today confined to basically Gondwanalandic loci: *Treubia, Sewardiella, Noteroclada, Verdoornia, Symphyogyna, Xenothallus, Allisonia, Hymenophytum, Podomitrium,* and *Phyllothallia*. A much smaller ensemble are purely Laurasian (Blasiineae, with the Carboniferous *Treubiites*, the modern genera *Cavicularia* and *Blasia*).

The Apparent Late Origin of Jungermanniales

First reports of Jungermanniales are from the Upper Jurassic and Lower Cretaceous (Krassilov 1970, 1973), from which *Cheirorhiza* and *Laticaulina* are described.[8]

Anyone who advocates a thalloid ancestry for the Hepaticae may be tempted to lean heavily on the foregoing facts. However, fossilization of Hepaticae under conditions in which abundant moss fossils were preserved may be singularly lacking (Steere 1942; Schuster 1966a)—and Steere cites four Jungermannialean genera (*Cephalozia, Calypogeia, Mylia,* and *Lo- phozia*) normally accompanying peat mosses on moors. Moor-inhabiting mosses are freely preserved, the presumably associated hepatics, never. Although this was noted for Pleistocene bryophytes (Schuster 1966a, p. 355) the "facts which conspired to prevent fossilization" of such leafy, moor- inhabiting Jungermanniales "during interglacial periods of the Pleistocene operated with equal relevance earlier." With perhaps as many as 7,000 extant species, *almost no Jungermannialean fossils are known prior to the period when modern types were preserved in (mostly Oligocene) amber deposits.* By contrast, the more robust Marchantiales, with perhaps under 200 species today, are abundantly fossilized in Mesozoic and later strata. Thalloid Metzgeriales and the Marchantiales are larger and often firmer organisms, hence preservation is more likely. Also, the mesophytic Metzger-iales often grow along streams, where silting and sedimentation may lead readily to fossilization; they also grow flat on the ground, thus more readily buried in thin layers of silt. Jungermanniales—often confined to areas with organic substrates (logs, peat, or tree bark) or rock faces, inhabit largely sites where fossilization is less likely.[9]

The Apparent Late Origin of Marchantiales and Sphaerocarpales

No reliable reports of Paleozoic fossils of Marchantiales exist. Town-row (1959) has described Triassic fossils from South Africa, pertaining to the Marchantiales, one, *Hepaticites cyathodes*, with supposed similarities to the modern genus *Cyathodium*.[10] Although several early Mesozoic fossils *may* belong to the Marchantiales (e.g., *Hepaticites rosenkrantzi*, *H. amauros*, and *H. wonnacotti*), their attribution remains uncertain (Lundblad 1954). Other dubious fossils known are discussed by Lundblad; the ambiguity of their provenance is such they do not warrant further consideration here. However, convincingly Marchantialean fossils are described by Lundblad (1954) from the Rhaetic-Liassic (*Ricciopsis florinii*, *R. scanica*, and *March-antiolites porosus*). Thus, by Triassic and Jurassic times at least a limited number of fossil Marchantiales are known, belonging apparently to the Ricciineae and Marchantiineae. From the Triassic we also know the erect, isophyllous leafy *Naiadita* (Harris 1938, 1939), whose sporophyte resembles that of the Order Sphaerocarpales. (I have studied the types of *R. florinii* and *M. porosus* at Stockholm, through the courtesy of Dr. Lundblad. She reports "branched" rhizoids from the former, supposedly "septate." In my opinion these represent admixed—and unattached—fungal hyphae. On one slide *all* attached rhizoids seen were unbranched, nonseptate, and of two diameters, which is consistent with modern Marchantiales. The "septae" seem to represent folds rather than cross-walls in the unbranched rhizoids. In *M. porosus* the pores are ring-elevated and, surely, are of the *Marchantia*-type. Air chambers are in one layer and the thallus anatomy corresponds to the *Neo-hodgsonia* model. These observations thus tend to "harden" the provenance of these very important fossil finds.)

Conclusion

The fossil evidence is too fragmentary to allow definitive conclusions. The only certain conclusion that can be derived is that a flora of limited diversity, pertaining to the Metzgeriales, existed by Devonian-Carbon-iferous times.

Since, on several bases, the Metzgeriales are to be regarded as already moderately advanced phylogenetically, one must almost assume from this that the hepatics *originated* simultaneously with other land plants. The fact that some real diversity (with the presence of Jungermanniales, Sphaerocar-pales, and Marchantiales) existed by Jurassic-Triassic times suggests that even though origin may have been very early, primary *diversification* seems to have occurred somewhat later. In particular, evolution of taxa with some physiological drought resistance (Marchantiales, some modern Jungerman-niales) seems to have occurred *relatively* late.

PALEOZOIC EVOLUTION: THE INITIAL CREATION OF SUITABLE SITES

The Initial Steps of Liverwort Evolution

Past ideas of evolution of the Hepaticae, chiefly in the nineteenth century, centered on the concept of a thalloid, prostrate ancestral type versus the twentieth century concept (Wettstein 1903–08; Evans 1939) of an erect, symmetrically leafy ("isophyllous") ancestral type. As already intimated, both such ideas are basically simplistic. Several recently discovered types are relevant: (1) *Phycolepidozia* Schust. (Schuster 1966c); (2) the *Haplomitrium intermedium* Berrie–*ovalifolium* Schust. complex (Schuster 1967b, 1971a); (3) *Takakia* Hatt. & Inoue (Schuster 1967b); (4) *Treubiites* Schust. (Schuster 1966a), a Devonian fossil. The following conceptualizations, thus, are based almost wholly on organisms, chiefly living, but in one case fossil, discovered only in the last two decades.

Initial steps in evolution involved only a limited suite of *relatively minor* alterations, once a highly generalized ancestral type existed. Such an ancestral type is conceptualized (Fig. 5.2[1]) as being *neither thallose nor leafy.* Thallose and leafy gametophytes thus may represent adaptations, ancient and modern, derived from a simple ancestor unencumbered by much sophisticated morphology. Such an organism possessed a tetrahedral apical cell, cutting off a helical sequence of derivatives that underwent limited secondary divisions, each to form eventually a segment, or *merophyte.* The naked radial polyseriate axis which ensued gave rise to four types of outgrowths based on single cells: (1) unicellular rhizoids, from undersurfaces of creeping or prostrate portions of the axis; (2) unicellular slime papillae, typically ephemeral, clavate cells that release mucilaginous matter, which overarched and protected the apical cell and its immediate derivatives; (3) gametangia, developed from the distal part of an initially unicellular protuberance, produced as an outgrowth from segments still posited very close to the shoot apex; (4) branches, probably produced from an epidermal cell, perhaps relatively mature, which dedifferentiates, then divides by three oblique divisions to cut off a new tetrahedral apical cell from which a new axis will proceed.

Coincident to the first tentative attempts at land invasion, the initially one-celled gametangia (perhaps, not unlike the plurilocular gametangia of algae, such as *Ectocarpus*) showed adaptation to intermittent desiccation by sterilizing a peripheral layer from which gametangial jackets were derived. Thus, pluricellular gametangia evolved and, furthermore, had differentiated

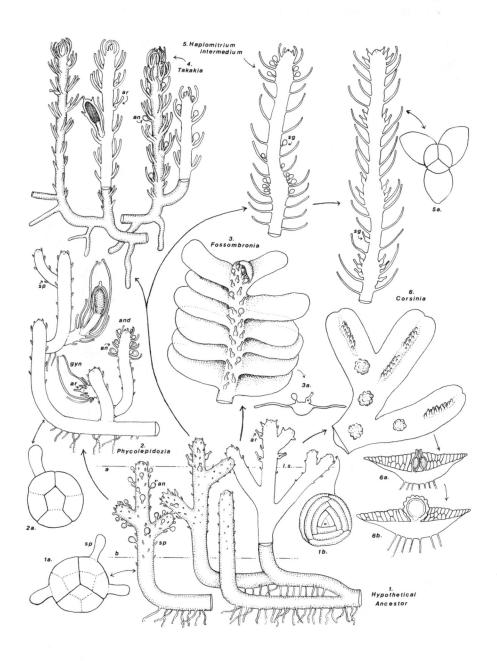

into archegonia and antheridia. (In early groups, the ontogeny of the gametangia was nearly identical; it is assumed that organisms with identical early stages in gametangial ontogeny are primitive in this respect. This idea will again emerge in the discussion of *Haplomitrium*.)

Such a gametophyte perhaps showed ill-defined division into creeping sectors and erect branches. It was, in fact, not much different from gametophytes of several kinds of Tracheophyta that still exist today, for example, *Psilotum*, *Tmesipteris*, *Lycopodium*, *Actinostachys*, and Ophioglossaceae, and various Gleicheniaceae. Only four features seem to have distinguished early types of bryophyte gametophytes from those of (at least extant) tracheophytes: (1) gametangia appear to have been cut off only near the shoot apex, in simple acropetal sequence; (2) mycorrhizal relationships, if they existed, were of limited significance; (3) vascularization was lacking, or, if present, not via tracheids; and (4) gametangia, evidently, had a significantly different history, *both* types being stalked, more elongated, and showing significant differences in ontogeny. (It *might* be significant that in

FIGURE 5.2. Some evolutionary pathways in Hepaticae. 1. Hypothetical ancestor, showing naked antheridia (*an*) and archegonia (*ar*) on a naked shoot bearing only slime papillae (*sp*); at *1a*, a cross section showing triradial symmetry, with primary segmentation of each merophyte drawn in (*stippled lines*), the origin of slime papillae indicated; at *1b*, apex of shoot, showing apical cell and helical segmentation; *ls* shows a sector in longisection; fluctuation of water level from level *a* to level *b* is believed to have occurred, facilitating aquatic fertilization (probably assisted by weak tidal effects). 2. Longisection of plant of *Phycolepidozia* (Jungermanniales); the leaves are reduced to slime papillae (*sp*), but remain well developed in androecial areas (*and*) and gynoecial areas (*gyn*); sterile shoot sectors approximate closely the hypothetical ancestor, but have achieved their similarity by reduction; at *2a*, a cross section of the axis showing reduction (lateral merophytes of two cells, side-by-side; ventral merophytes of two cells, superposed, one forming the medulla); cf. figure 5.4. 3. Dorsal view of *Fossombronia* (Metzgeriales), showing a bilaterally leafy type, with acropetal production of naked gametangia; as in *Phycolepidozia*, the gametophyte has become bisexual; at *3a*, a cross section showing juxtaposed an antheridium and an archegonium. 4. *Takakia* (Calobryales); a reconstruction of the ♂ plant, at right, and a fertilized ♀ plant, at left; it is assumed that the acropetally produced archegonia, occurring singly, *if* fertilized, would not inhibit further growth of the gametophyte (as shown in the lateral position of the developing sporophyte, encased in a shoot calyptra such as is known in *Haplomitrium*); cf. figure 5.5. 5. *Haplomitrium intermedium*, showing longisections of a ♂ shoot (left) and ♀ shoot (right) in which fertilization has not occurred; note the seasonal production of gametangia on *nodes*, with tendencies for scattered gametangia (at *sg*); at *5a*, a cross section showing isophylly and equally developed merophytes; cf. figure 5.6. 6. *Corsinia* (a primitive member of the Marchantiales), showing the bisexual, aplanate thallus, with acropetal development of sex organs (anteriorly, antheridia; posteriorly, gynoecia with developing sporophytes); at *6a*, cross section through androecial sector; at *6b*, cross section through gynoecial sector, showing immature sporophyte surrounded by the tuberculate calyptra.

See text for partial interpretations of the evolutionary patterns involved. In 1, 4, 5, the primitive radial symmetry and unisexuality persist; in 1, early ontogeny of gametangia is believed to be identical; in 4, ♂ gametangia are not known; in 5, early gametangial ontogeny is perfectly identical. Evolution of bisexuality (in 2, 3, 6) is believed to be advanced.

both the very primitive genus *Haplomitrium* and in *Psilotum* we find archegonial necks formed by four cell rows.) Significant characteristics of such early liverwort (and also moss) gametophytes include, among others: (1) No tendency toward losing *terminalization* of gametangium production—which soon proved to be a major limitation, as progression from amphibious to terrestrial sites occurred. (2) They made no attempt to "go underground"; in the Tracheophyta, some gametophytes remain surface-posited and chlorophyllose, while many others developed subterranean gametophytes in which "sex went underground." (There have been many recent discussions of the role played by fungi in making possible land plant invasion; such a role appears real in the case of many early tracheophytes, but not all [e.g., the paradoxical Sphenopsida]. Close mycorrhizal relationships between gametophytes of Bryophyta and fungi do *not* seem to have evolved at an early date. A subterranean symbiotic gametophyte, devoid of chlorophyll—has been evolved, *in modern times*, by a single genus of liverworts, *Cryptothallus* [Fig. 5.3]. There is no evidence that such symbiosis had ever been experimented with before.)

At the ultimate—perhaps indefensible—level of conceptualization, we can visualize this gametophyte as of ultimate simplicity: a naked, chlorophyllose axis, polyseriate, developed as a consequence of the activity of a single tetrahedral apical cell. Probably the *first* specialization, which must have evolved at a very early date, if not present in the ancestral algae, was the evolution of tubular anchoring devices, or rhizoids. Slime papillae may even derive from rhizoids arrested in initial stages of elongation, subsequently cut off by a cross wall and *converted* into ephemeral devices for mucilage secretion (Fig. 5.2[1–4]). Slime papillae, which play such a crucial role in protecting the apical cell and its derivatives, would *not* have been needed prior to inception of periodic, if very short, exposure to a drying atmosphere. *Perhaps* the ancestral polyseriate algae possessed unicellular gametangia, formed as similar outgrowths from epidermal cells, subsequently cut off by a cross wall. Such surface gametangia, exogenous in origin, characterize gametophytes of *all* "Archegoniatae"—except for the phylum Anthocerotophyta (Schuster 1977), in which endogenous gametangia occur which do not really qualify as archegonia—at least, in the traditional sense (they *lack*, in one sense at least, a jacket layer!).

Such very early gametophytes were not invested with much morphological debris. The simpler, less committed a gametophyte that we visualize (Fig. 5.2[1])—one in which essential new devices could be conceived as originating from simple preexisting devices (the rhizoid, probably already present in the ancestral algae), the closer we are likely to be to reality.

Such gametophytes have numerous potentials, not all realized: they can evolve leaves (from slime papillae); they can either foreshorten and ultimately lose erect sectors and compensate for corresponding loss of photo-

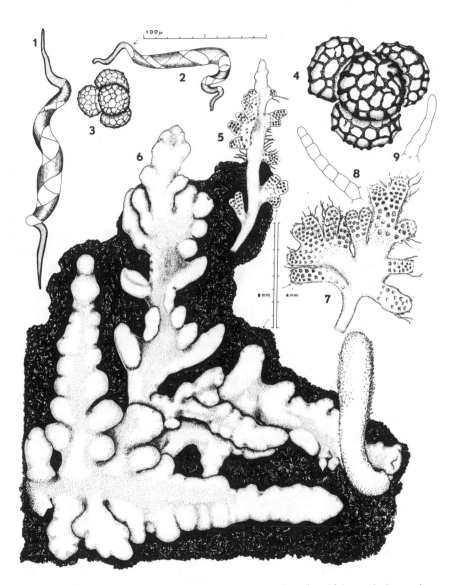

FIGURE 5.3. *Cryptothallus* Malmborg, the only bryophyte in which a colorless, subterranean gametophyte has evolved. Both gametophytic and sporophytic tissue is penetrated by fungal hyphae. Note the heterothallic nature of the gametophyte (figs. 5, ♂ and 6, ♀, are drawn to one scale, the 8-mm scale adjoined). 1–3. Elaters and a spore tetrad, all drawn to 100 μ scale. 4. Spore tetrad; the permanently adherent spores result in production, juxtaposed, of two ♂ and two ♀ thalli, from germination of a single tetrad. 5–6. ♂ and ♀ Plants, as seen after the 3 to 6 inches of peaty ground are removed; drawn against a background of peaty material. 7. ♂ Plant, to larger scale. 8–9. Cilium-like scales.

synthetic surface areas by planation of remaining creeping gametophytic sectors (Fig. 5.2[3]); they can go the route of evolving vascular strands—or, if they show rapid reduction and simplification, avoid evolving such devices; they can retain the initially radial axis, or experiment with flattening distal portions of axes (as in *Podomitrium*; Fig. 5.7), or show strong divergence between basal radial axes and erect axes consisting of a series of sterile telomes organized into structures that closely mimic megaphylls of some fern sporophytes (cf., e.g., the gametophyte of *Hymenophytum* [Figure 5.8] with the sporophyte of *Hymenophyllum*!). The potential for these departures—and of more—was basically inherent in the simple radial gametophytes postulated—even the potential for becoming foreshortened, massive, mycorrhizal, and subterranean—a potential realized in *Cryptothallus* (Fig. 5.3).

To try to document these various early steps in evolution is beyond the scope of this chapter. Figures 5.2 to 5.9 illustrate *some* of the more radical end points achieved, *all* derivable relatively rapidly *after* land was first invaded. The legends document some of the morphological adaptations which occurred. However, three key organisms deserve special emphasis: *Phycolepidozia* (Figs. 5.2[2]; 5.4), *Takakia* (Figs. 5.2[4]; 5.5), and *Haplomitrium intermedium* (Figs. 5.2[5]; 5.6). They illustrate what *some* of the earliest derivatives may have been like. A fourth type, *Podomitrium*, illustrates a response to a recurrent problem: how fertilization can go underground while the gametophyte remains photosynthetic. These examples are discussed in the following paragraphs.[11]

Phycolepidozia Schust. (Figs. 5.2[2]; 5.4)—A genus with major phylogenetic implications, though highly evolved, which suggests that leaf-like organs, at least in Jungermanniales, evolved by elaboration of slime papillae (Schuster 1966c). Other facts also suggest such a derivation: (1) When a radial leafy gametophyte becomes prostrate and bilateral, the third row of vegetative leaves, ventrally displaced, is often reduced to slime papillae, even though in gynoecial regions, which remain erect, all three rows of leaves remain identical, as in *Cephalozia*. (2) Young leaves or their divisions terminate in slime papillae (Greig-Smith 1958), although secondary, nonapical cell divisions may result in overtopping of the slime papillae. The concept of slime papillae–derived leaves receives major support from *Phycolepidozia*: here vegetative shoots are mere terete polyseriate axes growing by a tetrahedral apical cell (Fig. 5.4 and legend), and lateral merophytes each develop a slime papilla at the locus where, normally, a leaf is found; the ventral axis surface develops rhizoids. Vegetative shoots thus closely mimic the archetype earlier suggested for all Hepaticae!

Takakia Hatt. & Inoue (Figs. 5.2[4]; 5.5)—This unique genus shows a large number of primitive and/or generalized features. The chromosome number ($n = 4$ or 5) is the lowest known number in Bryophyta; it has been

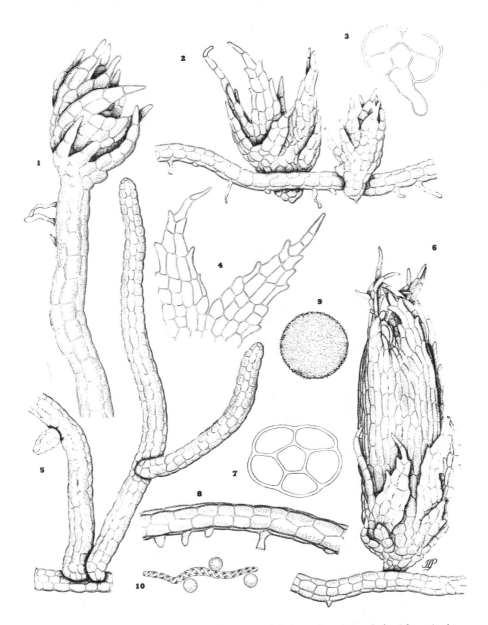

FIGURE 5.4. *Phycolepidozia exigua* (Jungermanniales), a reduced type derived from the *low* family Lepidoziaceae, in which sterile axes are naked, except for ephemeral slime papillae (which represent the reduced leaves). 1. ♂ Axis, showing few-celled leaves distally, each shielding an antheridium. 2. Sterile axis, with rhizoids, with ♂ branch at right, ♀ branch at mature archegonial stage at left. 3,7. Sections through sterile axes. 4. Leaf (♀ bract), from ♀ branch. 5. Part of the creeping, branched, sterile shoot system. 6. ♀ Branch with mature perianth, shielding enclosed sporophyte, arising from naked sterile axis. 8. Sector of sterile axis with rhizoids, lateral aspect; the ephemeral leaves (slime papillae) have already disappeared. 9. Spore. 10. Elater and spores. (From Schuster [1966c].)

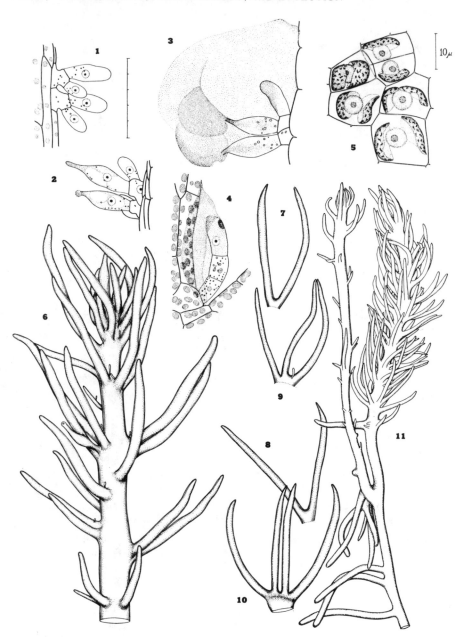

FIGURE 5.5. *Takakia lepidozioides* (Calobryales), in many ways the most primitive extant hepatic. 1–4. Slime papillae (1–2, the *beaked type*, 3–4, solitary nonbeaked type). 5. Cells near apex, showing paired plastids and nuclei. 6. Shoot apex; at arrow an archegonium; note slime papillae. 7–10. Leaflike appendages, formed of polyseriate segments. 11. Sterile shoot, showing rootlike organs. (From Schuster [1967b].)

suggested (Tatuno 1959; Schuster 1966a) that other numbers in Hepaticae, usually $n = 8$ or 9, rarely 10, may be derived by paleopolyploidy. The erect axis bears a weak central strand—perhaps a remnant of an earlier, more sophisticated one. The radial gametophyte produces polymorphic leaf-like outgrowths which may bear from one to four terete, polyseriate segments which mimic short shoots. The green and massive archegonia may, rarely, occur at apices of such segments, but normally occur, singly, in remote acropetal sequence, from epidermal cells.

Haplomitrium intermedium Berrie (Figs. 5.2[5]; 5.6)—This extraordinary plant (Schuster 1967b) bears remote affinities to *Takakia*. It is equally malleable and, indeed, "sloppy" in its morphology. Architectural sloppiness is presumed to be a sign of great age; a regular and closely regulated mode of growth, inversely, is a sign of evolutionary advancement. On any such basis, *H. intermedium* is extraordinary. Like *Takakia*, it may produce scattered archegonia (Fig. 5.6[1,6]), although they tend to be aggregated in gynoecia on weakly swollen shoot apices. Gametangium production does not inhibit further axial growth, so that, with lack of fertilization, gynoecia occupy the surface of rather swollen nodes, succeeded by sterile leafy regions, again succeeded by gynoecia; between such gynoecial regions the "sloppy" development of occasional scattered archegonia may persist. Antheridial production and localization are identical (Fig. 5.6[5]). The triradially leafy gametophyte has gone from simple acropetal production of sex organs to the point where these are more or less aggregated in zones (gynoecia and androecia)—but restriction to sexual regions is still imperfect, so that, in actuality, there is an acropetal sequence with partial sequestration of gametangia.

A central theme of gametophyte adaptation to land was the progressive lowering of gametangia to the ground surface, facilitating fertilization. (Even in very advanced organisms, like *Conocephalum*, archegonia occur on *archegoniophores* that are sessile or almost so at the time of fertilization; they are elevated only *after* fertilization.) Another adaptation was the evolution of acrogyny: development of a system in which, after a gynoecium forms, the erect shoot ceases to grow. In *H. intermedium*, an almost ludicrously improbable condition prevails: archegonia are found on shoots on which as many as four to five seasonal increments of sterile leafy tissue occur (Fig. 5.6[5,6]), each ending in a gynoecium—so that the archegonia eventually occur farther and farther away from ground level, and are less and less likely to get fertilized! If one were to purposely design an incompetent land gametophyte, one could hardly do better (or worse) than *H. intermedium*!

H. intermedium and *Takakia* already exhibit several concessions to land life: both bear root-like organs (Grubb 1970; Schuster 1967b) and vestigial vascularization; both are extraordinary mucilage producers (Fig. 5.5[3]). The nearly equally primitive *H. ovalifolium* Schust. (Schuster

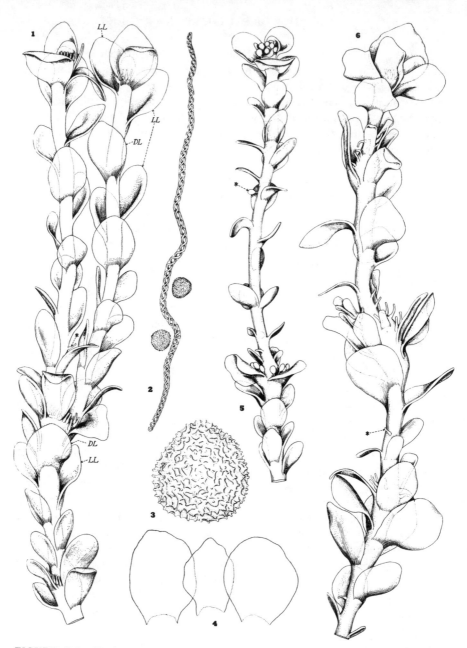

FIGURE 5.6. *Haplomitrium intermedium* (Calobryales), an exceptionally primitive and morphologically nonstandardized type. 1. ♀ Plant with unfertilized archegonia; at asterisk a solitary archegonium; *DL*, *LL*, dorsal and lateral leaves. 2. Spores and an elater. 3. Spore. 4. Cycle of leaves, the dorsal leaf in middle. 5. ♂ Plant; at asterisk a solitary antheridium. 6. ♀ Plant, with 4 unfertilized gynoecia at nodes; note solitary archegonium at asterisk. The anacrogynous and anacroandrous condition, with proclivities for scattered gametangia, is very primitive.

1971a) serves to somewhat connect them: it may have leaves divided to the base into two flat segments, or leaves that are all undivided. This organism also shows excessive sloppiness and even such an organ as the leaf is apt to show gross variation, often on a single stem. This recalls the malleability in leaf form and venation which Doyle and Hickey (1976) have demonstrated for early angiosperm leaves. The analogy may go back to a common factor: both types of organisms represent a level of organization in which a fixed morphology had not yet evolved.

Phycolepidozia, Takakia, and *Haplomitrium intermedium* agree in retaining radial growth patterns (cf. Figs. 5.2[1,2,4,5] with Figs. 5.4–5.6); all but ♀ axes of *Phycolepidozia* agree in retaining an acropetal sequence in production of sex organs. Environmental pressures, as stressed elsewhere, have led to derivation of a wide array of types in which dorsiventrality and bilaterality prevail. One may see simple planation, with retention of lateral leaves, and with retention of a tetrahedral apical cell and of an acropetal sequence in gametangium production (as in *Fossombronia,* Fig. 5.2[3]), or, more commonly, leaves are variously reduced or lacking. This may be linked with retention of an acropetal sequence in gametangium production (as in the otherwise rather specialized *Corsinia,* Fig. 5.2[6]), but more commonly, gametangia are organized into well-defined gynoecia and androecia, as in the following examples.

Podomitrium Mitt. (Fig. 5.7)—The three preceding types show, basically, radial organization, and the last two show evolution of leaves of sterile regions, even though leaves are primitive in form and (in *Takakia*) could be argued to be short shoots. In the *Takakia* and *Haplomitrium,* the naked gametangia, at best barely shielded by neighboring leaves, do not offer much with respect to reproductive efficiency. In *Podomitrium,* we see an example of where two striking new things have occurred: (1) the gametophyte has become dorsiventral; and (2) sex has gone underground. Gametangia occur on very short branches that issue from the terete, slender stipe-like portions of the sterile gametophyte. These sexual branches are weak, determinate in length, and occur somewhat below ground level (ground level here may be beneath the fibrous bark of tree ferns—the usual habitat of *Podomitrium*). Thus, fertilization is very considerably eased.

Hymenophytum Dumort. (Fig. 5.8)—This extraordinary organism combines several primitive features (the vascularized creeping rhizomes grow by means of tetrahedral apical cells and are radially symmetric) with others that are very highly advanced. Thus, the erect sector shows planation (*webbing*) and repeated pseudodichotomous branching, with growth via an apical cell with two cutting faces, simulating the megaphyll of certain Hymenophyllaceae! Equally, sexual branches are exceedingly reduced and occupy weak, determinate ventral branches on the lower face of the frond— a position not unlike that of sporangia on the megaphylls of certain ferns!

In *Podomitrium* and *Hymenophytum*, the gametophyte shows evolutionary excursions in directions early explored by tracheophyte sporophytes: vascularization; evolution of root-like organs; and evolution of differences between axis (rhizome), root, and photosynthetic organs (megaphylls). The implications of these analogies are extremely important.

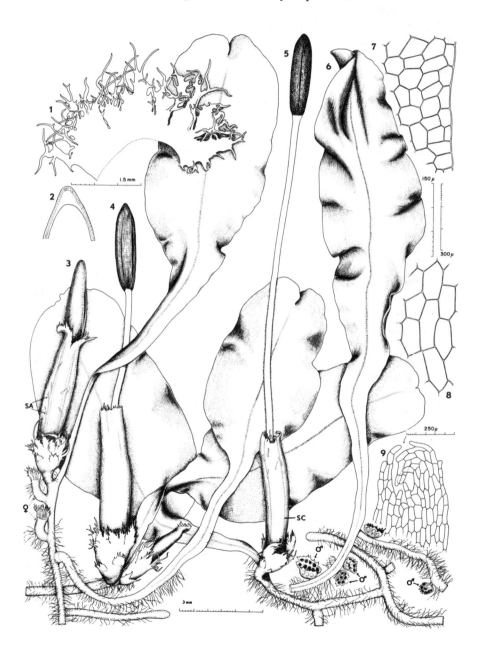

Such seemingly highly specialized features as vascularization, division into stipe (*stem*) and frond- or *megaphyll*-like sectors and evolution of geotropic terete axes (*roots*) appear to have occurred early in hepatic evolution: all but the last are already in evidence in the Devonian *Pallaviciniites*. Furthermore, even though such types of organization are associated usually with a thalloid configuration, all Pallavinioids show at least rudiments of leaves—in some species of *Pallavicinia*, reduced to mere sessile or stalked slime papillae in alternate sequence on thallus margins, but several other species of the same genus, as well as *Pallaviciniites*, retain few-celled alternate lateral *teeth*. The *leafy* to *thalloid* transition is marked in the extant related genus *Symphyogyna* (Fig. 5.9), in which some species show distinct lateral *leaves*, while in other (and sometimes the same) species there is transition to a perfectly thalloid form. Indeed, in *Symphyogyna*, transition occurs from a leafy condition, with independent lateral leaves inserted on an axis, to a condition with a deeply lobate thallus (see Fig. 5.9[2,3]; distal sectors of several shoots).

The series of examples here cited document that major differences in basic architecture are commonplace, sometimes even within the same genus. Examples illustrated in Figures 5.7 to 5.9 show *some* of the striking gametophytic changes in the single order Metzgeriales. These illustrate only a fraction of the total architectural variation: thus, we may also see bilaterally leafy types with retention of acropetal sex organ production (Fig. 5.2[3], *Fossombronia*)—as well as evolution of highly simplified, parenchymatous thalli—almost truffle-like in ecology and in simplicity of gross aspect—with sex organ restriction (Fig. 5.3; *Cryptothallus*). Similar sequences showing progressive simplification and eventual thallus formation occur in the Jungermanniales (Schuster 1966a). The extraordinary malleability in gametophyte form, in the Hepaticae as a whole, is probably a response to strong, long-continued and pervasive environmental stresses on plants with little or no cutinization, and with limited or no desiccation resistance.[12]

From the preceding one could conclude that an archetype not too dissimilar to vegetative axes of *Phycolepidozia* shows the greatest possible

FIGURE 5.7. *Podomitrium phyllanthus* (Metzgeriales), an advanced taxon, showing (*a*) restriction of gametangia to abbreviated branches from stipe-like bases, sunken in the substrate, thus, with *sex underground*; (*b*) simple, exposed, photosynthetic *fronds*. 1. Capillary ring of tissue surrounding archegonia, a device to facilitate fertilization. 2. Capsule apex, formed of coherent valves. 3–5. ♀ Plants, in part with unfertilized acrogynous gynoecia (♀), in part with them fertilized and developing a sterile sheath around the shoot calyptra (*SC*), which bears sterile archegonia (*SA*) on its face. 6. ♂ Plant, with androecia (♂); note production of cylindrical, root-like geotropic axes. 7–8. Marginal and intramarginal cells, showing the typically delicate cells of all Metzgeriales. 9. Apex of the sterile sheath (*pseudoperianth*) around fertilized gynoecia. This genus illustrates an attempt to cope with aquatic reproduction: the gametangia are on reduced branches, from buried, colorless, stipe-like axes that give rise to aerial, expanded, chlorophyllose fronds. Sterile plants of the Devonian *Pallaviciniites* are very similar.

conservation of primitive features (cf. Fig. 5.2[1 and 4]). Stresses were in the direction of: (1) bringing sex organs down as close as possible to ground surface or even (as in *Podomitrium* and even more so in *Cryptothallus*) below substrate surfaces; (2) protecting the gametangia; and (3) reducing sporophyte size and complexity, so that it would place minimal demands on

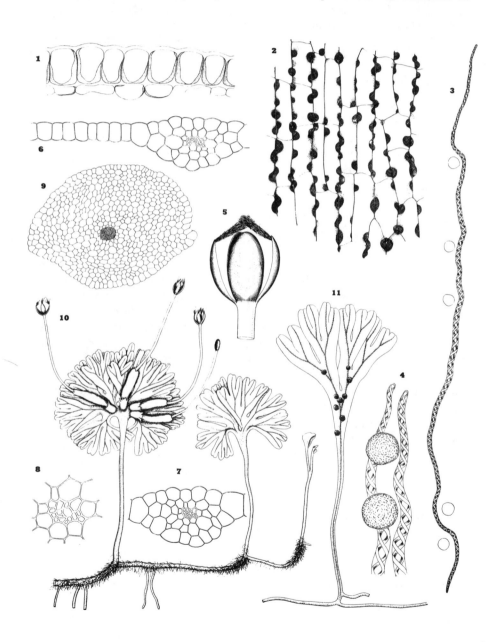

the resources of the perennial gametophyte. Much morphological and anatomical diversification has occurred in the Hepaticae to meet these needs. This diversification is, indeed, so great it tends to obscure the *basic* response. Figure 5.2[3–6], a diagram showing how *some* responses occurred to meet these needs, demonstrates that such responses in various groups of Hepaticae displayed the utmost diversity—while yet maintaining one exceedingly primitive feature—the simple, acropetal production of gametangia on the leading, gametophytic axis. In genera as diverse as *Fossombronia*, *Haplomitrium*, *Takakia*, and *Corsinia*—members of orders of both subclasses of Hepaticae—acropetal production of sex organs is maintained, even though everything else undergoes variation.

The Initial Creation of Suitable Environmental Niches

The survival strategies of early land plants can only be inferred. However, early *amphibious* candidates for transmigration had only two basic options: (1) progressive sporophyte simplification coincident with retaining the diploid within the modified archegonial wall for increasingly longer times; and (2) retention of the ancient pattern of relatively swift externalization of the embryo sporophyte, which was thus not locked into a pattern of continuous simplification. The nature of the archegonium dictates that there be *initial* sporophyte retention within gametophytic tissue, and the two options are merely extremes of a single basic type of life cycle. Evolution of the *degenerate* hepatic type of life cycle could thus be accomplished very rapidly from a more *normal* cycle involving rapid sporophyte externalization. A life cycle with rapid externalization and autonomy of the diploid would not preclude rapid selection of progressively taller diploids which would bear sporangia elevated further and further from ground level; hence, effective wind dispersal of spores, over long distances, is progressively

FIGURE 5.8. *Hymenophytum flabellatum* (Metzgeriales), a member of the Metzgeriales showing a mix of primitive traits (vascularization; terete creeping axes growing by tetrahedral apical cells) and highly advanced traits (evolution of megaphyll-like *fronds*; highly reduced sexual branches from ventral face of frond). 1. Capsule wall cross section, showing highly reduced condition with tapetum-like inner layer. 2. Epidermal capsule wall cells, showing secondary thickenings stiffening the cells. 3. Highly attenuated, *Haplomitrium*-like elater and spores. 4. Parts of fig. 3, highly magnified. 5. Capsule, dehisced, showing attached elaters at valve apices. 6–7. Part of ultimate segment of frond, showing part of wing, costa, and weak vascular strand. 8. Vascular strand of ultimate segment, in detail. 9. Cross section of stipe-like base, showing sophisticated vascular strand and cortex. 10. ♀ Plant, showing seasonal production of solitary *megaphylls*, the oldest with ventral sexual branches bearing sporophytes; note hymenophyllaceous growth pattern, with development of root-like axes. 11. Same, ♂ plant, the round bodies are highly reduced ♂ branches, reduced to a cushion of tissue into which antheridia are sunken. This plant illustrates one ultimate point in evolution, where the gametophyte approaches that of Hymenophyllaceae sporophytes in size and complexity.

FIGURE 5.9. A primitive species of the genus *Symphyogyna* Nees & Dumort. (Metzgeriales), showing maximal retention of lateral leaflike organs and the typical bilateral organization of all Metzgeriales. 1–3. ♀ Plants, showing perigynium (*p*) with unfertilized archeogonia (*a*) at apex; branches are both terminal (*tb*), some of which may remain dormant (*dtb*), and ventral intercalary (*vib*). 4. Cross section of shoot, a complete *wing* (*leaf*) at right; note conducting strands. 5. Same, the wing at left. 6. Part of axis in cross section to show vascular strand. Note the dorsal position of gynoecia, each with protective scales to the posterior (fig. 3), the archegonia in groups (gynoecia), acropetally produced.

facilitated. By contrast, on opting for progressively longer retention of diploid within haploid tissues, and eventual internalization of sporophyte within gametophytic tissue to the point where spores are mature prior to sporophytic extrusion, hepatics were automatically locked into an inferior life-style: with intense selection pressures to lower and/or reduce the gametophyte, the dependent sporophyte could not exceed the gametophyte in size or complexity. Thus, the selective pressures leading to reduction and simplification of the gametophyte were, in effect, also transferred to the sporophyte. Elevation of sporangia to a level where wind dissemination of spores is truly effective thus fails to occur, at least in early ground dwellers. Initial hepatic types had unisexual gametophytes. Combined with less effective wind dispersal of spores, this further emphasizes the intrinsic inferiority of the hepatic life cycle.

Implicit is progression from a plant type with alternation of (at least eventually) free-living haploids and diploids to the hepatic type of life cycle; this progression could be accomplished repeatedly and in a short time span. There should, therefore, not be anything surprising about the possibility that the Devonian explosion of early vascular plants may have been accompanied by rather rapid evolution of early bryophyte types.

Other things being equal, the tracheophyte-type life cycle is thus more primitive. On that basis alone one could assume that early vascular plants probably evolved marginally before early bryophytes. This further suggests that *before* early diversification of Bryophyta occurred, some environmental amelioration by the first vascular plants existed; early vascular plants, because of both their morphology and life cycle, could cope with more severe environmental conditions, such as might recur only very intermittently (perhaps once a century, or once a decade), than could early bryophytes. This hinges directly on the intrinsic superiority of a perennial sporophyte (able to produce spores each season in the absence of water) over a perennial gametophyte (limited in height and complexity because of the need to position gametangia close to ground level), which could reproduce only when abundant external water was available. The delicacy of gametangia versus the relative firmness and resistance to desiccation of sporangia is an obvious factor in this equation, but this is only a part of the inherent superiority, on land, of sporophyte versus gametophyte.

The durable sporophyte, erect in growth (or with erect components), of early vascular plants was in a sense preadapted for survival on land. As soon as adequate cutinization, vascularization, and evolution of stomatal mechanisms had occurred, it was—so to speak—fit for invasion of relatively hostile environments, *in spite* of the fact that the free-living (and presumably chlorophyllose) gametophyte was bound to aquatic reproduction.

One can outline a scenario, hypothetical, but surely recurrent, some 400 m.y. BP (million years before the present) that illuminates various facets of land invasion by sporophyte-dominant organisms: (1) Amphibious peren-nial gametophytes, under strong reduction pressure, after countless seasons of fruitless production of gametangia, *finally*, due to a fortunate juxtaposi-tion of gametangium maturity plus a long period of nearly constant rainfall, high cloud cover, and high humidity, produce *one* embryo which survives to form a mature sporophyte; it is assumed that thousands of tries were needed to accomplish a single fertilization. (2) The ensuing sporophyte, growing via a creeping and ramifying axis, showing gradual basal decay of older sectors, forms a *large and indefinitely perennial clone*; countless branches each produce sporangia, and these sporangia are produced (seasonally or contin-uously) over countless seasons; innumerable spores are produced each season. The cloning of the single sporophyte proceeds to the point where, after decades, large sporophyte stands are produced. (3) Since sporophyte stands produce potentially viable disseminules (spores) without reference to external factors—availability of water becomes noncritical, so long as water is available in sufficient quantity to allow vegetative survival. (4) The sporophyte stands are eventually extensive and dense enough to increase surface moisture (humidity) appreciably, to provide at least minimal amounts of shade, and the decay of older sectors of the clones creates at least a thin, moisture-retentive organic soil layer. (5) Evolution of a ground microenvironment that significantly ameliorates the scenario outlined under (1) follows. Subsequent spore germination, gametophyte development, fertilization, and embryo formation is thus eased. This scenario entails only *one* initial successful ecesis by a single bisexual gametophyte.[13]

Once ecesis on land succeeded, *progressive amelioration* occurred rapidly—and progressively further from the original marginal and semiam-phibious areas. Such a scenario partly explains *why* the Devonian tracheo-phyte explosion occurred. Once the initial problems of land invasion were solved and land organisms existed, the nature of the tracheophyte life cycle permitted *sporadic* invasion of *normally* relatively gametophyte-hostile environments, and, of course, anything that the gametophyte could tolerate could be tolerated by the sporophyte. *As soon as* such amelioration occurred, long-lived gametophytes became less and less necessary: a higher percentage of gametophytes survived, hence less reliance had to be placed on individual (long-term) gametophytic survival. Thus, an early pattern of evolution away from *gametophytic dependence* is visualized. This, of course, is *exactly the reverse* of the situation in the Hepaticae.

No such scenario can be visualized for early bryophyte types: with unisexual perennial gametophytes, on which the sporophyte had become dependent, even very intermittently limiting environmental problems, such as a persistent drought, would be sufficient to eliminate entirely the gametophyte—and with it the attached sporophyte. *If* fertilization does occur, the short-lived sporophyte, synchronously producing spores, *a single time*, allows only a "one shot" chance at survival. Liebig's law applies here with a vengeance: any intermittent catastrophe that would eliminate the gametophyte would, of necessity, also eliminate the gametophyte-dependent sporophyte. With the indefinitely persistent sporophytes of early vascular plants producing spores seasonally over many years (the basic *Lycopodium* life cycle), generation after generation of gametophytes could be wiped out, such organisms would still survive—so long as, even very rarely, *some* gametophytes could mature and produce embryo sporophytes.

I have perhaps overtaxed the obvious. But however obvious, the evolutionary and paleoecological implications have never been really carefully analyzed in a comparative fashion. It is, thus, evident from the natures of their life cycles that: (1) early vascular plants could more effectively (and more rapidly) colonize sites that were almost self-excluding for taxa with perennial gametophytes; and (2) possessing bisexual gametophytes, early vascular plants could—if outbreeding was impossible for spatial reasons—self-fertilize so that sporophyte formation remained possible; having unisexual gametophytes, early bryophytes had to have ♂ and ♀ gametophytes nearly juxtaposed so that fertilization could occur;[14] since often, under the harsh conditions of the early land surface fertilization was impossible, such gametophytes would have had to persist for years before sporophyte production became possible.[15] Hence, early bryophyte types would tend to be excluded from difficult loci where indefinite gametophyte survival was impossible. In essence, the combination of short-lived gametophyte plus long-lived sporophyte is enormously more fit for invasion of intermittently difficult sites than the opposite combination.

In a paleoecological sense, the above facts suggest that: (1) of necessity, primary invaders on land had to be early vascular plants; (2) with clone formation and aggregation of perennial sporophytes, early vascular plants could rapidly ameliorate initially harsh and limiting environments; (3) relatively sheltered microenvironments were, possibly rapidly, created to allow a higher percentage of gametophytes to survive for longer periods—hence, the environment was modified at least marginally to allow enhanced gametophyte survival—survival of gametophytes of the initial invader, and of other types of gametophytes, of both vascular plants and of bryophytes; and (4) by their very nature, therefore, the bryophytes had to be followers, rather than leaders, in the process of progressive invasion of increasingly difficult sites.

The inferiority of early bryophyte types dictated that both land invasion and early diversification had to come *after* preconditioning and amelioration of the land environment by the intrinsically superior early vascular plants. On paleoecological bases, therefore, the classical cliché that Bryophyta gave rise to vascular plants seems to be demonstrable nonsense.

The time lag between the Devonian proliferation of early tracheophytes and the first, Carboniferous, evolution of a series of recognizable hepatic types is thus not unexpected: the time lag is implicit in the preceding paragraphs. From the very beginning of their evolution, Hepaticae have been tracheophyte-dependent. Their diversification ultimately depended on *niche creation* by higher plants. Only by the end of the Devonian, when ferns and seed ferns (such as *Archeopteris*) began to play a significant role, can we visualize the formation in swampy lowlands of humid and sheltered niches in which organisms with perennial gametophytes could survive in considerable numbers and diversity.

Inferred in the preceding conclusions is that the first known (Devonian and Carboniferous) Hepaticae already represent extant groups in all of which we already, uniformly, find that the sporophyte is surrounded by gametophytic tissue(s) until spore maturity; it is assumed (Schuster 1977, 1979a) that such a life-style evolved at the very beginning of hepatic evolution. Implicit in the bryophyte life cycle is a situation where the subsidiary role of the sporophyte dictates that the latter *cannot, in any way, condition the environment of the next generation of gametophytes.* By contrast, the erect and ramified, indeterminately growing sporophytes of all tracheophytes, if aggregated in colonies, serve to ameliorate, in general, the environment for the next generation of gametophytes. Thus, marginally at least, the life cycle of even the earliest Tracheophyta had to be *superior* to that of early Bryophyta.

THE MESOZOIC EXPLOSION OF THE ANGIOSPERMS AND THE COEVOLUTION OF MODERN JUNGERMANNIALES AND METZGERIALES

Mesozoic Fossils

Fossil evidence for early Mesozoic bryophytes is scanty; Arnold (1947) noted that "fossil mosses in the Mesozoic rocks are almost nonexistent"; in the following quarter century this situation has hardly eased. Thus, Krassilov (1973) notes that—as regards fossil data—"Mesozoic mosses are even more scanty than hepatics," and goes on to describe two Upper Jurassic and Lower Cretaceous leafy Hepaticae (Jungermanniales: *Cheirorhiza* and *Laticaulina*). Halle (1913) had earlier described a Jurassic fossil from the Antarctic as *Schizolepidella gracilis*; this seems to be the first solid evidence of Jurassic Jungermanniales (Schuster 1966a). During this same time span, a number of Marchantioid types are known ("*Hepaticites*" *cyathodes* [Townrow 1959]; *Striatothallus*, *Aporothallus*, and *Riccia*-like forms [Krassilov 1973];[16] *Marchantiolites porosus*, *Ricciopsis scanica*, and *R. florinii* [Lundblad, 1954]). Perhaps, the best documented Triassic fossil is *Naiadita lanceolata* (Harris 1938, 1939)—a plant probably forming a suborder, the Naiditineae (Schuster 1966a, p. 385) of the Marchantioid order Sphaerocarpales. Thus, by the Jurassic, members of four orders of Hepaticae existed: Sphaerocarpales, Marchantiales, Jungermanniales, and Metzgeriales. Only the species-poor, small orders Monocleales, Treubiales, and Calobryales are unknown from Mesozoic (or any other) strata. By the Cretaceous, a considerable array of Marchantioid fossils are known (Schuster 1966a, Table 7, p. 358), as well as several supposed Jungermanniales (*Jungermanniites bryopteroides* Ball, *J. cretaceus* Berry, and *J. eophilus* [Cockerell] Steere).

In spite of the "*very few certain records of megascopic fossil remains*" Lundblad (1954) correctly notes [emphasis hers] "*representatives of living families—in cases even modern genera* [of Hepaticae]—*might be expected in Mesozoic floras.*" This conclusion is in agreement with the assumption made in the preceding section. However, I think *diversity* (at the family and genus level) remained relatively low into Mesozoic time. The wide array of liverwort types (about 8,000 to 10,000 species in seven orders and 69 families and 327 genera; see Schuster 1979a) represents a modern assemblage, and diversity has increased progressively and dramatically with accelerated angiosperm diversification. In pre-Cretaceous times, at any one time, there were surely no more than 1,000 species of Hepaticae; the eight- to tenfold increase in numbers has been in part a by-product of angiosperm evolution and diversification.

It would, however, be an error to place all responsibility for modern diversification of Hepaticae on the angiosperm explosion. The latter, in part, was surely induced not only by the usually assigned causes (perfection of the plant-insect relationship and evolution of progressively more effective dispersal devices employing passive methods, such as wind, as well as animal vectors), but by the fact that the time span of the Cretaceous into the Tertiary marks the period I refer to as the Great Tectonic Revolution. During this period, fragmentation of sialic plates and their mass migration resulted in the evolution of a world with a larger number of wholly isolated fragments than at any time since land plants evolved. Fragmentation and consequent isolation led to increased diversification (*provinciality*) in both groups.

Environmental Amelioration by Angiosperms

The scanty fossil evidence for Early Mesozoic Bryophyta *may* not be wholly due to imperfections in the fossil record. Some of the paucity in records of Hepaticae and Musci from this time may reflect: (1) lack of abundance of individuals and (2) lack of diversity. This, in turn, *may* reflect lack of a wide array of suitable microenvironments. I wish it were possible to show, in one way or another, whether the trunks of the Late Paleozoic Lepidodendroids were suitable loci for corticolous bryophytes; surely, the fallen logs and resultant humus must have been invaded. However much one tries to extrapolate from knowledge of *existing* groups, it would seem that one prime series of *present-day* microenvironments was *not* utilized by Hepaticae at the start of the Mesozoic: the bark and leaves of Cordaites and other early gymnosperms. Probably, at the start of the Mesozoic, the ancestors of today's predominantly epiphytic suborders Porellineae and Radulineae existed, if at all, as genus- and species-poor generalized saxicolous types. Probably 90 to 95 percent of the members of the five most (only?) successful groups of Hepaticae, the Lejeuneaceae (perhaps over 2,000 species in about 90 genera), Jubulaceae (over 400 species in about five to six genera), Radulaceae (about 250 species in one genus), Porellaceae (more than 100 species in two to three genera), *Plagiochila* (600 to 1,200 valid taxa; numbers highly uncertain) have evolved not only in angiosperm-conditioned and dominated habitats, but *as epiphytes on angiosperms.*

Several observations and deductions are involved: (1) In spite of the high productivity of Carboniferous forests, the constituent taxa do not seem to have formed forests with a density corresponding to today's angiosperm-dominated forests: shade was, perhaps, less extensive and humidity in lower strata was possibly lower. All reconstructions of Carboniferous forests seem to agree that largely microphyllous trees, casting less shade, existed and that these were in less dense stands. (2) No credible evidence exists that the derivative Radulaceae, Porellaceae, Jubulaceae, and Lejeuneaceae existed prior to the start of the Mesozoic.[17] (3) The bark of tree ferns and of most gymnosperms is mostly unsuitable for the ecesis of Hepaticae, aside from certain taxa (*Podocarpaceae* and *Agathis* in Gondwanalandic areas and *Thuja, Juniperus, Chamaecyparis,* and *Cryptomeria* in Laurasian areas). (4) Permo-Carboniferous glaciations *may* have severely compressed the forested, largely frost-free zones where taxa of these families are predominant today.

Bark and leaf characters of primitive angiosperms were presumably suitable for ecesis of Hepaticae of the families cited previously. One finds numerous taxa on bark and leaves of extant primitive taxa, for example, Magnoliaceae, Illiciaceae, Winteraceae, Annonaceae, Lauraceae, and *Laurelia*. The generally smooth and relatively porous, water-retentive bark of these families is suitable for ecesis of liverworts. Also, numerous taxa (especially of genera like *Siphonolejeunea, Austrolejeunea,* and *Cololejeunea*; all Lejeuneaceae) may invade leaf surfaces of taxa of Winteraceae (especially *Tasmannia, Pseudowintera,* and *Drimys*) and Magnoliaceae (especially evergreen taxa of *Magnolia*). Thus, as soon as such primitive families of angiosperms existed, a host of new microenvironments came into being which were freely exploited by certain leafy Hepaticae—of the previously cited families, as well as to a lesser extent by taxa of *Riccardia* (Aneuraceae), *Metzgeria*, and *Austrometzgeria* (Metzgeriaceae), *which constitute the most modern and only species-diverse, successful families of the Metzgeriales*. Thus, diversity and abundance, the key criteria by which one may gauge relative success, *both* seem linked with evolution of new microenvironments suitable for Hepaticae. The series of microenvironments created is much larger than often realized: one set of taxa occurs in high-humidity sites on lower tree trunks; others, on the upper trunks of trees; still others undergo ecesis only on twigs and small branches high in the crown or only on the surfaces of leaves. As many as 10 to 15 species, belonging to five to six genera and two to three families, may undergo ecesis on a single leaf!

Angiosperm Habitats and the Acceleration of Rates of Evolution

A corollary of the fact that ancestors of the Hepaticae (and almost surely those of other Bryophyta as well) were unisexual is the frequent failure of sexual reproduction, due to formation of isolated unisexual clones. In effect, *if* sexual reproduction occurs at all, it is infrequent, so that the effective duration of a single generation may be measured in terms of decades, if not centuries. Reliance on asexual propagation results in a gametophytic "holding pattern." Such a life cycle is simply an impossibility in epiphytic taxa: the habitat is too impermanent. Bark may exfoliate; trees die; and succession and associated lowered light intensities may change the habitat perimeters so that initial invaders are soon supplanted. Living leaves, especially, form a unique problem: they may impose a life cycle of a maximum of 2 to 3 years on the invading Hepaticae.

Certain taxa have adapted themselves to angiosperm-created microenvironments: bisexuality has become frequent and in some groups (the advanced and species-rich subfamily Cololejeuneoideae of the Lejeuneaceae, for example) predominant; spatial distance between ♂ and ♀ gametangia are reduced—often to millimeters when not to microns; and asexual reproductive devices have become commonplace. One, or the other, of these devices—and sometimes both—are joined, resulting in modern reproductive strategies. In the predominantly epiphytic Radulaceae, Lejeuneaceae, Jubulaceae, and Plagiochilaceae, we find either frequent bisexuality (the first three families) or evolution of diverse and often very sophisticated methods of asexual, gametophytic propagation (all four families). In effect, strategies have evolved to cope with the impermanent nature of the angiosperm habitat; these, in general, have guaranteed free and frequent sexual reproduction (and sporophyte production: hence, meiosis and all it implies). Thus, starting in the Cretaceous, new niches evolved whose nature was such that selection pressures operated to evolve taxa of Hepaticae with short generations. A by-product of this was greatly enhanced rates of evolution.

All students of the Hepaticae are painfully aware of the fact that genus limits in families like Jubulaceae, Lejeuneaceae, and Plagiochilaceae are ill defined, that species are plastic and complex, and gaps between species may be slight. All of this diversity, and the problems involved, can thus be blamed on coevolution with the angiosperms. This occurred during a time frame which carries with it other implications, next discussed.

Effects of the Cretaceous-Tertiary Fragmentation of Laurasia and Gondwanaland

One by-product of the greatly accelerated breakup of sialic mega-continents—Laurasia and Gondwanaland—was the rupture of former widespread floras into isolated segments, which, as a consequence, underwent divergent evolution (Schuster 1976). Divergence was especially accelerated when plate migrations resulted in radical climatic alteration. Such climatic alterations could result from migrations from south to north (India and Australasia), or from migrations in any direction where plate collisions elevated mountains (thus, simultaneously creating both arid, rain-shadow environments, exploited by Marchantiidae, and rain-forest environments, exploited by Jungermanniidae). Migration of sialic plates, particularly in Cretaceous–Early Tertiary times, fragmented formerly uniform floras and also induced the creation of stresses when these plates moved from one climatic zone to the next. These stresses enhanced selection pressures and rates of evolution. The Cretaceous-Tertiary explosion in the angiosperms is largely conditioned by these facts. Equally, the associated Hepaticae underwent a roughly similar history. In short, the massive Cretaceous-Tertiary increase in diversity of the Hepaticae was in part induced by creation of new microhabitats—but, in part, was due to enhanced selection pressures resulting from sialic plate migrations and enhanced fragmentation. The specifics of these effects and the quantitative differences between them cannot be gauged today; collectively, they were enormous and account for a great deal of the diversity observable today. The high endemicity on *continental fragments*, such as New Caledonia and New Zealand, reflects these processes.

Timing, as well as magnitude, of the tectonic disruptions are exactly in phase with both modern floristic diversification of angiosperms and of modern types of Jungermanniales and Metzgeriales (of the genera *Metzgeria* and *Riccardia*), and—to some extent—of some Marchantiales. (Due to the fact that spores of the latter are large, long retain their viability, and can be transported in soil, as on the feet of birds, the latter retain a relatively high level of *cosmopolitanism*.) Numerous examples of speciation—especially the creation of species pairs and genus pairs—could be cited in support of this thesis; a large number are cited and mapped in Schuster (1979b). *Some* effects of this disruption in the Southern Hemisphere are documented by Figures 5.10 to 5.12 (see legends for details). Timing of the migrations of the relevant sialic plates suggests that, in many instances, time intervals of the order of magnitude of 60 to 80 m.y. (m.y. equals million years) may have served to establish species pairs in some cases, genus pairs in others (Schuster 1979b). Although subject to rather wide margins of error, because of the ambiguity of much of the data, this represents a first attempt at trying to establish slow type evolution rates in Hepaticae, using tectonic data as a base. (The data are derived almost wholly from taxa with the critical constellation of unisexuality plus lack of asexual propagative devices.) Timing, thus, is an essential factor in this equation; the following rough chronology is adequate for present purposes: (1) Extant fragments of the Campbell Plateau, including New Zealand, New Caledonia, Auckland, and Campbell Islands (or land areas now eroded and/or foundered that existed in that general area) were isolated by about 80 m.y. BP from Australia, but may have retained tenuous contact with Antarctica. (2) South America was isolated from Africa by 80 to 90 m.y. BP, and the "filter" (through the now widely extended Caribbean Arc?) allowing interchange with North America was ruptured by the end of the Cretaceous; by about 60 m.y. BP, westward migration attenuated and ruptured the Scotia Arc, hence cut South America off from Antarctica. (3) North America, with the opening of the North Atlantic, became isolated from Europe by 55 to 60 m.y. BP; it must then have been isolated from the now juxtaposed Asiatic Plate for much of the Early Tertiary. The Beringian Bridge probably did not exist until the middle of the Tertiary or later—by which time, rotation of the westward moving North American Plate approximated present-day Alaska to northeastern Asia. (4) By about 45 m.y. BP, Australia (and then existing and attached fragments of New Guinea) was isolated from Antarctica—and had already been isolated from the continental fragments still extant that represent the foundered Campbell Plateau. (5) By the Late Cretaceous, if not earlier, India had begun migrating from its Gondwanalandic origin toward Eurasia; it remained isolated until about 40 to 45 m.y. BP.

Such a very abbreviated chronology leaves out of consideration many significant but lesser events: creation of new land areas along midoceanic rifts, such as the Tristan Group and Kerguelen; isolation of small but old continental fragments, such as Réunion, the Seychelles, and, especially, Madagascar; the complex and fluctuating picture associated with vulcanism, in turn associated with plate-interactions, which created fragments of present-day Central America and, in part, the Antilles; elevation of the Andes, effectively cutting off the present-day Amazonian seaway, which, with sedimentation, became narrowed so that the major northern and southern plates in South America became biotically joined; the creation of a new major migratory route from southeastern Asia into present-day Australia. All these events, concentrated into *a relatively short time*, played a critical role.

The period from about 90 m.y. BP to 20 m.y. BP represents a Great Tectonic Revolution; it created the present-day world and made possible the extraordinary present-day biotic diversity. Fragmentation, assemblage of heterogeneously originating sialic fragments, creation of wide ocean barriers, creation of new mountain barriers, creation of new volcanic *target areas* for vagrant diaspores—all conspired to not only stimulate the modern diversity in angiosperms, *but had the same effect on the Hepaticae*. By the Middle Tertiary—some 30 to 35 m.y. BP—there was a world with maximal diversity. Fragmentation was surely a large factor, but creation of barriers to migration was another. Thus, starting with the almost precipitous decline in world temperatures in Oligocene times, forests and forest communities became extinct in Antarctica, with the result that the physical isolation of Antarctica served to create a major barrier in the cold Antipodes where formerly a major source for mass migrations had existed (Schuster 1976, 1979b).

ON MALLEABILITY AND PERSISTENCE: SURVIVAL OF PALEOZOIC AND EARLY MESOZOIC RELICT TYPES IN SEGMENTS OF FORMER GONDWANALAND

Critical recent investigations (Schuster 1959, 1961a, 1961b, 1963a, 1963b, 1964a, 1964b, 1964c, 1964d, 1964e, 1965a, 1965b, 1965c, 1965d, 1966b, 1967a, 1967b, 1967c, 1968, 1970, 1971a, 1972c, 1974; Schuster and Scott 1969; Schuster and Engel 1973, 1974; Fulford and Hatcher 1959; Fulford and Taylor 1960) have shown that the cool Antipodes possess a uniquely rich and largely relict flora, chiefly of stenotypic types (families often with only two to nine species; genera with one to three species; suborders with one to three families). This demonstration, partly outlined in tabular form in Schuster (1979b),[18] leads to the conviction that what we see is probably persistence of very early types (primitive Calobryales; all Monocleales; most primitive families and genera of Jungermanniales; all Treubiales; and many genera and several families of Metzgeriales) to a highly asymmetric degree; cf. Figures 5.10 to 5.14. The phylogenetically widely spaced, stenotypic remnants we see are probably the last survivors of the late Paleozoic Gondwana or *Glossopteris* flora—which, as regards the vascular plants, has long been extinct. *If* this hypothesis can be substantiated with fossil data *then* the nature and constitution of the hepatic flora of the cool Antipodes becomes of pervasive significance. At the least, one must assume that many of these relict taxa represent survivors from earliest Mesozoic time. The evidence is circumstantial: the taxa are nearly all cool adapted and probably evolved in Late Paleozoic time as periglacial and alpine elements, as a response to the widespread climatic changes that arose in Late Paleozoic time. Extensive glaciation of Gondwanaland in the Late Paleozoic displaced the *Glossopteris* flora far to the north—isolated glossopterids have been shown to occur north as far as present-day Mexico. Prior to the Late Paleozoic glaciations, conditions in Gondwanaland were presumably such that little basis existed for evolution of a diversified cool- or cold-adapted flora, involving numerous genera (by a conservative estimate about 50 to 60 genera of Hepaticae in 20 to 25 families survive today!);see Figures 5.10 to 5.13. Such a copious and diversified relict flora could have evolved as a response to glaciations and coincident climatic deterioration of *either* Late Paleozoic or Late Tertiary–Pleistocene, but the remote phylogenetic lines between many groups of Gondwanalandic taxa can only be explained by evolution during the earlier period of climatic deterioration. It is impossible to visualize origin of Treubiales (Fig. 5.13), Monocleales, Calobryineae (Haplomitriaceae; Fig. 5.14), Phyllothalliineae, Lepidolaenineae (Fig. 5.11), Ptilidiineae, Hymenophytaceae, Verdoorniaceae, *Noteroclada*, Balantiopsidineae, Herbertineae (Fig. 5.10), and many other isolated taxa in the period since the start of the Tertiary. Thus, the *Glossopteris* land flora is not wholly extinct: at a time prior to the origin of the dinosaurs, the direct ancestors of the aforementioned taxa of Hepaticae existed, probably little changed from extant types. This startling assertion seems to stand in direct contradiction to data in the preceding section; hence, there appear to be strikingly different evolution rates.

Rapid versus Slow Rates of Evolution

Stress was earlier placed on the fact that certain modern taxa (e.g., Radulaceae, Jubulaceae, Lejeuneaceae, and the genus *Plagiochila* of the Plagiochilaceae) exhibit massive and modern speciation and/or genus formation, largely but not exclusively associated with warm to tropical climates. Analogous modern speciation occurs, to a minor extent, in some families and/or genera found chiefly in cold to frigid climates (e.g., Lophoziaceae in cold parts of Laurasia). However, in much of the Antipodes, one finds, chiefly, persistence—often of single species in ancient genera, of monogeneric or digeneric families, of suborders including few genera and species, with relict and disjunct ranges (Figs. 5.10–5.12), suggesting there has been *some* survival, through long time periods of *old* and generalized, malleable taxa. Of the 13 to 14 suborders of Jungermanniales, generalized or primitive stenotypic genera remain in some 10 to 11 of these (Schuster 1972c, 1979b); generalized representatives of all but one of these 10 to 11 suborders are found today exclusively or very largely in sectors of former Gondwanaland; see Figure 5.13. Persistence in this region of an extraordinarily large

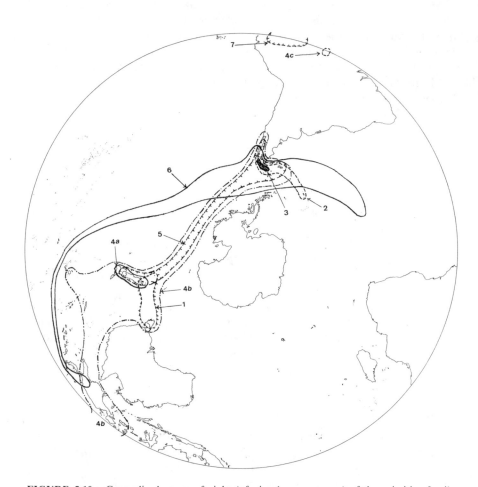

FIGURE 5.10. Generalized range of eight (of nine known genera) of the primitive family Blepharostomataceae (Pseudolepicoleaceae; Jungermanniales). Most primitive are the monotypes, *Pseudolepicolea* (2), *Temnoma* subg. *Eotemnoma* (4a). With the 60 to 80 m.y. period in which taxa were physically isolated, plus the destruction of Antarctica as a refugium, starting by Oligocene times, there was creation of species-pairs and genus-pairs. Taxa are all strongly mesophytic and often paludicolous and do not tolerate frequent desiccation. Only two species (of *Temnoma* subg. *Temnoma*) have extended their range significantly beyond the boundaries shown: *T. setigerum* and *T. quadripartitum*. 1. *Isophyllaria* Hodgs., monotypic. 2. *Pseudolepicola* Fulf. & Tayl., monotypic (allied to *Lophochaete* Schust., which underwent *rafting* on the Indian Plate; see Schuster 1976). 3. *Herzogiaria* Fulf. & Tayl. and *Fulfordiella* Hässel, both monotypes. 4. *Temnoma* Mitt.; *a*, the monotypic subg. *Eotemnoma* Schust.; *b*, subg. *Temnoma* with about ten species; *c*, subg. *Blepharotemnoma* Schust., a monotype known only from Venezuela. 5. *Archeophylla* Schust., with three species 6. *Archeochaete* Schust., with three species. 7. *Chaetocolea* Spruce, monotypic. Cf. Figures 5.11 and 5.12.

FIGURE 5.11. Generalized range of Jungermanniales: of all four genera of the Lepidolaenaceae, 1–4, and of the monotypic, isolated Goebeliellaceae, 5. 1. *Gackstroemia* Trev., with six species (four South American, two Australasian). 2. *Lepidolaena* Dumort., with six species, all in Australasia. 3. *Lepidogyna* Schust., a species-pair of South America and Australasia. 4. *Jubulopsis* Schust., forming a monotypic subfamily Jubulopsidoideae Hamlin; subantarctic islands south of New Zealand to southern New Zealand. 5. *Goebeliella* Steph., monotypic; restricted to fragments of the former Campbell Plateau, from New Caledonia to New Zealand.

number of primitive taxa (Schuster 1979b) is true not only of the Jungermanniales cited, but of Monocleales and Calobryales (the two most primitive species of *Haplomitrium* occur in Australasia); see Figure 5.14. Why such disproportionate survival of generalized types on former Gondwanaland, with the notable exceptions of Africa and Antarctica?[19]

Such persistence of generalized taxa probably reflects three principal factors: (1) Persistence of favorable climates through long time spans: with rainfall patterns, such as occur in south Chile, Fuegia, western and especially southern New Zealand, western and southern Tasmania, parts of eastern Australia, and—above all—on various subantarctic islands (e.g., Campbell Island and Auckland Islands), there is persistence today of climates which must have long characterized much of the former circum-Thallassan basin. (2) Persistence throughout relevant geological time (from Late Devonian to Early Mesozoic) of a large supercontinent, Gondwanaland (Bambach et al. 1980), *some* of which was located in appropriate latitudes through almost all relevant geological time. By contrast, prior to the assemblage of Pangaea, the plates that eventually formed Laurasia existed principally as relatively small, rather freely migrating sectors. Of the 13 to 14 suborders of Jungermanniales, only one (Antheliineae) seems clearly of non-Gondwanalandic origin (it may have evolved in Baltica or Laurentia, prior to the assemblage of Laurasia). (3) Slow, if not sluggish rates of evolution linked with *somatic malleability*. The generalized, morphologically malleable—I am tempted to use the inelegant but expressive term "sloppy"—taxa that form the *base* of the 10 to 11 suborders of Jungermanniales, and all taxa of Calobryales and of Monocleales are relict in range and often endemic to small areas. They represent very stenotypic groups. (Thus, the Calobryales include only about nine to ten species of *Haplomitrium* and two taxa of *Takakia*; Monocleales, two taxa of *Monoclea*; the primitive subisophyllous taxa of the 10 to 11 suborders of Jungermanniales cited each include, on an average, no more than one to four species per genus and one to six genera per family. For details see Schuster 1979a.) Linkage of stenotypy, relict range, unisexuality, and lack of asexual propagative devices collectively dictates: (1) ineffective dispersibility; and (2) slow rates of evolution—often the taxa occur in clones of a single sex. Many genera (e.g., *Grollea* Schust., *Herzogiaria* Fulf., *Anisotachis* Schust., *Isophyllaria* Hodgs., and *Eoisotachis* Schust.), known from a single sex, never go through the sexual cycle; lacking sporophytes, recombination and meiosis—with all their implications—have been lost. Such taxa, often limited to difficult niches where competition is low or almost nonexistent, represent the last recognizable remnants of the old *Glossopteris* flora (Figs. 5.10–5.12).

With elimination of fertilization and meiosis only intrinsically malleable taxa would survive: they are usually no longer able to migrate; hence, their theme of existence must be: persist or perish. Most, surely, have perished— but in the cool to cold Antipodes, a significant number persist, and it is from the environmental demands and tolerances of these taxa that we must extrapolate backward in time, to attain some idea of the paleoecology of Paleozoic Hepaticae.

Most such taxa are cited in tabular form in Table I in Schuster (1979b). Leaving out several rather advanced genera (Tuyamaelloideae, *Anastrophyl-*

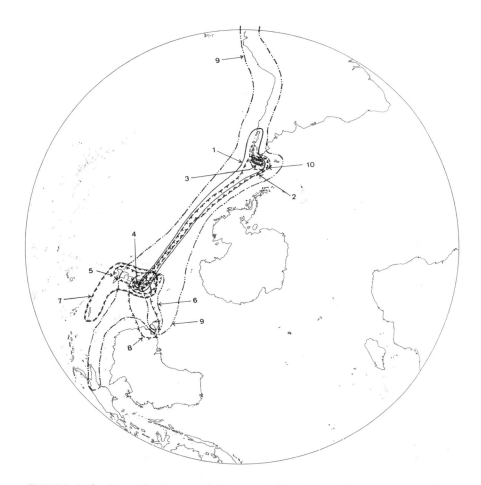

FIGURE 5.12. Generalized range of 11 genera of Jungermanniales. 1. *Austrolophozia* Schust., with two species in South America, one in New Zealand. 2. *Acrolophozia* Schust., with a species pair only. 3. *Grollea* Schust., monotypic. 4. *Neogrollea* Hodgs., monotypic and *Megalembidium* Schust., monotypic (both Lepidoziaceae). 5. *Lembidium* Mitt., two species, only in New Zealand. 6. *Isolembidium* Schust., monotypic, from Campbell I. to Stewart I. and Tasmania. 7. *Chloranthelia* Schust., two or three species, from New Zealand to New Caledonia. 8. *Kurzia* subg. *Dendrolembidium* (Herz.) Grolle, two species. 9. *Pseudocephalozia* Schust., six species, three Australasian, three American (one of which has extended its range in the Andes to Colombia, Venezuela, and Costa Rica). 10. *Pleurocladopsis* Schust., monotypic.

Figures 5.10 to 5.12 show the ranges of some 23 generic units, out of about 45 to 50 that could be mapped, *all* showing the identical pattern of circum-Antarctic range, and/or reduction to isolated monotypes. It is assumed that ancestors of nearly all were present in Antarctica, probably to at least Oligocene times, and prior to 80 m.y. BP, the ranges were essentially continuous. If the ranges were plotted on a Permian reconstruction, they would show a basic identity with those of *Treubia* (cf. Fig. 6 in Schuster 1972a).

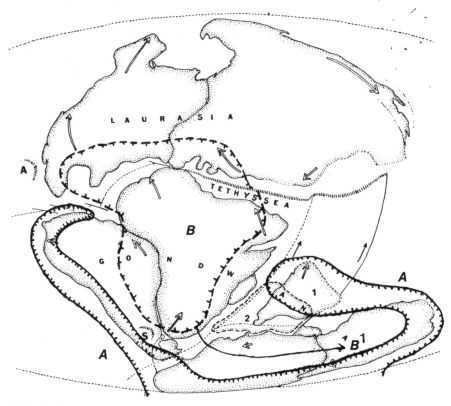

FIGURE 5.13. Generalized distribution of: A. Primitive taxa of orders with furrowing-type meiosis (Calobryales, Treubiales, Jungermanniales, Metzgeriales, Monocleales; cf. Figures 5.10 to 5.12, in which the actual distribution of some 23 genera with peripheral-Gondwanalandic ranges is plotted. B. The continental range of most taxa of orders with meiosis that is followed by cell plate formation (Sphaerocarpales and Marchantiales). The latter group is not known to have occurred prior to the Jurassic; two species of *Riella* (Sphaerocarpales) are known from Australia; arrow, B¹. Base map from Schuster (1976, based on Dietz and Holden 1971; position of Indian Plate uncertain, probably somewhere between 1 and 2).

lum, Adelanthus, and *Wettsteinia*), the remaining 29 genera contain an average of only three to four species each; the average relevant family includes fewer than four genera. *If* we include the remainder of primitive Gondwanalandic taxa, not cited in Schuster (1979b), including Vetaformaceae (one genus, one species), Chaetophyllopsidaceae (two genera, two species), Lepicoleaceae (one genus, seven to eight species), *Grollea* (one species), *Triandrophyllum* (four to five species), Trichotemnomaceae (one genus, one species), *Pachyglossa* (four species), Hymenophytaceae (one genus, one species), Verdoorniaceae (one genus, one species), *Noteroclada* (one species), etc., *then* a picture of stenotypy at the generic and species level obtains suggestive of both age and persistence.

Detailed analyses of these distributions (Schuster 1969, 1972a, 1976, 1979b), allow several pertinent generalizations: (1) The ranges of these stenotypes are not only basically Gondwanalandic, but—where showing slight expansions beyond the borders of former Gondwanaland—there is a tendency for these ranges to extend only along the fringes of the ancestral Pacific, or Panthallassa.[20] Thus, *some* of these ancient ranges are purely Gondwanalandic, with current distribution confined to the oceanic fringes of former Gondwanaland; the ranges of other taxa involve oceanic, moist sectors of other parts of the borders of the ancient Pacific. By the time the events here discussed occurred, there presumably had been assemblage or collision of the fragments (Angara, Cathaysia, etc.) that eventually formed significant portions of Laurasia. (2) The ranges seem to reflect remnants of former more considerable ranges which were basically transantarctic (Schuster 1979b).

This summation excludes pertinent ecological data. A consideration of the ecology of the cited taxa suggests that: (1) Many taxa have solved the problem of water loss by growing in or at the margins of rills or cascades (*Herzogiaria, Eoisotachis, Pachyglossa,* and *Monoclea*), or on boulders or rock walls fringing streams where regular submersion must occur (*Vetaforma* and *Triandrophyllum*), or on rock faces where rain water in sheets is a frequent factor (*Pleurocladopsis, Pachyglossa, Grollea, Isotachis,* and *Herberta*), or they have (secondarily?) reinvaded aquatic sites, such as moory rills (*Eoisotachis, Anisotachis,* and *Pseudolepicolea*). (2) Angiosperm *dependence* is low, or lacking—these taxa inhabit niches that occurred, presumably, in much the same fashion in pre-Cretaceous times. (3) These ecological niches are *in phase* with assumptions that land invasion by the Bryophyta was from estuarine-riverine sites, by migration, upstream so to speak, from original loci of invasion. By contrast, *no generalized* genus today inhabits microhabitats that can be derived from lake margin habitats. There is *nothing* in the ecology of primitive taxa of Hepaticae that would support the assumption that land invasion occurred from shallow waters of closed bodies of fresh water.

CONCLUSIONS

Striking differences exist between Bryophyta, Anthocerotophyta, and Tracheophyta in aspect, reproductive patterns, and strategies (Tracheophyta show a fundamental reliance on the sporophyte as the durable reproductive unit versus the gametophyte as the long-lasting reproductive unit in Hepaticae and other Bryophyta), sexuality (the Tracheophyta gametophyte seems to be genetically bisexual, at least in all relevant extant taxa, versus unisexual in not only the Hepaticae, but in Musci as well), and survival strategies. In Tracheophyta, experimentation with both chlorophyllose and chlorophyll-free gametophytes was repeatedly initiated; in many groups the critical gametophyte stage showed early tendencies to become subterranean and mycorrhizal, in others, such as the ferns, short-lived. In Hepaticae and Anthocerotae, reliance has been on persistent, chlorophyllose surface gametophytes (submersion occurred in only a single genus of Hepaticae, *Cryptothallus*). I assume that taxa with green surface gametophytes, as in modern ferns, became successful chiefly after forest environments were created, starting with Late Devonian time, but principally in Carboniferous times. Experimentation with heterospory occurred early in Tracheophyta; not in Bryophyta. These and other differences suggest that the history of the Bryophyta, Anthocerotophyta, and Tracheophyta has been a distinct one since the very beginning of land invasion: it may be that the first represents a group with basically unisexual gametophytes that showed very early tendencies to evolve gametophytic leaves, the last group, a sequence that showed basically bisexual gametophytes that never evinced *any* tendency for gametophytic leaf formation. In the first, perhaps, clearly limited symbiosis with fungi led to retention of the initial reliance on gametophytic photosynthesis and thus strictly limited the gametophyte to land surfaces (hence, exposed to the dangers of desiccation and to limited availability of capillary water needed for sexual reproduction). In the vascular plants, we find apparently early adoption of diverse gametophytic strategies: chlorophyllose surface gametophytes *and* subterranean mycorrhizal types of considerable longevity, as well as green, short-lived types. Groups with gametophytes of the last two types, obviously better adapted for the at least intermittently harsh land environment than taxa with long-lasting photosynthetic surface gametophytes, *may* have been the initial land invaders.

Only *after* such groups modified or prepared the initial, raw land environments, perhaps through clonal formation of large and relatively dense stands of sporophytes that ameliorated the land environment, is it likely that groups with persistent, delicate green surface gametophytes could really succeed on land. Hence the Bryophyta, and, especially, the desiccation-sensitive, primitive Hepaticae, as well as the Anthocerotae, and perhaps some groups of Tracheophytes with green surface gametophytes, probably were secondary invaders of the land surface.

The *inadequacy* of Paleozoic Hepaticae may, in large part, derive from another factor: the dominant, indefinitely perennial gametophyte, being haploid, is genetically *inferior* to the dipolid sporophyte. There are clear inadequacies in a life cycle where the long-lived haploid, with a single genome (hence, with even *mildly* deleterious alleles freely expressed), is exposed to the exigencies of an environment which, during a matter of decades, may repeatedly approach the margin of tolerance of that gametophyte. The permanently attached diploid, intrinsically genetically superior (among other things, because recessive deleterious genes are masked), is, so to speak, damned by the limitations of its *host*. By contrast, in evolutionary

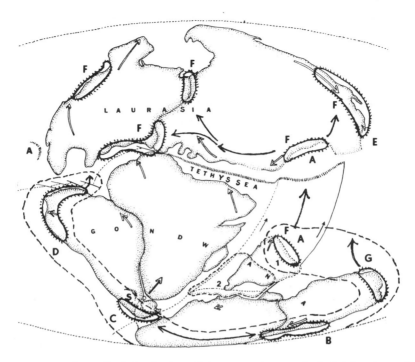

FIGURE 5.14. Generalized distribution of the constantly mesophytic Calobryales, *s. lat.*, superposed on a reconstruction (based on Dietz and Holden 1970, modified by Schuster 1976) of the world about the time Pangaea began its breakup at the end of the Triassic (about 180 m.y. BP); 1 indicates the position of India (as per Dietz and Holden), but a position nearer 2 seems more likely at that time. Distributions as follows: *A*, the distributional center of *Takakia* (both *T. lepidozioides* and *T. ceratophylla* here; it is uncertain whether the original range was Gondwanalandic, on the Indian Plate, or on the north fringe of the Tethys, in Laurasia); here also are found two to three other species of *Haplomitrium* (*H. indicum*, *H. hookeri*—the two are perhaps identical—and an additional, poorly known species), so that the greatest concentration of Calobryales is at A. *Question*: Is this due to the juxtaposition of Laurasian and Gondwanalandic elements? B, range of two primitive *Haplomitrium* species (*H. intermedium*, *H. ovalifolium*, plus a third, somewhat more advanced species, *H. gibbsiae*). C, range of *H. chilensis*, clearly allied to *H. gibbsiae*; the two, probably, once constituted a single taxon with transantarctic dispersal (*via arrows*). D, range of *H. andinum*, now extended to the Antilles (*arrow*). E, range of *H. mnioides* (note presence there also of F, *H. hookeri*). F, range of *H. hookeri* (this may derive from an Indian Plate infusion into Laurasia, spread in mid-Tertiary times, and subsequent extinction in intervening areas; current range includes coastal Greenland, Spitsbergen, northern Europe (central F), eastern North America, central Europe (lower-median F), Pacific North America, Japan, and northern India. G, *H. blumei*, allied to E, *H. mnioides*; range extends across present-day Wallace's Line (*arrow*). Note the restriction, basically, to the fringes of Pangaea, with some spread around the eastward-extending sinus between (present-day) North America and Europe. The most primitive complex in *Haplomitrium* includes *H. intermedium* and *H. ovalifolium* (the latter with some similarities to *Takakia* in leaf form!); this complex and the rather primitive *H. gibbsiae-chilensis* complex are today cool-Gondwanalandic. The origin of *Haplomitrium* may have been in the area B–C. The rather primitive *H. hookeri* may be a sibling of the primitive *H. ovalifolium*, carried north on the Indian Plate and then dispersed (*arrows*) throughout parts of Lauarasia. Note the general absence from continental areas (exception: *H. hookeri* is known from Colorado!); the absence from Africa is noteworthy. The general area of the original range of *Haplomitrium* may have encompassed the area within the heavy broken line.

sequences with short-lived gametophytes and perennial sporophytes, survival of the inadequate gametophyte has to be over only short time spans. With countless spores produced by long-lived sporophytes, the opportunity exists for sporadic survival of the admittedly inadequate gametophytes, *if* these are repeatedly produced during a wide spectrum of environmental extremes over long time spans. Since the sporophyte is indeterminately perennial and can repeatedly, if not continuously, produce spores over long time spans, gametophytes can be produced in large numbers also over long intervals—thus, those lost due to environmental stresses, such as inadequate growing seasons, can be replaced, year after year.

Associated with these differences is one final, important difference between the Hepaticae and Anthocerotae, on one hand, and the various groups of early vascular plants on the other: basic chromosome numbers. In hepatics, the original haploid base number was apparently $n = 4$ or 5; this still occurs only in the primitive *Takakia*. In Anthocerotes, $n = 6$ (reports of $n = 5$ by Proskauer [1948] have never been confirmed by other workers; see Newton [1971]). There appears to have been early gametophytic paleodiploidy in the Hepaticae, where n today is normally 8, 9, or 10, but polyploidy beyond this point ($n = 18, 27$, or 36) is infrequent to rare. In the Anthocerotae, a single case of gametophytic diploidy is known. There thus appears to be a high level of cytological conservatism in Hepaticae and Anthocerotae. By contrast, as is well documented, in living members of primitive vascular plants (e.g., *Psilotum*, Ophioglossales, various modern ferns, lycopods), high levels of genomic duplication are the norm. Thus, deleterious recessive alleles are masked to a much higher extent in the gametophyte generation. I cannot assay the *degree* to which these differences may have contributed to the relative lack of success of Hepaticae and Anthocerotes versus the early vascular plants. There *seems* to be some relationship, however, between cytological conservatism, as expressed in lack of extensive polyploidy, and lack of success, as measured in both number of taxa and degree of dominance in the vegetation. Thus the Anthocerotae, cytologically most conservative, include only five to six genera with probably under 100 valid species. The Hepaticae, cytologically a little less conservative, include over 350 genera with perhaps 8,000 valid species. The mosses, in which polyploid series are abundant and many cases of aneuploidy are documented, also seem to show several different base numbers ($n = 5, 6, 7, 8$, and 9) and polyploid derivative numbers of $n = 10, 11, 12, 13, 14, 15, 16, 20, 21, 22, 24, 27, 28, 36$, and 56 occur frequently. Also, gametophytic polyploidy may occur in the same taxon in the same locality (thus, *Hypopterygium rotulatum* may have $n = 9, 18$, about 27, and 36; Ramsey 1967). The relatively high levels of *genetic adventurism* in the mosses are perhaps linked with the relatively higher number of taxa (550 to 600 genera; perhaps over

15,000 species) and the general dominance of mosses in those vegetational strata where bryophytes predominate. One is thus tempted to hypothesize that there is an intimate link between the other *inadequacies* of Hepaticae and Anthocerotae, and their genetic conservatism.

We thus arrive at the general conclusions that: (1) Hepaticae, or at least their primitive and stenotypic extant types, represent an isolated unit that preserve not only conservative cytological and genetical systems, but also a primitive morphology and an ecology that seems hardly changed from that of ancestral types. (2) The combination of unisexuality (hence, enhanced difficulty in fertilization, leading to an identical difficulty in sporophyte production), large spores (leading to difficulty in dispersal—hence, in positing ♂ spores next to ♀ spores, and vice versa), and inadequate gametophytic reproduction (hence, inability to disperse in the gametophytic state) are a *deadly* combination—leading, at best, to very slow genetic change. (3) Ramifications of the preceding point lead to the conclusion that taxa with this combination are believed to retain ecological perimeters today that accurately reflect those of their ancestors. Many taxa are *living fossils*—taxa whose present-day ecology reflects ecological limitations and needs that probably have come down little changed from the Late Paleozoic; the frequency of highly disjunct and relict distribution reflects this phenomenon. (4) In striking contrast, a series of modern families and genera (*Rwdula*, *Plagiochila*, *Frullania*, *Porella*, and Lejeuneaceae—all modern Jungermanniales, often with high desiccation tolerance) show the *angiosperm syndrome*: explosive evolution—with poorly demarcated species and/or genus limits, linked with rapid evolutionary rates and short effective life cycles— sometimes with a duration of a year or less, analogous to the modern *annual* angiosperm. Such groups give us no valid basis for deducing anything about the paleoecology of the Hepaticae, since they clearly have coevolved with their frequent, or normal hosts, the angiosperms.

The Hepaticae are thus a complex group to which easy generalizations are difficult to apply; they include both ancient groupings that seem to have evolved in the *shadow* of the earliest vascular plants, and a limited suite of truly modern types, whose diversification—if not evolution—was conditioned clearly by the Mesozoic and Cenozoic shift from a gymnosperm-dominated to an angiosperm-dominated world.

ACKNOWLEDGMENT

I wish to acknowledge the assistance of my wife in all phases of the preparation of this chapter, and in the typing and editing of the several versions this manuscript went through.

NOTES

1. Parts of such an imaginary evolutionary progression are shown in Smith (1938, 1955), in which *Anthoceros* is portrayed as evolving into a Psilophyte!

2. Conceptual bases for this assertion rest in part on internal evidence derived from study of primitive Hepaticae.

I refrain from using biochemical data because, to date, such data are ambiguous. Biochemical evolution and evolution of larger taxa do not seem well correlated. Thus, even though analysis of the fine anatomy of the spermatozoid suggests that Calobryales are very primitive (Carothers and Duckett 1979), a conclusion I had arrived at from analysis of the entire spectrum of data then available (Schuster 1966a, 1967b, 1971a), recent biochemical data have been interpreted to suggest the opposite. Similarly, biochemical study of evolution within Marchantiales supported my classification into suborders (Schuster 1966a)—but this initial analysis has not been supported by other biochemists. The ambiguous and fragmentary nature of this type of data is also shown in this work: Cooper-Driver and Swain (vol. 1, chap. 4) show that flavonoid chemistry of the Hepaticae suggest they are more advanced than the mosses— even though fossil evidence fails to support this. On the other hand, liverworts possess sesqui- and di-terpenes which are identical stereochemically to those of fungi—and are fundamentally different from those of vascular plants.

The biochemist will have to face the fact that as many opportunities for evolution of biochemical parallelisms and convergences exist in Hepaticae—and other plants—as for such parallelism and convergence in characters that are expressed phenotypically. Hence, biochemical data, by definition, carry no more profound implications than data derived from other sources.

3. The Anthocerotes *may* represent a relatively late attempt at land invasion, subsequent to the period when a localized, rather close vegetation had evolved. Conceivably, their ancestors did not migrate to land until late in Devonian times. The relatively sophisticated capsule wall, with residual air chambers and stomata, and the indurated (*cutinized*) capsule wall, plus the supportive columella, jointly suggest the sporophyte was once more complex. From this one could argue that the sporophyte once attained eventual independence from the gametophyte, and that the bryophyte life cycle *evolved relatively late*, much as in the fern *Actinostachys* (Bierhorst 1971). Once a ♀ gametangium with a sterile outer layer evolved, the potential for permanent epiphytism of sporophyte on gametophyte arose; that it was exploited in Bryophyta, Anthocerotophyta, and *Actinostachys*, independently, suggests it may have arisen also in other groups, now all extinct. Reconstructions of Rhyniales with creeping axial gametophytes, to which erect, furcate sporophytes are permanently attached, and of *Sporogonites*, with permanently attached sporophytes arising from a *thallus* (Andrews 1960), are thus suggestive.

4. Variations from one niche to the next may be significant. Thus, Schade (1917) showed that tufts of *Webera nutans* and *Mylia taylori*, posited only 50 m apart, showed differences in average yearly temperatures of 17°C. Clausen (1952) showed that in mats of *Cephaloziella byssacea*, the temperature in August (in Denmark) may rise to 55°C; the same species on exposed rock faces in coastal North Carolina occurs under significantly higher temperatures— yet in windy and exposed loci in west Greenland it survives temperatures of well below −50°C. *Mannia fragrans* had thallus surface temperatures of 109°F and probably over 115°F; these plants did not survive temperatures of −39°C without prior cold conditioning or desiccation— yet intermingled *Bryum argenteum* easily survived *both* temperature extremes (Schuster 1957). In the related *Mannia dichotoma*, desiccated plants survived temperatures of 94°C.

5. There are several other instances of supposed Paleozoic Hepaticae: *Thallites willsi* Walton (1925), initially compared to *Riccia*, later (Walton 1928) to *Riccardia*. Also *Thallites lichenoides* (Matthew) Lundblad was suggested by Lundblad (1954) to perhaps represent a xeromorphic liverwort allied to *Riccia*. This is most unlikely (Schuster 1966a, p. 354). These are

the earliest references to *Riccia*-like taxa (Marchantiales) from the Upper Paleozoic. Early Mesozoic (Rhaetic-Liassic) spores described as "*Ricciisporites*" were compared to *Riccia*, but later (Lundblad 1959), such a comparison was admitted to be invalid. In short, there is not a shred of credible evidence that Subclass Marchantiidae existed prior to the Mesozoic.

6. *Treubiites* Schust. (1966), based on *Hepaticites kidstoni* Walton (1925, 1928), was suggested by Müller (1939-40, p. 176) and Lundblad (1954, p. 405) to show a possible affinity to *Treubia* (Treubiales). Study of Walton's slides at Glasgow convinced me the plant had succubous leaves and that the *dorsal scales* that suggest an affinity to *Treubia* are, in actuality ventral in insertion. Affinities are to the living genera *Blasia* and *Cavicularia* (Blasiaceae, Blasiineae; Metzgeriales). *Treubiites*, from Greenland, and the purely Laurasian genera *Blasia* and *Cavicularia* appear to form a phylogenetically and phytogeographically discrete, isolated unit for which Douin's Order Blasiales should perhaps be resurrected.

7. For this I propose the form genus *Metzgeriothallus* Schust., gen. n. Plants similar to the modern genus *Metzgeria* in being thallose, branching furcately, bearing narrow (under 1.5 mm wide) thallus segments divided into an apparently unistratose wing and median polystratose costa. *Type: Hepaticites metzgerioides* Walton (1925). Perhaps allied to *Hepaticites* (*type: H. Langi* Walton, 1925), but this form genus exhibits no distinction between thallus wings and a costa, and is thus clearly more similar to *Riccardia*.

8. Krassilov (1973, p. 97) suggests an affinity of the former to Lejeuneaceae; the scattered rhizoids effectively exclude *Cheirorhiza* from *any* affinity to the Porellineae (to which Lejeuneaceae belong). Judging from both diagnosis and figures, I cannot recognize any affinity to any modern suborder.

Laticaulina is even more ambiguous as to affinity: similarities *may* be to *Noteroclada* or *Symphyogyna*-like Metzgeriales, *if* the structures Krassilov interprets as underleaves are really not underleaves. I would regard *only* *Cheirorhiza* as surely a member of the Jungermanniales.

9. Also, cutinization, or its lack, may be a factor. In Metzgeriales, and at least most Jungermanniales, true cutinization is lacking; Ziegenspeck (1942) demonstrated a cuticle in several living Marchantiales. Lundblad (1954) also showed that the epidermis of *Riccia* spp. may survive ordinary maceration techniques used for fossil leaf cuticles (treatment with nitric acid plus potassium chlorate, followed by ammonium hydroxide). (It is, however, incorrect to assume from this, as did Lundblad [1955] that "there is a true cuticle in many Liverworts.")

10. Medwell (1954) also describes Lower Jurassic fossils from Australia which seem Marchantioid in the thallus with two rows of ventral scales; the disposition of the material remains ambiguous (Lundblad 1955). Lundblad (1955) describes a Lower Cretaceous plant from South America as *Marchantites hallei*. Thus, by the early Cretaceous, Marchantialean fossils appear to be known from South Africa, Australia, and South America.

11. These organisms are particularly pertinent to the discussion, in part to counteract the "brainwashing" that results from reading most modern elementary botany and plant morphology texts. Even if the reader is familiar with them, the new interpretations of them given here are relevant.

The figures were also selected to show the paleobotanist some examples of generalized Hepaticae so that, hopefully, attribution of paleozoic fragments may be attempted.

12. By contrast, Musci, with their strong desiccation tolerance, show relatively minor departures from a single basic growth pattern: the ability of even their most primitive members to cope with strong variations in moisture conditions seems, in part at least, to account for their "conservative" morphology.

13. This scenario derives principally from two studies, of 14 and 20 years, respectively, of reproductive strategy of several species of *Lycopodium* (*L. obscurum*, *L. annotinum*, *L. tristachyum*, and *L. complanatum*) in Conway and Hadley, Mass. During the two decades, two of these species have extensively spread through a pattern of cloning. Evidence of new establishment of sporophytes, from embryos (hence, evidence of spore germination followed by gametophyte production), was seen in three cases (*L. obscurum*, in two cases; *L. annotinum*, in

one case). Several acres of almost pure *Lycopodium* cover, in an area of about 70 acres, may have derived from no more than 60 to 90 original gametophytes! The *Lycopodium strategy* probably applied with equal force to many Early Devonian vascular plants. During the 14-year period cited, the number of spores produced by the numerous clones studied must have exceeded 1,000 billion per year. No evidence that more than four successful life cycles were passed during this time is available. If we assume that out of 14,000 billion spores at least 1 million germinated, only four out of the million formed mature gametophytes that, in turn, produced externally visible sporophytes! The whole point is that a tracheophyte cycle that shows such a ghastly failure rate is yet, ultimately, successful.

Relevant is another fact: this *failure rate* is occurring today in sites where arborescent vegetation provides considerable amelioration and produces a water-retentive and nutritionally adequate organic soil layer. We must assume an even higher failure rate under Early Devonian conditions, yet we know that some, at least, of these *inefficient* organisms survived! Even more relevant: such sporophytic persistence, linked with sporophytic cloning, clearly works today, and presumably must have worked under Silurian-Devonian conditions, while a pattern of gametophytic persistence linked with short-lived epiphytic sporophytes cannot be visualized as successful under the outlined scenario.

14. Unpublished observations suggest that level surface distances of 30 to 50 cm are the maximum that sperm in *Polytrichum commune* will travel; on slopes, fertilization down hill—so to speak—over long distances remains a possibility.

15. The vigorous, widespread unisexual *Bazzania trilobata* (Jungermanniales) illustrates this point: under conditions *today*, with environmental amelioration by a vascular plant cover, sporophyte production is extraordinarily rare: I have seen it only three times, even though thousands of clones of this species have been investigated in the last 35 years. This case may seem exceptional, but it is almost *normal* in taxa that are unisexual and lack asexual propagula. Thus, among about 24 species of the unisexual genus *Plagiochila* in North America, sporophytes (hence fertilization) have been seen in only two (Schuster 1959-60). The ancestors of these plants, under Early Paleozoic conditions, surely found it even more difficult to complete their life cycles.

16. The firmer, structurally more complex Marchantiales clearly fossilize more readily than Jungermanniales and Metzgeriales; fragmentary material is also readily placed to the correct order. Hence, undoubtedly, numerically the *relative* frequency of Jungermannioid and Marchantioid fossils is skewed off, artificially. No conclusions as to time of origin, or diversification, can be made from such data.

17. Stotler and Crandall-Stotler (1974, p. 150) claim the extant *Bryopteris trinitensis* (Lehm. & Lindenb.) Lehm. & Lindenb. (Bryopteridoideae, Lejeuneaceae) arose at least 160 to 180 m.y. BP. This assumption is clearly wildly wrong (Schuster 1979a); it is based on, i.a., an erroneous reading of the timing of Indian plate movement. The first *real* evidence for the existence of taxa of the four cited families does not occur prior to the Oligocene.

18. The sequence in cited localities at the head of Table I in Schuster (1979b) was, unfortunately, reversed by the printer; to read correctly, the localities must thus be reversed.

19. For long relevant time spans when a supercontinent, Gondwanaland, existed in the Southern Hemisphere, much of Africa occupied a central and thus *continental* position with, presumably, continental as opposed to oceanic climates; hence, groups like the Jungermannii-dae simply did not find many suitable niches for colonization (Schuster 1972a). By contrast, Antarctica *once* surely was a significant segment of the *central core* where significant parts of the Gondwanalandic hepatic flora evolved (Schuster 1979b); those taxa that were not wiped out, starting probably with the Oligocene, were, of course, *forced* to migrate northward.

20. The Panthallassan range of several primitive subfamilies of otherwise advanced and subcosmopolitan families is notable. Thus two *low* and stenotypic subfamilies of the highly specialized Lejeuneaceae are: the Nipponolejeuneoideae (Schuster 1963; only two species of *Nipponolejeunea*; oceanic eastern Asia only) and Tuyamaelloideae (Schuster 1963; about 12 to

13 species in five genera; oceanic eastern Asia, Madagascar, Australasia to Java and Borneo, South America-Juan Fernandez Islands). Similar examples could be cited.

REFERENCES

Andrews, H. N., Jr., 1960. Notes on Belgian specimens of *Sporogonites. Palaeobotanist* 7:85–89.

Arnell, H. W. 1922. Die schwedischen Arten der Gattungen *Diplophyllum* und *Martinellia.* Göteborg, pp. 1–80.

Arnold, C. A. 1947. *An Introduction to Paleobotany.* New York: McGraw-Hill, pp. 1–433.

Bambach, R. K., C. R. Scotese, and A. M. Ziegler. 1980. Before Pangaea: The geographies of the Paleozoic world. *Amer. Scient.* 68:26–38.

Banks, H. 1968. The early history of land plants. *Symposium on Evolution and Environment* (E. T. Drake, ed.). New Haven: Yale University Press, pp. 73–107.

Banks, H. P. 1969. *Crenaticaulis,* a new genus of Devonian plants allied to *Zosterophyllum* and its bearing on the classification of early land plants. *Amer. J. Bot.* 56:436–49.

Bierhorst, D. W. 1971. *Morphology of Vascular Plants.* New York: Macmillan, pp. i–xi, 1–560, figs. 1–28.

Bold, H. 1973. *Morphology of Plants.* 3rd ed. New York: Harper and Row, pp. 1–668.

Campbell, D. H. 1925. The relationship of the Anthocerotae. *Flora* N.F. 18–19:62–74.

Carothers, Z. B., and J. G. Duckett. 1979. Spermatogenesis in the systematics and phylogeny of the Hepaticae and Anthocerotae. In Systematics Association Special Volume No. 14 *Bryophyte Systematics* (G. C. S. Clarke and J. G. Duckett, eds.). New York: Academic Press, pp. 425–45.

Caspary, R. 1906. Die Flora des Bernsteins und anderer fossiler Harze des ostepreussischen Tertiärs. *Abh. Geol. Landesanstell.* (Berlin) N.F., Heft 4:1–181, pls. 1–30.

Clausen, E. 1952. Hepatics and humidity, a study on the occurrence of hepatics in a Danish tract and the influence of relative humidity on their distribution. *Dansk. Bot. Ark.* 15:1–80.

Dietz, R. S., and J. C. Holden. 1970. The breakup of Pangaea. *Sci. Amer.* 223:30–41.

———. 1971. Pre-Mesozoic oceanic crust in the eastern Indian Ocean (Wharton Basin?). *Nature* 229:309–12.

Doyle, J. A., and L. J. Hickey. 1976. Pollen and leaves from the mid-Cretaceous Potomac Group and their bearing on early angiosperm evolution. In *Origin and Early Evolution of Angiosperms* (C. B. Beck, ed.). New York: Columbia University Press, pp. 139–206.

Evans, A. W. 1939. The classification of the Hepaticae. *Bot. Rev.* 5:49–96.

Fulford, M., and R. Hatcher. 1959. *Triandrophyllum,* a new genus of leafy Hepaticae. *Bryologist* 61:(1958):276–85.

Fulford, M., and J. Taylor. 1960. Two new families of leafy Hepaticae: Vetaformaceae and Pseudolepicoleaceae, from southern South America. *Nova Hedwigia* 1(1959):405–22.

Gottsche, C. M. 1886. Über die im Bernstein eingeschlossenen Lebermoose. *Bot. Centralbl.* 25:95–97;121–23.

Greig-Smith, P. 1958. Notes on Lejeuneaceae. III. The occurrence of hyaline papillae. *Trans. Brit. Bryol. Soc.* 3(3):418–21.

Grolle, R. 1969. Review: Rudolf M. Schuster. The Hepaticae and Anthocerotae of North America. *Nova Hedwigia* 16(1968):539–42.

Grubb, P. J. 1970. Observations on the structure and biology of *Haplomitrium* and *Takakia*, hepatics with roots. *New Phytol.* 69:303–26, figs. 1–8.

Halle, T. G. 1913. The Mesozoic flora of Graham Land. Wiss. Ergebn. Schwed. Südpolar-Exped. 1901–1903 unter Leitung v. Dr. O. Nordenskøld. Bd. 3, Lief 14. Stockholm.

———. 1916. A fossil sporogonium from the Lower Devonian of Röragen in Norway. *Bot. Notiser* 1916:79–81.

———. 1936. Notes on the Devonian genus *Sporogonites*. *Sv. Bot. Tidskr.* 39:613–23.

Harris, T. M. 1938. The British Rhaetic Flora. pp. 1–84, figs. 1–26, pls. 1–5. London: British Museum of Natural History.

———. 1939. *Naiadita*, a fossil bryophyte with reproductive organs. *Ann. Bryol.* 12:57–70, figs. A–G.

Jarzen, D. M. 1979. Spore morphology of some Anthocerotaceae and the occurrence of *Phaeoceros* spores in the Cretaceous of North America. Pollen et Spores 21(1–2):211–31, pls. I–VI.

Krassilov, V. 1970. Leafy hepatics from the Jurassic of the Bureja Basin. *Paleont. J.* 3:131–42.

———. 1973. Mesozoic bryophytes from the Bureja Basin, Far East of the USSR. *Palaeontographica* (Abt. B) 143:95–105, pls. 41–51.

Leitgeb, H. 1879. Untersuchungen über die Lebermoose. 5. Heft. Die Anthoceroteen. Graz: Leuschner and Lubensky, pp. 1–50.

Lundblad, B. 1954. Contributions to the geological history of the Hepaticae. Fossil Marchantiales from the Rhaetic-Liassic coal mines of Skromberga (Prov. of Scania), Sweden. *Sv. Bot. Tidskr.* 48(2):381–417, figs. 1–5, pls. I–IV.

———. 1955. *Ibid*. II. On a fossil member of the Marchantiineae from the Mesozoic plant-bearing deposits near Lago San Martin, Patagonia (Lower Cretaceous). *Bot. Notiser* 108(1):22–39, pls. 1–3.

———. 1959. On *Ricciisporites tuberculatus* and its occurrence in certain strata of the "Höllviken II" Boring in S.W. Scania. *Grana Palynol.* 2(1):1–10, pl. 1.

Medwell, L. M. 1954. A review and revision of the flora of the Victorian Lower Jurassic. *Proc. Roy. Soc. Victoria* N.S. 65(2):63–111.

Mehra, P. N., and O. N. Handoo. 1953. Morphology of *Anthoceros erectus* and *A. himalayensis* and the phylogeny of the Anthocerotales. *Bot. Gaz.* 114:371–82.

Mielinski, K. 1926. Über die Phylogenie der Bryophyten mit besonderer Berücksichtigung der Hepaticae. *Bot. Arch.* 16:23–118.

Müller, K. 1939–40. Rabenhorst's Kryptogamen-Flora. VI. Lebermoose. Ergänzungsband. Fasc. 1:1–160, figs. 1–128(1939). Fasc. 2:161–320, 26 figs. (1940). Leipzig.

Newton, M. 1971. Chromosome studies in some British and Irish bryophytes. *Trans. Brit. Bryol. Soc.* 6(2):244–57.

Niklas, K. J. 1979. Simulations of apical development sequences in bryophytes. *Ann. Bot.* 44:339–52.

Proskauer, J. 1948. Studies on the morphology of *Anthoceros*. I. *Ann. Bot.* N.S. 12:237–65, figs. 1–9, pl. V.

Ramsey, H. P. 1967. Intraspecific polyploidy in *Hypopterygium rotulatum* (Hedw.) Brid. *Proc. Linn. Soc.* N.S.W. 91:220–30.

Renzaglia, K. S. 1978. A comparative morphology and developmental anatomy of the Anthocerotophyta. *J. Hattori Bot. Lab.* 44:31–90.

Schade, F. A. 1917. Uber den mittleren jährlichen Wärmegenuss von *Webera nutans* und *Leptoscyphus taylori* im Elbsandsteingebirge. *Ber. Deutsch. Bot. Gesell.* 35:490–505.

Schopf, J. M. 1970. Gondwana paleobotany. *Antarctic J. U.S.* 5:62–66.

Schuster, R. M. 1957. Boreal Hepaticae, a manual of the liverworts of Minnesota and adjacent regions. II. Ecology. *Amer. Midl. Nat.* 57(1–2):203–99, figs. 17–23.

———. 1959. Studies on Hepaticae. I. *Temnoma. Bryologist* 62(4):233–42.

———. 1959–60. A monograph of the nearctic Plagiochilaceae. *Amer. Midl. Nat.* 62(1):1–166; 62(2):257–395; 63(1):1–130.

———. 1961a. Studies on Hepaticae II. The new family Chaetophyllopsidaceae. *J. Hattori Bot. Lab.* 23(1960):68–76, figs. 1–2.

———. 1961b. Notes on Nearctic Hepaticae. XIX. The relationships of *Blepharostoma*, *Temnoma*, and *Lepicolea*, with descriptions of *Lophochaete, Chandonanthus* subg. *Tetralophozia*, subg. nov. *J. Hattori Bot. Lab.* 23(1960):192–210, figs. 1–2.

———. 1963a. An annotated synopsis of the genera and subgenera of Lejeuneaceae. I. *Nova Hedwigia.* 9:1–203.

———. 1963b. Studies on Antipodal Hepaticae. I. Annotated keys to the genera of antipodal hepaticae with special reference to New Zealand and Tasmania. *J. Hattori Bot. Lab.* 26:185–309.

———. 1964a. Studies on Hepaticae. XIV. The genus *Austrolophozia* Schust. *Bryologist* 67(2):179–86.

———. 1964b. Studies on Hepaticae. XVII. *Trichotemnoma* Schust., gen. n. *J. Hattori Bot. Lab.* 27:149–58.

———. 1964c. Studies on Antipodal Hepaticae. IV. Metzgeriales. *J. Hattori Bot. Lab.* 27:183–216.

———. 1964d. Studies on Hepaticae. XXII–XXV. *Pleurocladopsis* Schust., gen. n., *Eoisotachis* Schust., gen. n., *Grollea* Schust., gen. n., with critical notes on *Anthelia* Dumort. *Nova Hedwigia* 8(3–4):275–96.

———. 1964e. Studies on antipodal Hepaticae. VI. The suborder Perssoniellinae. *Bull. Torrey Bot. Club* 91:479–90.

———. 1965a. Studies on Hepaticae. XXVI. The *Bonneria-Paracromastigum-Pseudocephalozia-Hyalolepidozia-Zoopsis-Pteropsiella* complex and its allies–a phylogenetic study (Part I). *Nova Hedwigia* 10(1–2):19–61.

———. 1965b. Studies on antipodal Hepaticae. II. *Archeophylla* Schust. and *Archeochaete* Schust., new genera of Blepharostomaceae. *Trans. Brit. Bryol. Soc.* 4(5):801–17, figs. 1–5.

——. 1965c. Studies on antipodal Hepaticae. VII. Goebeliellaceae. *J. Hattori Bot. Lab.* 28:129–38, figs. 1–2.

——. 1965d. Studies on Hepaticae. XXVII. *Xenocephalozia* Schust. *J. Hattori Bot. Lab.* 28:139–46, figs. 1–2.

——. 1966a. The Hepaticae and Anthocerotae of North America. Vol. I. New York: Columbia University Press, pp. i–xvii, 1–802, figs. 1–84.

——. 1966b. Studies in Lophoziaceae. 1. The genera *Anastrophyllum* and *Sphenolobus* and their segregates. 2. *Cephalolobus*, gen. n., *Acrolophozia*, gen. n. and *Protomarsupella*, gen. n. *Rev. Bryol. et Lichén* 34:240–87, figs. 1–4.

——. 1966c. Studies on Hepaticae. XXVIII. On *Phycolepidozia*, a new, highly reduced genus of Jungermanniales of questionable affinity. *Bull. Torrey Bot. Club* 93(6):437–49, figs. 1–2.

——. 1967a. A memoir on the family Blepharostomataceae. I. and II. Candollea 21(1):59–136, figs. 1–21; 21(2):241–355, figs. 22–50 (1966).

——. 1967b. Studies on Hepaticae XV. Calobryales. *Nova Hedwigia* 12(1966):3–64, figs. I–XII.

——. 1967c. Studies on antipodal Hepaticae. IX. Phyllothalliaceae. *Trans. Brit. Bryol. Soc.* 5(2):283–88.

——. 1968. Studies on antipodal Hepaticae. X. Subantarctic Scapaniaceae, Balantiopsidaceae and Schistochilaceae. *Bull. Natl. Sci. Mus.* (Tokyo) 11:13–31, figs. 1–3.

——. 1969. Problems in antipodal distribution in lower land plants. *Taxon* 18:46–91, maps 1–24.

——. 1970. Studies on antipodal Hepaticae. III. *Jubulopsis* Schuster, *Neohattoria* Kamimura and *Amphijubula* Schuster. *J. Hattori Bot. Lab.* 33:266–304, figs. 1–6.

——. 1971a. Two new antipodal species of *Haplomitrium* (Calobryales). *Bryologist* 74(2):131–43, figs. 1–29.

——. 1971b. On the genus *Pleurocladopsis* Schuster (Schistochilaceae). *Bryologist* 74:493–95.

——. 1972a. Continental movements, "Wallace's Line," and Indomalayan-Australasian dispersal of land plants: some eclectic concepts. *Bot. Rev.* 38(1):3–86.

——. 1972b. Studies on Cephaloziellaceae. *Nova Hedwigia* 22(1971):122–265, pls. 1–25. *Ibid.* II. *Cylindrocolea madagascariensis. Nova Hedwigia* 22:266a–c.

——. 1972c. Phylogenetic and taxonomic studies on Jungermanniidae. *J. Hattori Bot. Lab.* 36:321–405, figs. 1–11.

——. 1974. Studies on antipodal Hepaticae. XI. The Chaetophyllopsidaceae: Their taxonomy, phylogeny and phytogeographic affinities. *Bull. Natl. Sci. Mus.* (Tokyo) 17(2):163–180, figs. 1–2.

——. 1976. Plate Tectonics and its bearing on the geographical origin and dispersal of angiosperms. In *Origin and Early Evolution of Angiosperms* (C. B. Beck, ed.). New York: Columbia University Press, pp. 48–138, figs. 1–45.

——. 1977. The evolution and early diversification of the Hepaticae and Anthocerotae. In *Beiträge zur Biologie der niederen Pflanzen* (W. Frey. H. Hurks, and F. Oberwinkler, eds.). Stuttgart and New York: Gustav Fischer Verlag, pp. 107–115.

———. 1979a. The phylogeny of the Hepaticae. In Systematics Association Special Volume No. 14 *Bryophyte Systematics* (G. C. S. Clarke and J. G. Duckett, eds.). London and New York: Academic Press, pp. 41–82.

———. 1979b. On the persistence and dispersal of transantarctic Hepaticae. *Can. J. Bot.* 57:2179–2225.

Schuster, R. M., and J. J. Engel. 1973. Austral Hepaticae. II. *Evansianthus,* a new genus of Geocalycaceae. *Bryologist* 76(4):516–20, fig. 1–9.

———. 1974. A monograph of the genus *Pseudocephalozia* (Hepaticae). *J. Hattori Bot. Lab.* 38:665–701, figs. 1–17.

Schuster, R. M., and G. A. M. Scott. 1969. A study of the family Treubiaceae (Hepaticae, Metzgeriales). *J. Hattori Bot. Lab.* 32:219–68, figs. 1–12.

Smith, G. M. 1938. *Cryptogamic Botany.* Vol. I. New York: McGraw-Hill, pp. 1–545, figs. 1–299.

———. 1955. *Ibid.* Vol. II. 2nd ed. New York, pp. 1–399, figs. 1–254.

Steere, W. C. 1942. Pleistocene mosses from the Aftonian interglacial deposits of Iowa. *Papers Mich. Acad. Sci.* 27(1941):75–104, pls. 1–5.

Stotler, R. E., and B. Crandall-Stotler. 1974. A monograph of the genus *Bryopteris* (Swartz) Nees von Esenbeck. *Bryophytorum Bibliotheca. Band* 3:1–159, figs. 1–219.

———. 1977. A checklist of the liverworts and hornworts of North America. *Bryologist* 80:405–28.

Tatuno, S. 1959. Chromosomen von *Takakia lepidozioides* und eine Studie zur Evolution der Chromosomen der Bryophyten. *Cytologia* 24:138–47, figs. 1–5.

Townrow, J. A. 1959. Two Triassic bryophytes from South Africa. *J. S. African Bot.* 25:1–22, figs. 1–4, pl. 1.

Walton, J. 1925. Carboniferous Bryophyta. I. Hepaticae. *Ann. Bot.* 39:563–72, pl. XIII.

———. 1928. *Ibid.* II. Hepaticae and Musci. *Ann. Bot.* 42:707–16, pl. XII.

Wettstein, R. von. 1903–08. Handbuch der systematischen Botanik. II. Band. Leipzig and Vienna: Franz Deutiche, pp. 1–577, figs. 127–496.

Ziegenspeck, H. 1942. Die Spaltöffnungen der Marchantiaceae. *Feddes Repert.* 131:94–120.

Ziegler, A. M., K. S. Hansen, M. E. Johnson, M. A. Kelly, C. R. Scotese, and R. Van der Voo. 1977. Silurian continental distributions, paleogeography, climatology, and biogeography. *Tectonophysics* 40:13–51.

6

DIVERSITY AND MAJOR EVENTS IN THE EVOLUTION OF LAND PLANTS

Bruce H. Tiffney

INTRODUCTION

In this chapter, a synthetic review is presented of the major events in the evolutionary history of land plants as derived from paleo- and neobotanical data and from evolutionary principles. The initial adaptation of plants to land involved a host of characters which became integrated over time and which evolved primarily in response to the relatively desiccating land environment. While the sporophyte and gametophyte plants were on an equal competitive basis during the earliest stages of this evolution, moisture requirements eventually dictated the dominance of the sporophyte. Together with considerations of energetics, this in turn led to the appearance of mega- and microspores and gametophytes, and ultimately to the evolution of the seed habit. The latter was perhaps the single most important event in the evolution of land plants. Simple (gymnospermous) seed plants retained dominance through the latter portions of the Paleozoic and most of the Mesozoic, but were ultimately displaced by the more complex seed plants, the angiosperms. Much as with the origin of land plants, the angiosperms achieved success through a complex of adaptive characters, which presumably accumulated over a long period of time under the influence of a flexible developmental system.

The recognized patterns and process of evolution have been delineated through both paleontological and neontological inquiry. In each field, individual investigations range in scope from an examination of one or a few organisms to broad syntheses involving a wide range of taxa, a wide range of time, or both. To date, the study of the fossil record of plants has been confined largely to the description of fine scaled transitions involving one or a few taxa. Such data are then often used to support or modify broad-scale hypotheses erected by students of the comparative morphology and anatomy of modern plants. Thus, in many cases (but not all; cf. Florin 1951; Zimmerman 1952; Doyle and Hickey 1976), concepts of the nature of evolutionary trends in plants are derived more from modern organisms than from the succession of past organisms. In part, this apparent myopia is understandable. The paleobotanical record is often sketchy, and at times nearly nonexistent. Plants frequently do not grow in environments where they have a good potential for being preserved; even when they do, exacting morphological and anatomical techniques are required to identify the organism and even partially comprehend its biology. As a result, the paleobotanist is often too busy looking at the trees to see the forest.

The same thinking characterized paleontology as a whole up to the time when Simpson (1944, 1953) commenced an examination of the larger patterns involved in the evolution of animals. Since that time, vertebrate and invertebrate paleontologists have become increasingly aware of their potential contribution to the framing of evolutionary theory. One of the central aspects of this inquiry in the field of invertebrate paleontology has been that of the track of diversity in the fossil record (Raup 1972, 1976a, 1976b; Valentine 1973). Such studies can be pursued from the perspective of overall diversity through time or can be subdivided, permitting one to follow the diversity tracks of individual lineages. These tracks can be compared to one another, or can be tested for correlation with known perturbations in the physical world.

More recently, awareness of the evolutionary significance of paleontological data has spread to paleobotany, with the result that a growing number of workers are addressing questions concerning evolution as recorded in the fossil record. Knoll, Niklas, and Tiffney (1980) have specifically looked at patterns of diversity in land plants in terms of total diversity, individual lineage diversity, and community diversity. In addition to providing direct information on diversity, these patterns also serve to highlight critical or interesting periods in the evolution of land plants (e.g., effects at the Permian-Triassic boundary, or in the Early and middle Cretaceous). On a broader view, the patterns in diversity provide the framework within which to pursue the question of processes in land plant evolution, processes which might well be different from those operating in animals, particularly in view of the different speciation mechanisms occurring in plants and animals in the present day.

While sufficient data have been gathered (discussed later) to permit the construction of these land plant diversity curves, they represent a first approximation, as new data is continually being gathered. Similarly, the data presently on hand are still being analyzed from several perspectives, and new perspectives and methods of data analysis will undoubtedly be applied in the future; all of which is to say that our understanding of the full significance of land plant diversity is still in a state of flux. However, in the context of this chapter, the patterns of overall and individual lineage diversity serve in particular to focus our attention upon the critical phases of evolution in land plants, both as individual events and as parts of a continuum. Thus, it is attempted to provide a scenario which fits within the observed diversity variation, while also conforming to our knowledge of the fossil record, the biology of the organisms involved (as inferred from both fossils and modern relatives), and the interactions of the diversity curves of the individual lineages. In addition, attention has been paid to the variation of geological, geographical, and climatological factors as they contribute to influence the environment of growth. Some elements of this scenario are not original, having been initially elucidated by others, but often these are presented here with refinements. What is most important is that these published viewpoints join with original observations in a broad outline of the continuous adaptive history of the evolution of land plants, an outline within which more restricted studies can be placed.

BACKGROUND

The diversity patterns that led initially to the examination of the evolution of land plants in a broad context are based on more than 11,500 species citations gathered from the primary literature, largely from North America. Each citation records the presence of one species at one geographical locality at one time. In gathering this data, attention was paid to the host of potential biasing factors which occur on several levels. Efforts have been taken to account for synonymy in the literature, presence of several organ genera of one plant in a single deposit, and a group of other biases unique to paleobotany. In addition, the more generalized paleontological biases, such as those discussed by Raup (1972, 1979), have also been dealt with. Of all of these biases, only one will be considered in the present context, that is the effect of available sediment area. The reader is referred to a previous work (Niklas, Tiffney, and Knoll 1980) for a full discussion of the other biases.

The actual patterns of diversity may be discerned in one of three ways. The most simple of these is a direct plot (time standardized) of species versus time, which is presented for both all land plants (Fig. 6.1) and for specific lineages of land plants, including thallophytes (Fig. 6.2), pteridophytes (Fig. 6.3), gymnosperms (Fig. 6.4), and angiosperms (Fig. 6.5). Each graph includes curves for net profit and loss and for turnover, which jointly influence the third curve, that for diversity. The latter is the most immedi-

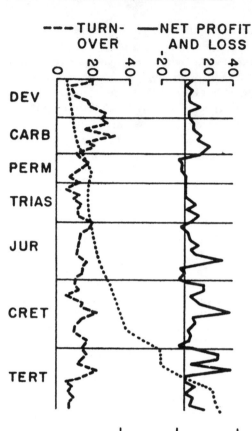

FIGURE 6.1. Cumulative species citation data for all land plants, as analyzed through curves for net profit and loss, turnover, and overall diversity. The scales for each curve indicate the number of taxa per epoch. These conventions apply in Figures 6.2–6.5 and 6.9, as do Dev = Devonian, Carb = Carboniferous, Perm = Permian, Trias = Triassic, Jur = Jurassic, Cret = Cretaceous and Tert = Tertiary.

ately interesting, and comparison of the diversity curves for each of the four groups reveals parallels as well as differences.

However, previous analyses (Raup 1972, 1976a, 1976b) indicate that these curves may be severely biased by the amount of sediment available for sampling from each geological period. Figure 6.6 is a time standardized histogram of number of species per million years versus geological time, while Figure 6.7 is a similar histogram of North American nonmarine

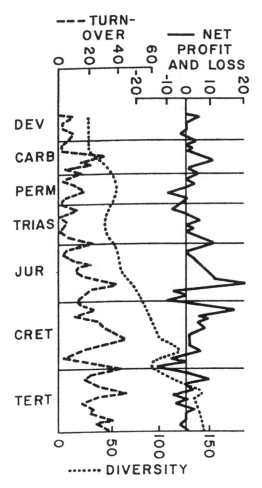

FIGURE 6.2. Species citation curves for Thallophytes.

sedimentary outcrop per million years versus geological time. Clearly, the two are quite similar, indicating a close correlation between the amount of exposed rock outcrop and the diversity of fossils for a given period. A linear regression of diversity against nonmarine outcrop area (Fig. 6.8) from the Devonian through the Tertiary yields a correlation coefficient of 0.868. If attention is restricted to the period of the Carboniferous through the Jurassic, the correlation coefficient becomes 0.947. However, as is visually evident in Figure 6.8, certain periods (the Permian, Triassic, and Jurassic) fall below the regression line, indicating that they possess a lower diversity than available outcrop area would lead one to expect, while other periods (the Devonian, Cretaceous, Tertiary, and to a lesser degree, Carboniferous)

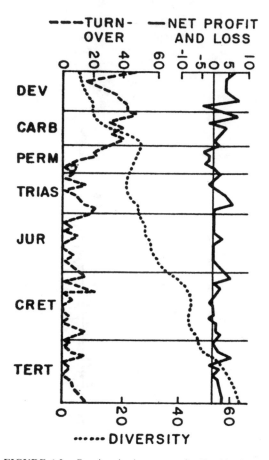

FIGURE 6.3. Species citation curves for Pteridophytes.

fall above the line, indicating that they possess a greater diversity than would be predicted by the available outcrop area. The reality of these positive residuals is supported by two features of the overall and individual lineage diversity plots. First, those periods featuring positive residuals often correspond to the times of major increases in diversity. Second, and more importantly, while the individual lineage diversity plots are all based on the same rock outcrop area data, the actual diversity curves vary considerably in their shape. This suggests that, while sediment area may exert a broad control over diversity patterns, biological factors are still detectable and of importance.

A broad estimate of diversity changes independent of sediment area may be obtained by examining the change of within-community diversity through time. Given the assumption that the grouping of species found in an

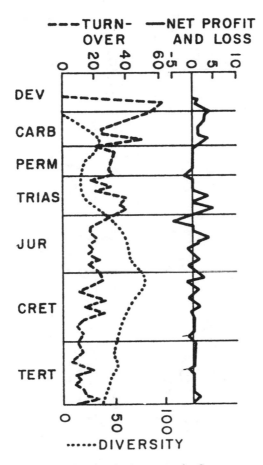

FIGURE 6.4. Species citation curves for Gymnosperms.

outcrop or a closely associated series of outcrops represents a sample of a contemporaneous community, it is possible to tabulate changes in community diversity through time. As evident in Table 6.1, there are two major times of community diversity increase (excluding that associated with the initial appearance of land plants, which is somewhat artifactual). The first of these falls in the Late Devonian—Early Carboniferous, corresponding to the advent of the seed and arboresence. The second falls between the Lower and Upper Cretaceous, and corresponds to the appearance of the angiosperms.

Again, while it is possible to argue that much of the observed variation is artifactual, or is due to bias, we believe that much of it is not, but rather reflects actual biological events. The previous paragraphs defend this viewpoint in a general manner, and the reader is again referred to Niklas, Tiffney, and Knoll (1980) for a more complete consideration. At present, the

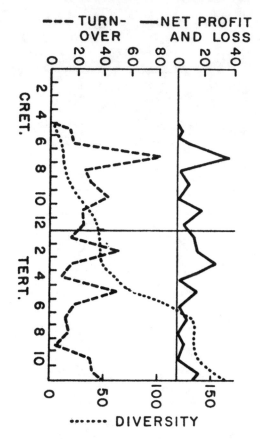

FIGURE 6.5. Species citation curves for Angiosperms.

diversity curves will be accepted as representative of biological events, and will be used as the impetus for an examination of significant events in the evolution of land plants.

THE TRANSITION TO LAND

The first major change involved was the very appearance of land plants, that is, their evolution from an algal group to a group of organisms which completed their life cycles in a subaerial environment. The initial exploitation of the land environment by a presumably green algal lineage (ignoring apparently unsuccessful attempts by other lineages, including the brown algae), was mediated by a number of vegetative adaptations. Most immediately, these would have to involve protection from desiccation by means of

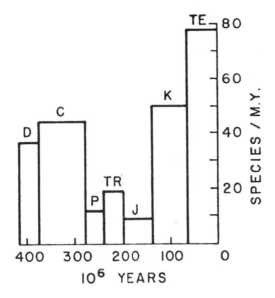

FIGURE 6.6. Histogram showing species per million years as a function of geologic period. D, C, P, Tr, J, K, Te respectively, Devonian, Carboniferous, Permian, Triassic, Jurassic, Cretaceous, and Tertiary.

water-impermeable substances, and simultaneously, cuticular perforations of lesser or greater morphological differentiation for the exchange of gases. As indicated by Swain and Cooper-Driver (1981) chemical evolution would play a significant role at the time, both in the origin of impermeable cuticular substances, and in the strengthening of the cell walls against increased compressive forces. Additionally, some degree of morphological differentiation into aerial portions and anchoring devices would be required to stabilize the plant in its environment and permit efficient photosynthesis.

Perhaps the most coincidental and least immediately adaptive feature of the earliest land plants would have been their mechanisms of reproduction. Thus, vegetative adaptations permitted the initial steps of the invasion of the land habitat, while reproductive mechanisms followed along by chance. The life cycle of the ancestor of the earliest land plants could have taken one of three forms. It could have been diplobiontic, possessing alternating haploid and diploid generations of a similar morphology and stature (as in many modern algae); haplobiontic haploid, where a dominant haploid generation results from zygotic meiosis (as in modern bryophytes); or haplobiontic diploid, wherein the dominant diploid generation has sporic meiosis. In all three cases, the haplobiontic-diplobiontic distinction is employed in an ecological sense. However, as soon as the organism possessing one of these life cycles becomes established upon land, certain features of reproduction

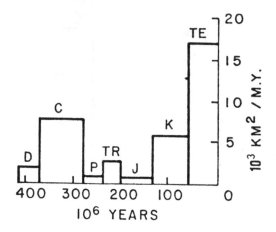

FIGURE 6.7. Histogram showing North American non-marine sedimentary outcrop per million years as a function of geologic period. Conventions as in Figure 6.6.

would assume a particular adaptive value. This is attested to by the fact that modern land plants are öogamous, archegoniate, and ecologically haplobiontic diploid. Clearly, these features are at the very least selectively neutral, or much more likely, advantageous.

Given a desiccating environment, it is only logical that öogamy would be favored, as it reduces by approximately one-half the potential for gametic dehydration; only one gamete must navigate the hostile environment outside the plant, rather than two. With the establishment of öogamy, three consequences follow and reinforce the öogamous condition. First, the developing zygote is retained, at least in its initial stages, within the protective environment of the egg-bearing gametophyte. Second, without the necessity for motility, the egg may become enlarged and thereby increase the nutritive base available to the developing zygote. Third, the zygote has the potential to draw further resources directly from the gametophyte, adding to those available in the egg. These advantages are synergistic, and would favor the development of an enlarged egg held within a complex protective structure. The archegoniate condition thus can be seen as a logical adaptation following the invasion of the land.

The distinction between gametophytic and sporophytic plants which is involved in the early stages of the archegoniate condition would appear to lend itself to an ecological bipartitioning of the environment within a single species, much as occurs with the holometabolous insects (on the basis of developmental stage), or in some island species of birds (on the basis of sex). However, the archegoniate condition, together with the inability of individual plants to migrate, dictates that the two stages could achieve an isomorphic alternation of generations only by continued asexual reproduction.

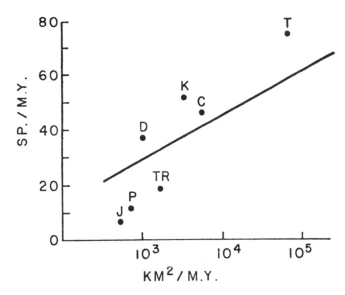

FIGURE 6.8. A regression of species per million years (Fig. 6.6) versus non-marine outcrop area per million years (Fig. 6.7) The solid line indicates the predicted relationship between species number and outcrop area, while the dots locate residual diversity values for individual periods. Those falling above the regression line possess greater diversity than predicted by outcrop area, while those falling below the line possess less diversity. Conventions as in Fig. 6.6.

However, the importance of genetic recombination involved in sexual reproduction apparently imposes the constraint that fairly strict asexual reproduction, while viable in an ecological time scale, is not viable in an evolutionary time scale. Certainly those groups of land plants in which asexual reproduction comprises the major means of genome maintenance (sphenopsids, lycopsids, and bryophytes) have generally been outpaced by sexually reproducing groups. In the absence of zygotic motility, sexual reproduction has the immediate consequence of sporophytic overgrowth, and ultimately elimination, of the gametophyte.

Given the advantage of sexual reproduction, the protoarchegonium of these early land plants would be the site of fertilization, and subsequently, either the site of either zygotic meiosis or the development of a multicellular sporophyte. These two possibilities correspond respectively to the antithetic and homologous concepts of the alternation of generations (Celakovsky 1874; Bower 1935). The protective advantage of the archegonium would select for the retention of the zygotic product on the gametophyte for some period of time in either case. However, in the hindsight of the modern flora, the homologous solution appears to have been the most favorable, in that all modern land plants possess a diploid multicellular phase developed in situ. Subsequent factors would favor a limitation of either the gametophyte or the

TABLE 6.1: **Species Diversity of Fossil Vascular Plant Floras**

	No. Floras	Max. Sp./Flora	Av. Sp./Flora
Late Silurian	1	1	1
Early Devonian	7	11	5
Middle Devonian	4	9	5.3
Upper Devonian	9	17	8.2
Early Mississippian	7	14	10
Late Mississippian	2	26	22.5
Pennsylvanian	24	57	35
Permian	—	—	—
Triassic	11	48	19
Jurassic	10	50	27.5
Early Cretaceous	15	46	21.5
Late Cretaceous	7	67	43
Tertiary	22	123	51

Source: Taken from Niklas et al. 1980.

sporophyte, engendering the dominance of one generation over the other. This relationship is reflected in either the sporophyte or the gametophyte attaining a determinate habit, while the other member of the cycle becomes relatively indeterminate.

In modern bryophytes, the determinate sporophyte is of limited size and terminates the vertical growth of the gametophyte, while the gametophyte possesses indeterminate lateral growth and can spread over a large area. Pteriodophytes, on the other hand, possess an indeterminate sporophyte and a gametophyte which, while not necessarily determinate, is generally restricted in its growth and is normally supplanted by the growth of the sporophyte. Presumably, neither the sporophyte nor the gametophyte of the earliest land plants had a clear dominance, but selective advantages would exist for tall sporophytes, which would have a greater probability of dispersing spores over a wide area than short sporophytes. As a result, it is possible to envision a small determinate sporophyte, similar to that of modern bryophytes, growing taller and taller through evolutionary time, and coming to dominate the gametophyte by insensible degrees. In its later stages, such an organism would not look all that different from *Cooksonia* or *Rhynia*, which, while considerably larger than bryophytic sporophytes, are still determinate in the sense that apical growth is terminated by sporangial production.

This scenario would suggest that the transition from ancestral land plants to bryophytic and pteridophytic lineages was gradational, commencing with a gradual extension of the duration of determinate sporophytic growth. As Schuster (1969; Chap. 5 in this volume) has pointed out, the extremely low rate of success of sexual reproduction in modern bryophytes suggests that a substantial selective pressure would exist for longer lived and larger sporophytes. This pressure would further select for a transition from determinate sporophytes, capable of producing only one crop of spores, to indeterminate sporophytes which could yield successive crops of spores over a long period of time. This latter step presumably followed the restriction of the production of reproductive structures on lateral branches, freeing the terminal meristem to grow in an indeterminate manner; a simple developmental shift. As the sporophyte dominated lineage of pteridophytes grew more organized and successful, competitive displacement would insure the gradual disappearance of intermediate forms and the delineation of the group presently recognized as the bryophytes.

The dominance of the sporophytic stage in the pteridophytes is further enforced by the moisture requirements of the sperm. In both bryophytes and pteridophytes, the gametophytes must grow in a situation which provides sufficient moisture for swimming gametes in order to achieve successful fertilization. This limits the gametophytes to a semiaquatic situation, or to one where they grow close to ground moisture. Apparently the selective advantage of sexual reproduction has reinforced the low stature of the gametophytes, much as it selected for the increase in the stature of sporophytes. While both gametophytic and sporophytic plants might possess the capacity for the production of vascular tissue (and attention should be called here to the hydroids and leptoids of mosses), only in the case of the indeterminate sporophyte did the organism attain sufficient height to result in the evolution of specialized vascular tissue recognized as xylem and phloem. Additionally, either the genetic capability for the synthesis of lignin was present only in the lineage leading to the pteridophytes, or if present in both bryophytic and pteridophytic lineages, it manifested itself only in the lineage with indeterminate sporophytes.

The foregoing provides a potential explanation for the early divergence of pteridophytic and bryophytic lineages, based on a differential optimalization of their sporophytic and gametophytic characters. In terms of evolutionary flexibility, success clearly lay with the group possessing indeterminate sporophytes (i.e., pteridophytes). However, while the ultimate dominance of sporophytes resulted in large part from the selection for sexual reproduction, the resulting lineages were not initially particularly efficient at sexual reproduction. Hence, strong selective pressure undoubtedly favored those sporophytic lineages which could maintain themselves by vegetative reproduction, while the occasional successful spore served to colonize more distant areas. Early Devonian evidence for this is seen in both the low community diversity (Table 6.1) and the frequent occurrence of large slabs of rock covered by one taxon. The community involved can be visualized as consisting of a whole series of large, turfed in clones of plants, each occupying its own area, much as sphenopsids or lycopsids dominate an area in the present day.

With the passage of time, sexual reproduction increased in efficiency with the result that communities slowly became more diverse as vegetative growth decreased in significance. This diversification was further aided by the appearance of larger and larger plants, at first in the herbaceous layer, then as shrubs, and finally as trees. The effect of this increased verticality of the community was to permit the distinction of separate layers (herb, shrub, canopy tree), such that more organisms could be encompassed within one community. The attainment of true arboresence followed upon the continued refinement of vascular tissues and the origin of the vascular cambium, together with a redirection of energy expended originally in a lateral growth plane into a vertical one. In one sense, since a tree is more of a determinate organism than is a horizontally creeping herb, the achievement of tree status could only follow the development of moderately efficient sexual reproduction. Certain selective pressures undoubtedly helped push the development of arboresence, including competition for light among the members of the community, and the dispersal advantage to be gained from a tall sporophyte. Additionally, the Upper Devonian witnessed the appearance of a significant, but wingless group of plant predators, the true insects. It has been suggested that arboresence may have also served as the first defensive move on the part of land plants against insects (Kevan, Chaloner, and Savile 1975), a move that was successful until the insects discovered flight.

THE SEED HABIT

While the development of a multilayer canopy is of importance in considering the Late Devonian–Early Carboniferous increase in both overall and within habitat diversity, perhaps the single most important factor involved in that increase was the evolution of the seed. Indeed, this may well be the single most important evolutionary event in the history of land plants, one which represents in its most refined aspects an ecological approximation of the haplobiontic diploid life cycle similar to that of terrestrial vertebrates.

The development of the seed habit commenced with the advent of heterospory. Heterospory may be defined in one of two ways: either morphologically, in which case the spores differ in size, or physiologically, in which case separate spores produce separate antheridiate and archegoniate gametophytes. Morphological heterospory does not *a priori* dictate physiological heterospory, but all known morphologically heterosporous forms are physiologically heterosporous. For this reason, it is commonly accepted that morphological heterospory implies physiological heterospory in interpreting the fossil record. On the other hand, the reverse is not true; physiological heterospory need not imply morphological heterospory, as for instance in some extant species of *Equisetum* which are physiologically, but not clearly morphologically, heterosporous.

The fossil record gives no indication as to which form of heterospory evolved first, although it does show clearly that the condition of morphological heterospory has evolved independently in several separate lineages (e.g., bryophytes, lycopods, sphenopsids, pteridophytes, and progymnosperms). For this reason, a logical exploration of the evolutionary trends which would favor one or the other form of heterospory, or indeed both, would be profitable. In both cases, the assumed starting point is that of morphological and physiological homospory.

If we postulate that spore size change preceded the appearance of unisexual gametophytes, then the initial step would involve the production of a range of spore sizes within one sporangium. Such a condition could result in a random manner from the influence of exogenous stimuli (e.g., fluctuations in temperature, moisture, or photoperiod) on the presumed weakly homeostatic developmental systems of early land plants. Alternatively, spore size variation might have been a fairly common occurrence within a given species, reflecting poorly canalized endogenous genetic control. In either case, different sized spores would be produced and be exposed to the forces of selection. It is not unreasonable to assume that the larger spores would possess greater powers of dormancy and would give rise to larger and more vigorous gametophytes than would the smaller spores. These large gametophytes would thus be expected to survive for longer periods of time, harbor larger eggs, and give rise ultimately to larger sporophytes. In short, the larger spores would ultimately provide the most successful archegoniate members of the gametophytic population.

Initially, the gametophytes produced by differentially sized spores would be bisexual and capable of selfing. In modern homosporous plants, there is a strong tendency for the production of antheridia and archegonia to be temporally, and often spatially, separated upon one gametophyte. Frequently, it is the archegonium which matures prior to the antheridium, establishing a system favorable to outcrossing among the gametophyte population. The combination of such a temporal separation of gametangial production with the differential growth rates envisioned to result from the early stages of heterospory would lead by degrees to a situation of operative unisexuality. Large gametophytes might tend to establish archegonia more quickly than small, and to retain them for a longer period of time. Combined with the energetic advantage of the sporophyte borne upon a large gameto-phyte, this would serve to favor a consistently *archegoniate* function for the gametophytes born of large spores. Conversely, those born of small spores would find themselves more often the provider, rather than the recipient of sperm. Additionally, their sporophytes would be at a consistent energetic disadvantage. As a result of these factors, together with the value of outcrossing and genetic and energetic efficiency, selection would favor the gradual restriction of the antheridiate and archegoniate conditions to small and large gametophytes, respectively.

Once this size-related trend was established among gametophytes and sporophytes, it is only predictable that it generated a reciprocal pressure which initially favored the further enlargement of megaspores. Subsequently, reproductive and energetic efficiency would favor the establishment of discrete size classes of mega- and microspores, with the intermediate sizes being eliminated as a result of competitive displacement. In a historical perspective on morphological heterospory, it should be possible to observe an initial stage wherein the diversity of spore sizes increases in a continuous manner, followed by a period when intermediate sized spores disappear, thereby providing a clear distinction between micro- and megaspores. Independent of their complementary microspores, megaspores are normally recognized as those spores which exceed 200 μm in diameter. Heterospory could have logically arisen either by the continuous enlargement upon a small initial homosporous condition (as assumed in the foregoing discussion), or by a simultaneous increase and decrease from a median spore size condition. The fossil record (Chaloner 1967, 1970) indicates that the former case actually obtained. Additionally, these data support the sequence of events offered in the above scenario, as they are consistent with a progressive increase in spore size through the Lower and Middle Devonian, followed by a reduction in the diversity of intermediate size classes, and the distinction of micro- and megaspores.

If we reverse the approach and presume that physiological heterospory precedes morphological heterospory, then it is necessary to assume that endogenous or exogenous factors would give rise to a unisexual gametophyte from a bisexual ancestor. This established, the metabolic needs of the archegoniate plant, as contrasted with those of the antheridiate one, would select for larger archegoniate gametophytes relative to antheridiate ones. This in turn would provide a positive selective pressure for any larger spores which appeared by chance. In short, the same selective pressures which favored morphological heterospory favor physiological heterospory. The increased success of the sporophytes borne on larger gametophytes, together with the selective advantage of outcrossing and considerations of efficiency, dictate that morphological and physiological heterospory are closely intertwined. The value of this statement might be questioned in light of the occurrence of physiological heterospory in the absence of morphological heterospory as seen in modern *Equisetum*. However, the latter is probably a special case. While by and large morphologically homosporous, it does display some variation in spore size and more importantly, it is characterized by a particularly successful rhizomatous growth habit. On the one hand, this latter feature may remove considerable pressure from its sexual reproductive mode, but on the other, it is significant that the sphenopsids are presently among the most limited and moribund groups of vascular plants. In this light, their physiological heterospory does not appear to be fully integrated, and is apparently carried in large part by the vegetative success of the system.

Heterospory is represented initially in the fossil record by the Middle Devonian plant *Chaleuria* (Andrews, Gensel, and Forbes 1974), which possessed a range of spore sizes within a single sporangium. *Barinophyton* and *Protobarinophyton* (Brauer 1978), of a slightly younger age, are clearly heterosporous, bearing micro- and megaspores within one sporangium. By the Upper Devonian and Early Mississippian, seed-bearing plants had become clearly established.

The significance of the seed in the evolutionary history of land plants cannot be overestimated. It releases the gametophyte from its dependence on an aqueous environment for fertilization, and thereby permits the sporophytic plant to grow in a wider array of environments than possible previously. Its appearance represents the end of the bipartitioning of the physical environment by the sporophytic and gametophytic generations, uniting them in one ecologically haplobiontic diploid organism, even while remaining technically diplobiontic. This latter step, while a major advance, would not have been possible without the great size differential resulting from the earlier differentiation of the sporophyte and gametophyte, a size differential which permitted the sporophyte to bear the gametophyte. Similarly, the seed habit represents a strong commitment to sexual recombination, and its evolution caps a trend toward the increasing efficiency of sexual reproduction among its pteridophytic forebearers.

The concept of the seed plant embraces a series of structural and functional conditions, including: (1) the production of a single megaspore per sporangium; (2) the diminution of the megaspore wall; (3) the development of modified megasporangial structures, including the integument(s) and micropyle; (4) the retention of the megagametophyte upon the sporophyte for part or all of its development; (5) the appearance of microgametophyte capture mechanisms (salpinx, pollen drop, and stigma); and (6) elaboration and modification of the male gametophyte (distal germination and pollen tube), such that it can effect fertilization. Thus the seed habit involves both mega- and microsporophyllar conditions which permit the endosporic germination of the haploid phase and the initial development of the succeeding diploid phase within the ovule. Variations upon this theme are possible, as, for example, in *Lepidodendron*, wherein the endosporic megagametophyte was retained within a modified megasporangium, and the diploid embryo possibly developed while the whole structure was still on the megasporophyll. However, the presence of a clearly defined spore wall and the lack of a micropyle and integument directly homologous with those of other seed plants renders the lycopod solution morphologically distinct, although ecologically convergent upon the seed habit. Given more time, the lycopod "seeds" might well have evolved to such a degree as to have become morphologically indistinguishable from early gymnospermous seeds. The apparent morphological uniformity of Early Carboniferous seeds suggests their derivation from a limited phylogenetic stock, presumably evolving under highly circumspect and directed selectional forces.

The initial step in the evolution of the ovule, and ultimately, the seed, involved a reduction in the number of megaspores in each megasporangium from many to a few and ultimately to one. In effect, this eliminated competition for resources within the megasporangium and resulted in a substantial increase in the energy resources of the remaining megaspore(s). This trend was extended by a similar reduction in the number of sporangia borne within a megasporangial branch system or truss. Again, presumably some increase in the energy reserves of the megaspores in the remaining sporangia would take place, resulting in increased advantage to the gameto-phytes and their ensuing sporophytes. Simultaneously with this reduction in the numbers of megaspores and concomitant increase in size, selective pressures would favor the endosporal development of the gametophyte, and the retention of the spores within the megasporangium, much as is seen in modern *Selaginella* where endosporangial development is facultative. Both the spore wall and that of the sporangium afford protection from predation and desiccation; thus, the retention of the megagametophyte upon the sporophyte through the time of fertilization would release the gametophyte from its previous restriction to relatively moist areas. The retention of the megagametophyte within the megasporangium would select for microspores which could be more widely distributed in a horizontal and vertical sense, and which were adapted to being "caught" (e.g., spines and hooks), by megagametophytic or megasporangial structures. Germination and growth of the male gametophyte would then be restricted to that necessary to produce swimming gametes. As a result, while microspore size did not alter greatly during this time, the microgametophyte became increasingly simpli-fied and reduced in size, until such time as germination and microgameto-phyte growth could occur within the confines of the microspore proper.

A similar trend toward a reduction in size also took place with regard to megaspores. Megaspores initially grew in size as a consequence of selective pressures in favor of increased metabolic resources coupled with the self-sufficiency enforced on them by the spore coat. With the retention of the megaspore within the megasporangium, the protective advantage of the megasporal exine is lost. A mutation resulting in its disappearance would not be deleterious, but rather would result in a saving of energy. More importantly, with the demise of its exine, the mature megaspore would be able to derive direct nutritive support from the sporophyte. This would result in a decrease in megaspore size, inasmuch as it would no longer be totally self-dependent, but rather would be converging on a "parasitic" relationship with the sporophyte. While such a reduction would initially favor an increase in dispersibility, subsequent evolution would include considerable variation in seed size resulting from various ecological pressures outside the scope of this discussion.

As a sidelight, it may be observed that the same selective pressures (the success of the megagametophyte and ensuing sporophyte, together with considerations of energy efficiency), operating under different circumstances (origin of heterospory *versus* origin of the seed), have in both cases resulted in fluctuation of the size of the archegonium-bearing gametophyte. Initially, the megametophyte increased in size as the heterosporous condition developed, but with the advent of the seed habit, it decreased in size.

To date, we have pursued the evolution of the megagametophyte within the megasporangium. However, only following the development of an integument and of suitable structures concerned with pollination can this megasporangial structure be considered a seed. The origin of the integuments and of the pollen-trapping structures has been treated by a number of authors (Andrews 1963; Schweitzer 1977), resulting in the selection of a morphological series of fossils, including *Genomosperma kidstoni, G. latens, Salpingostoma dasu, Physostoma elegans, Eurystoma angulare,* and *Stamnostoma huttonense* (cf. Andrews 1963, figures 4–8) as representative of the possible intermediate stages in the evolution of the integument. While these taxa represent a phenetic, rather than a phylogenetic sequence, their occurrence in a limited stratigraphic interval and their overall uniformity of structure suggest a rapid and highly directional transition from an open to an integumented megasporangium. Of these, *G. kidstoni* possessed a ring of eight sterile telomes, which surrounded the base of the megasporangium, but did not invest it. They flared outward and remained distant, with the result that no micropyle was present. In *G. latens,* some degree of basal fusion of these telomes is observed, while in *Salpingostoma* and *Physostoma,* the telomes were fused to about one-half their length, forming a rudimentary integument and micropyle. In *Eurystoma* and *Stamnostoma,* the telomic fusion was complete almost to the apex, with a concomitant reduction in the lobed appearance of the integument. Several of these seeds possessed other specializations. The basal grooves on the exterior of the integument of *S. dasu* were lined with unicellular hairs. While these were absent on the distal exterior surface of the integumentary lobes, finer hairs reappear on their inner faces. The distal portions of the nucellus in several of these Early Carboniferous forms was modified into a salpinx or trapping structure for the microspore. In *S. dasu* and *P. elegans,* the proximal portion of the salpinx was modified into a pollen chamber.

The question remains as to the nature of the selectional pressures and the intermediate stages involved in the evolution of the seed. From a logical perspective, the trend in reduction in the numbers of sporangia in a truss would be expected to be continued through the slow reduction of the length of the resulting telomes (old sporangial stalks) until they disappeared entirely, all for energetic reasons. However, as the telomes shrank, they would come to invest the axis bearing the remaining sporangium (sporangia), and the fact that these telomes failed to disappear suggests that they assumed some further function in this process. Present wisdom suggests that this function would be protection, but at this stage, the telomes are too far apart to protect the ovule from predation by small insects, and would be of value in this respect only as diversionary forage. Similarly, the spacing of the telomes dictates that they would provide only limited protection from desiccation. However, and more importantly, these telomes would form a series of baffles which would disrupt windflow and cause airborne microspores to impact upon the ovule or the telomes, forming a microspore-trapping device. Further development of this entrapment mechanism would be directed by the importance of the telomes being near the tip of the megasporangium, selecting for apical fusion. In its latter stages, this trend would result in the telomes suddenly achieving a very important function as protective agents, as they came to completely invest the ovule. This function would supplant their initial significance, and would force the evolution of new pollination mechanisms such as pollen drops, and ultimately, the stigma.

The sequence of events described here involves two transfers of function. First, the initial reduction of the telomic truss for energetic reasons acquired a new pollination function as the resulting sterile telomes surrounded the remaining megasporangium. Second, as selection for increased pollination efficiency drew the telomes about the megasporangium, the telomes fused and assumed a protective function. The close stratigraphic occurrence of the morphologic examples of this sequence indicates that the selective pressures involved oscillated rapidly between the two optima and that the whole process occurred quickly.

A PHYSICAL INTERLUDE: CARBONIFEROUS-CRETACEOUS

With the appearance of the seed, land plants achieved something of a plateau, a condition which was maintained for approximately the next 200 million years. Although new lineages and adaptations appeared during this time, they were primarily concerned with vegetative variations on the common gymnospermous reproductive theme. Those events which occurred, and which are reflected in both the diversity and the species composition data, did not involve major evolutionary trends, but rather were limited to direct responses on the part of land plants to environmental stimuli.

The Carboniferous was a time of fairly equable climes and diminishing provinciality, and is featured by high but fluctuating rates of turnover accompanied by a gentle rise in overall diversity to a peak in the Late Carboniferous and Early Permian (Fig. 6.1). This may be attributed to one of several factors, including the development of a series of multilayered communities which were dominated by pteridophytes, gymnosperms, or a combination of the two. Within each group, differentiation may have been spurred by the appearance of flight among herbivorous insects in the Namurian (Niklas 1978). This new-found adaptation negated one of the earlier advantages of arboresence, an occurrence which forced plants to seek other avenues of physical, and perhaps chemical, defense. During this time, the major events of speciation recorded in the lowland coal swamp forests may well have occurred in response to those cyclical variations in environment which are recognized in North America by the presence of cyclothemic patterns of deposition. Such repetitious environmental fluctuation may have led to a taxonomic relay such as postulated by Boucot (1978) in his concept of diacladogenesis. This hypothesis suggests that environmental perturbation causes one species to give rise to a second, ecologically equivalent, species, which normally replaces the first, and so on with successive perturbations through time. This would account for continuous turnover coupled with the slow rise in diversity.

The Late Carboniferous–Early Permian peak in diversity is followed by a distinct decline, which is more pronounced in some groups than in others (cf. Figs. 6.1–6.4). This parallels closely the pattern of diversity in the marine biosphere for the same time period (Raup 1972, 1976a, 1976b; Valentine 1973). Presumably, the drop in land plant diversity reflects both a decrease in provinciality and an increase in the extent of continental climates resulting from the coalesence of Pangaea. The biological reality of this recorded diversity drop can be tested in a simple manner. Assuming that the spread of continental climates was a central environmental factor during the Permian and Triassic, one can search for significant differences in the relative survival of major groups of plants which are more (gymnosperms) or less (pteridophytes) well adapted to continental environments. Preliminary results indicate that a significantly larger proportion of gymnosperm taxa survived the Permian-Triassic periods than did pteridophyte taxa. A more detailed examination of these time periods is under way, and will be the subject of a future work (Niklas, Tiffney, and Knoll, in manuscript).

ANGIOSPERMS

While several lineages of gymnosperms evolved and diversified from the Late Devonian through the Early Cretaceous, each of which was successful in its own way and time, none really involved an improvement on the central feature of the group, the seed. No similar vegetative or reproductive character can be selected as characterizing the angiosperms. Rather, their success is attributable to a consortium of morphological, anatomical, and ecological characteristics which are united by a common developmental theme. Individually, the adaptive value of each of their characters ranges from moderately significant to debatable, and no single character appears to account for their success. Among the more significant of these characters are the production of ovules which are (to varying degrees) enclosed in an ovary, a reduced, nonarchegoniate gametophyte, the production of polyploid endosperm through the process of double fertilization, tectate columellate pollen possessing a laminated endexine, the production of a pollen tube by the microgametophyte, and the possession of specialized conducting elements, including vessels and sieve tubes with companion cells, as well as more ecological factors, such as the employment of relatively predator-specific plant toxins, together with a tendency for coevolutionary relationships with animals. While these characters commonly coexist in angiosperms, they do not necessarily form a consistently cohesive unit, but may be expressed in a multivariate manner. Thus, *Austrobaileya* has vessels associated with the phloem of gymnosperms, while the Winteraceae lack vessels, but possess sieve tubes and companion cells. Certain taxa (e.g., Amentiferae) have reverted to wind pollination, and others carry their ovules within unsealed carpels (*Degeneria*). Furthermore, many of these characters are not limited to angiosperms, but may be found individually in a host of other plant groups.

Thus, we can observe that a carpel-like structure appears to have evolved in several presumably unrelated gymnospermous groups of the early Mesozoic, including such taxa as *Dirhopalostachys* (Krassilov 1975), *Leptostrobus* (Krassilov 1977), and *Irania* (Schweitzer 1977). This would suggest that angiospermy in an etymological, but not necessarily a taxonomic, sense had evolved by the early Mesozoic, presumably as an adaptive response of several phylogenetically separate groups to a common environmental pressure. Of these, only one group assembled the host of other adaptive traits which led to its ultimate recognition as an angiosperm. In addition to its protective function, the carpel may also have selected for increased intermicrogametophytic competition as suggested by Mulcahy (1979). While this may have favored increased microgametophytic efficiency in the angiosperms, it should be pointed out that similar pressures may have operated on the microgametophytes associated with the pre-Cretaceous carpels mentioned previously, and that pollen tubes, and the associated potential for such competition, are encountered in other groups (conifers and seed ferns).

The angiosperm vessel presents a similar case, as analogous cells are found in gymnosperms, lycopods, and ferns. It is interesting to note, however, that the arrangement of sieve tubes and companion cells appears to be unique with the angiosperms, although its adaptive significance is not immediately clear.

The character of tectate-columellate pollen is a more complicated case, as this morphology possesses the potential of serving more than one purpose simultaneously. Additionally, the alliance of such pollen in the fossil record to a whole biological organism is rendered especially difficult by the inherent dispersibility of pollen. The tectate-collumellate structure functions in modern angiosperms both for the retention of germination recognition factors and to cause pollen grains to clump as an aid to insect pollination. Either of these two adaptations could have preceded the other in evolution-ary time and have been involved in a transfer of function; alternatively, the two could have arisen simultaneously. However, neither of these two adaptations is necessarily restricted to the angiosperms. It is becoming increasingly clear that insects have played some role in the reproduction of earlier groups (Taylor and Millay 1979; Meeuse 1979) and recognition systems are utilized both by other plant groups and by animals. Although the tectate-columellate structure is restricted presently to angiosperms, there is no reason why it should not have evolved in one or several distantly related gymnospermous groups prior to the Early Cretaceous. By way of example, Cornet (1977) has reported pollen grains of Triassic and Jurassic age which possess a tectate-columellate structure that could have been produced by an extinct sister group to the angiosperms, or possibly even by a direct angiosperm ancestor. However, in the absence of further data, it is equally possible that these grains were produced by an extinct gymnosper-mous group operating under similar selective pressures as the early angio-sperms, but totally unrelated to them.

The reduction of the megagametophyte to an egg plus a few associated nuclei is nothing more than a logical extension of the trend toward megagametophyte simplification which was initiated with the appearance of the seed. This process follows from the increasing physiological dependence of the gametophytic generation on the sporophytic generation, and estab-lishes a situation favorable to the accidental appearance of double fertiliza-tion through the fusion of a second sperm with an available egg sac cell. On the basis of hybrid vigor, this endosperm would be presumed to grow at a rapid rate, and perhaps to achieve a greater size and nutritive value than the original gymnospermous endosperm. Selection would then favor the gradual transfer of metabolic commitment on the part of the parent plant from a gymnospermous to an angiospermous mechanism of angiosperm formation.

Such a trend would be accentuated by the fusion of the residual sperm with two egg sac nuclei, as the resultant polyploid condition would lead to further enhanced vigor and rapidity of growth. This process also provides the sporophyte with a mechanism to conserve metabolic energy, as endosperm is not formed until fertilization has occurred. This situation is in contrast to the gymnospermous condition where the sporophyte places an energetic "bet" in the form of endosperm that pollination will take place. Additionally, the whole foregoing sequence of events would serve to speed up the life cycle of the angiosperms relative to that of the gymnosperms.

Perhaps the most important character of the angiosperms is one which, while not specific to the group, is the single character which links all the other features into a successfully functioning whole. The character in question is that of their progenetic developmental system, initially alluded to by Takhtajan (1972, 1976) and elaborated upon by Doyle (1978a). The value of this developmental system lies in its plasticity and its relative freedom from rigorous canalization, a characteristic which permitted the group to explore a variety of adaptive solutions in a relatively brief time and to integrate these adaptive solutions into a functioning whole. This plasticity is reflected in the relatively fast growing, rapidly reproducing, opportunistic nature of the group in the present day, features which approximate an ecologically "r" type of life history strategy.

Recognition of the importance of this developmental aspect of the nature of the angiosperms brings two other nebulous adaptive traits of the group into focus; their plasticity of growth habit and their chemistry. Modern angiosperms may be described as possessing at least 23 different patterns or models of forming a tree, based on branching strategy, position of buds, and other factors (Hallé, Oldeman, and Tomlinson 1978). This is in contrast to the four models recognized presently in modern gymnosperms, and more specifically to the two in modern conifers. This variability of growth model is a character which crosses taxonomic lines (Hallé, Oldeman, and Tomlinson 1978), and is a further reflection of the progenetic nature of the group. While the exact ecological significance of the various growth models has yet to be demonstrated, their distribution does show a broad correlation with world environment, suggesting their adaptive importance. It is improbable that all these models were present in the angiosperms from the time of their inception; rather it seems more likely that this diversity of form has accumulated since the Early Cretaceous. If these various growth forms are adaptively significant, then it is probable that their presence conferred a competitive advantage over contemporaneous gymnosperms with their limited number of models.

There is a second architectural distinction between angiosperms and gymnosperms which is of potentially even greater importance, and that is in their differing capacity for reiteration. Reiteration is an architectural adjustment on the part of the tree in response to damage incurred from physical or biological (predation and parasitism) sources (Hallé, Oldeman, and Tomlinson 1978). Often, this adjustment takes the form of a repetition of the original growth model of the tree, which fills in the damaged area and permits the tree to continue to occupy its place in the forest. Conifers generally adhere quite closely to their growth model, and are incapable of replacing effectively the damaged area: as a result, they often maintain a damaged appearance throughout the rest of their lives. On the other hand, most angiosperms have a well-developed capacity for reiteration (Hallé, Oldeman, and Tomlinson 1978) and can thereby respond to damage in a much more successful manner than gymnosperms. In a mixed forest of angiosperms and gymnosperms suffering equal amounts of damage, the competitive advantage lies clearly with the angiosperms.

The difference in reiterative ability between angiosperms and gymnosperms is perhaps correlated with a second distinction between the two groups: the nature of their respective chemical defenses against one of the prime sources of damage, insects. Gymnosperms are characterized normally by large quantities of relatively unspecialized broad-spectrum insect toxins, evenly distributed throughout the plant body (Niklas 1978). Presumably this is the culmination of a long coevolution with the insects in which the insects were never significantly more than pests. The upheaval of the developmental system involved in the postulated progenetic origin of the angiosperms may be presumed to have been more than skin deep, and to have involved equally the biosynthetic pathways of the group. This would permit the angiosperms to redesign and redeploy their chemical defenses in a different manner with several results. By differential redeployment, the group could select for the insects to visit one portion of the plant (flower), while continuing to maintain toxicity in the remainder of the plant body. Such an arrangement would enhance greatly the potential for insect pollination, and indeed this redeployment of toxins in the angiosperms may underlie their coevolutionary success.

On the other hand, while angiosperm toxins are generally quite powerful, they are also often limited in their effectiveness to one or a few groups of plant pathogens or predators, thereby conforming to an "r" type of life history strategy (Niklas 1978). Such a specificity of defense leaves the group subject to extreme predator pressure on those occasions when a pest emerges which is resistant to the toxins in question. This fact has two implications in the present context. First, as noted by Niklas (1978), a rapid interplay between the evolution of insect predators and plant chemical defenses, together with the fluctuation in angiosperm population sizes resulting from differential predator pressure, would stimulate the rate of angiosperm speciation. While this effect may have been limited, it might serve to increase the rate of evolution of the early angiosperms relative to that of contemporary gymnosperms. However, this also implies that the early angiosperms might generally be subject to a greater amount of insect damage than would contemporary gymnosperms. In this respect, the reiterative ability of the angiosperms complements their chemical plasticity, in that it provides a means of recovery in the face of increased insect damage, which is in turn a direct and unavoidable consequence of plant-insect coevolution resulting from the redeployment of chemicals.

This view of angiosperms as being defined by an interactive multitude of characters, rather than by a single character, suggests a series of perspectives upon the group. First among these is that their evolution was additive; that is, that the group of characters found in modern angiosperms did not evolve simultaneously, but were accumulated over a long period of time. This implies that, even within the direct ancestors of the angiosperms, several angiospermous characters could have evolved before the group was recognized as a whole. What is perhaps less obvious is the complementary probability that the group continued to accumulate characters after its recognized first appearance; after all, the operative definition of an angiosperm is based solely on extant representatives of the group. Such an additive model of evolution further means that the recognition of the origin of the angiosperms will be as difficult as the selection of the first land plant. Single morphological characters, while perhaps angiospermous in nature, cannot be used to define the presence of the group, inasmuch as its "spirit" lies in an interactive plexus of characters. Only at that time when evidence is available for said plexus operating as an ecological and evolutionary unit can it be assumed safely that the group is present. Attempts to relate isolated pre-Cretaceous morphological structures to the angiosperms will be complicated further by convergent evolution in response to similar environmental pressures. Given the common selective forces faced by the separate groups of early Mesozoic gymnosperms, the appearance of such convergent structures is, as Stebbins (1967) has indicated, only to be expected in view of the limited number of possible adaptive modes possessed by plants. At present, the first undeniable evidence for the angiosperms is that provided by Doyle and Hickey (1976; Hickey and Doyle 1977) for the floras of the Lower Cretaceous Potomac Group of eastern North America. The described plexus of leaf and pollen fossils is representative of an organism possessing a distinct ecology and occupying a distinct niche.

Given the additive model of angiosperm origin, it is most likely that the full integration of each new character took a considerable amount of time. Any nondeleterious feature in the early angiosperms or angiosperm ancestors which favored competitive parity with contemporaneous gymnosperms would permit the group to "explore" their progenetic developmental system. With the passage of time, several (although not necessarily all) of these individual features would become integrated to a degree, and lead to the attainment of an adaptive threshold which would be recognized as the first appearance of the group. However, there is no reason to believe that the angiosperms had assembled all their characters by this time, or that those characters which were present had achieved their most efficient interrelationship. Rather, it seems more probable that the group spent a good portion of the Cretaceous consolidating their abilities. This perspective would suggest that the group did not undergo an explosive radiation in the Early Cretaceous, but rather achieved a slow and steady diversification as they underwent an integration of their characters. Local dominance might then be expected in the later Cretaceous, and world dominance perhaps as late as the Early Tertiary. The diversity data appear to bear this out (Fig. 6.5), as the angiosperm curve shows a slow increase of diversity through the Late Cretaceous, with explosive diversification taking place only in the Tertiary. Thus, the initial portions of this curve are the same as those for the early stages of any other group; the angiosperms are not special in their diversification.

Finally, the fact that the success and the very nature of the angiosperms is based on the integration of a set of separate features, no one of which possesses an overriding importance, leads to the view that the angiosperms are only a specialized group of gymnosperms, rather than a totally new and unique development in land plants. The central feature of the group is still the seed, and many of their other adaptations are only the logical end products of trends, often involving reduction, initiated earlier in the history of land plants. However, even if they are "only" advanced gymnosperms, their diversity in the present day bespeaks the success of their adaptive complex of characters.

As previously stated, the angiosperms did not achieve a great diversity immediately, but rather, expanded their domain by slow degrees. The initial record of the group lies in sediments representative of unstable environments along the margins of streams and rivers, as befits their presumed weedy, "r"-selected nature (Doyle and Hickey 1976), although they may have evolved at an earlier date in arid uplands as suggested by Stebbins (1974). Both the morphology of the leaves and the pollen of these early angiosperms, together with the pattern of their diversification, have been admirably elucidated by Doyle and Hickey (1976; Hickey and Doyle 1977). This involves their gradual transition from riparian plants to first shrubs and then trees of disturbed sites, and ultimately to their participation in the canopy of the stable forest community. During this period, the diversity of leaf morphology is seen to achieve almost modern levels, although the diversity of pollen lags behind (Doyle and Hickey 1976; Hickey and Doyle 1977), an early expression of mosaic evolution in the group.

The slow expansion of the angiosperms is paced by a gentle decline in gymnosperm diversity (Fig. 6.4), which commenced in the Early Cretaceous.

This relationship between the diversity of the two groups serves to both support the idea that the first appearance of the angiosperms coincides fairly closely with their actual origin, and that angiosperms did not overwhelm the gymnosperms in an explosive burst of evolution. Indeed, the Cretaceous forests appear to have contained a wide range of gymnosperms, including Caytoniales, Nilssoniales, Bennettitales, Cycadales, Ginkgoales, and Coniferales. The last group in particular was fairly diverse, and may have occupied a wider range of habitats than do its descendents of the present day (Alvin, Spicer, and Watson 1978). Thus, it is possible that the angiosperms of the middle and later Cretaceous were generally organisms of the lowland flood plains and braided river systems, while the gymnosperms continued to dominate the relatively more stable upland habitats. During this time, it may be presumed that the angiosperms encroached upon the gymnosperm communities, but seriously supplanted them only in the Late Cretaceous, when many of the more archaic gymnosperms (e.g., Caytoniales, Bennettitales, and Nilssoniales) became extinct (Krassilov 1978). It is quite probable that this extinction was not a chance event, but one resulting from the competitive pressure from the increasingly efficient angiosperms, an expression of the Red Queen's hypothesis (Van Valen 1973a). This suggests that, in an environment, increased competitive efficiency in one group or lineage forces all other co-existing groups to adjust to some degree in order to maintain parity. Failing to do so, they ultimately become extinct. The competitive aspects of the angiosperm-gymnosperm interrelationship may have been exacerbated by still larger scale factors, perhaps including those postulated to be responsible for the Late Cretaceous–Paleocene boundary extinctions of marine organisms and dinosaurs (Russell 1979). However, it should be noted that the competitive ability of some gymnosperms remained undiminished, as evidenced in particular by the Pinaceae, which evolved to a considerable degree in the Cretaceous and Tertiary (Miller 1976).

While the angiosperm diversity curve advances at a slow, but steady, rate in the Cretaceous, the turnover curve for the same period exhibits two major peaks (Fig. 6.5). The first may correspond to Doyle's (1977) suggestion that the angiosperms achieved a maximum evolutionary rate at this time, thus reflecting an initial flush of adaptive success, which was followed by a period of consolidation. The second peak in turnover might correspond to the attainment of canopy status by the angiosperms, although it falls at a slightly later date than postulated for this event by Doyle and Hickey (1976; Hickey and Doyle 1977). However, both these peaks also correlate with fluctuations in diversity parameters of insects for the same period (Fig. 6.9). The first angiosperm turnover peak corresponds closely to a peak in insect turnover, while the second angiosperm turnover peak corresponds to a Late Cretaceous peak in the diversity of insects. The present level of resolution of

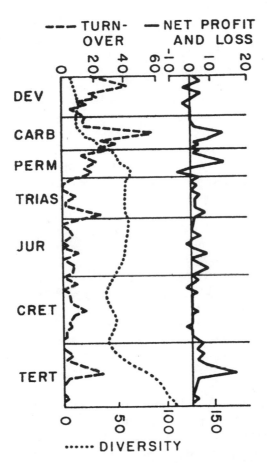

FIGURE 6.9. Species citation curves for presumable mandibulate terrestrial arthropods and *bona fide* insects.

the diversity data does not permit a clear delineation of cause and effect. However, to the degree to which such a synergistic system can be arbitrarily subdivided, we concur with Doyle (1978a) that the turnover rates and diversity variation of insects are probably in response to the evolution of their angiosperm hosts, rather than the cause of that evolution. This pattern finds a parallel in the increased diversity of ascomycetes witnessed during the same period (Tiffney and Barghoorn 1974), which is also presumed to reflect the increased diversity of angiosperm host plants.

While some of the earliest angiosperms may have been wind pollinated (Dilcher 1979), others already possessed some degree of coevolutionary linkage with the insects as attested by the morphology of early angiosperm pollen (Doyle 1978a, 1978b). However, both the angiosperm diversity data and the relation between it and the insect diversity data suggest that these coevolutionary links did not immediately dominate the evolution and success of the group; rather, these relationships slowly gained in their importance and complexity through the Cretaceous and Tertiary (cf. Crepet 1979a, 1979b). It may also be assumed that coevolution took place between angiosperms and dispersal agents in a similar manner during this period. Schuster (1976) has observed that the phylogenetic position of an angiosperm taxon (primitive versus advanced) often correlates with the degree of adaptation of its fruits to animal dispersal. Thus, phylogenetically primitive genera tend to possess woody or papery disseminules, while more advanced taxa tend to have disseminules directly adapted to animal dispersal. The significance of this parallelism is strengthened by the fact that the two trends (phylogenetic advancement and dispersibility) were recognized independently. The success of the angiosperms in establishing coevolutionary pollination and dispersal relationships with animals is probably in large part due to their progenetic origin. In each case, the variety of adaptations present and the rapidity with which they appeared reflect the developmental plasticity inherent in the group.

The Cretaceous-Tertiary boundary is recognized generally as a time of major extinction among marine invertebrates and terrestrial tetrapods (Russell 1979). Yet, whatever factors were involved apparently had little effect on terrestrial plants, as only the thallophytes (Fig. 6.2) exhibit any response at this time. Among the angiosperms, the Cretaceous-Tertiary boundary was a time of relative stability, as diversity was maintained at a plateau level from the Late Cretaceous through the Late Eocene. What rearrangements which did occur among the angiosperms were primarily phytogeographical. The latest Cretaceous of the Northern Hemisphere was dominated by two palynologically defined floras, the east Asian–western North American *Aquilapollenites* province and the eastern North American –European *Normapollenites* flora (Muller 1970). With the retreat of the North American Mid-Continental Seaway and the weakening of the barrier created by the west Asian Turgai Straits in the latest Cretaceous (Cox 1974), the distinctions between these two provinces began to disappear. The exact nature of this transition is clouded in an absence of data; apparently some elements of each flora went extinct, while others simply intermingled. At present, nothing suggests that this was a time of major evolutionary change (Hickey 1977a; Tschudy 1977).

This apparent stasis among the angiosperms is broken in the Late Paleocene and Early Eocene when a high turnover peak is witnessed against a backdrop of little change in diversity (Fig. 6.5). This indicates a major rearrangement in the taxonomic composition of the angiosperm community, and bears out the observations (Hickey 1977b, personal communication;

Muller 1970; Tiffney 1977) that the Paleocene-Eocene was the time of disappearance of many archaic taxa, and of the appearance of a diverse array of modern families and genera. This new assemblage spread across the entire Northern Hemisphere in the Eocene (Wolfe 1975), and contained a rich array of arborescent, shrubby, liane, and aquatic taxa, many of which are now restricted to southeastern Asia. The driving force behind this taxonomic upheaval is not identified presently. It could be a lag effect resulting from the factors which dominated the Cretaceous-Tertiary boundary in the zoological realm, a result of increased rates of hybridization and evolution following the mixing of the taxa of the *Aquilapollenites* and *Normapollenites* provinces, a function of the cooler temperatures of the Paleocene (Hickey 1977b), a result of undetermined forces, or a result of some combination of any of these factors. The high rate of turnover coupled with the lack of diversity change suggests that these newly evolved taxa were directly invading niches established by the archaic taxa which they were replacing, rather than evolving to fill newly created niches.

From the Eocene-Oligocene boundary, the angiosperm diversity curve rises sharply to a brief mid-Miocene plateau, and then rises again to the present. This second rise may be largely dictated by the bias introduced by the "pull of the recent," as it is during this period that modern species begin to appear. The Oligocene-Miocene increase, on the other hand, appears to be largely biological in nature, as it correlates with a major worldwide deterioration of climate (Wolfe 1975, 1978; Kennett 1977; Buchardt 1978), and with the appearance of nonaquatic angiosperm herbs (Scott, Barghoorn, and Leopold 1960; Tiffney 1977). The coincidence of the climatic deterioration and the advent of herbs is not surprising, as the annual and biennial reproductive cycle of the latter is well adapted to the seasonality provided by the former. The appearance of the herbs had a striking effect on angiosperm diversity because they both represented new species and formed an almost entirely new stratum in the community. Prior to the Oligocene, evidence suggests that the angiosperm community was primarily composed of trees and shrubs; with the appearance of the herbs the spatial diversity of this community was strongly enhanced (Table 6.1)). The impact of herbs on angiosperm diversity can also be sensed in the present day, in that herb species outnumber tree species three to one, and outnumber shrub species two to one in the modern flora (Van Valen 1973b).

While the evolution of the herbs looks like a new and distinct chapter in the history of land plants at first glance, it is in truth a recapitulation of the nature and history of the angiosperms. The early members of the group owed a large measure of their success to their progenetic developmental system and to the individual adaptive characters which it permitted to evolve and become integrated. Perhaps the most distinct outward expression of this developmental syndrome lay in the small habit of the early angiosperms.

This habit is inferred both from their initial habitat (unstable river margins favoring a rapid life cycle), and from the absence of angiospermous wood of any size from the deposits containing the earliest members of the class (Doyle and Hickey 1976; Hickey and Doyle 1977). This is further supported by the apparently herbaceous nature of the earliest known monocotyledonous remains, which do not appear to necessarily have been reduced aquatics (Doyle 1973), and may have been plants of seasonally dry areas, as suggested by Stebbins (1970). These lines of reasoning, together with the generally accepted close relationship of monocotyledons and dicotyledons, whether as lineal descendents of one another or as sister groups (Cronquist 1968; Doyle 1973; Stebbins 1974; Meeuse 1976), suggest the absence or weak development of secondary wood in their common ancestor. The presence of both monocots and dicots in the Early Cretaceous Potomac Group (Doyle and Hickey 1976; Hickey and Doyle 1977) suggests that such a common ancestor would have to be sought in earlier sediments, although not necessarily much earlier. It is unfortunate that the suggestion of the small size of the early angiosperms carries with it an all too facile argument for the inability to find the earliest members of the group. The central fact which emerges from all this is that the earliest angiosperms were, at most, plants of a shrubby stature, and perhaps were smaller.

Given that the earliest angiosperms were small, the subsequent history of the group provides a classic example of Cope's Rule, wherein a lineage enters a new adaptive zone by virtue of its small size and attendant ecological and evolutionary malleability, and then increases in size through evolutionary time to an optimal size appropriate to the utilization of the resources of that zone (Stanley 1973). The small size of the first angiosperms was a direct result of their derivation through reduction from some preexisting, presumably pteridospermous, lineage, and was central to the establishment of their progenetic nature. This initially suited the angiosperms to ephermeral environments which would be found most consistently in areas of continuous disturbance, such as along the margins of streams. The disturbed and contiguous nature of this environment permitted the maintainence of critical population sizes, while also affording some degree of allopatric speciation. As described by Doyle and Hickey (1976; Hickey and Doyle 1977), with the passage of time angiosperms expanded from this environment to take advantage of clearings in the gymnosperm forest, and then evolved to greater and greater sizes until they were in direct competition with the gymnosperms in most stable habitats, perhaps forming dominantly angiosperm forests in more unstable habitats. While the boreotropical forest of the Eocene (Wolfe 1975) is the first unequivocal evidence for a fully angiospermous forest, it is likely that this stage was achieved by the Cenomannian as evinced by such assemblages as the Dakota flora.

The central and most interesting aspect of this evolution is that even as the angiosperms grew larger in size and converged on the arborescent habit of the gymnosperms, they retained the essence of their progenetic developmental system. Thus, with the onset of seasonality initiated by the Eocene-Oligocene boundary climatic deterioration (Wolfe 1975, 1978), the group responded with the evolution of herbaceous derivatives, a reiteration of their progenetic heritage. It is presumed here that this reiteration was a feature of those temperate latitude regions most strongly affected by the climatic deterioration, a hypothesis which requires the test of more data from low latitude floras of Tertiary age. It is a measure of the inherent importance of progenesis to the advent (and almost to the definition) of angiospermy that when these herbs appeared, they did not evolve from one or a few closely associated arborescent taxa, but from a wide range of orders and families (Cronquist 1968). Angiosperm herbs are thus not a special condition, but rather are an expression of the nature of the group as a whole. In this sense, progenesis can be seen as a plesiomorphic (primitive) character shared by the group as a whole.

Some of these herbaceous lineages were more successful than others, as witnessed by their relative rates of evolution, diversity, and level of taxonomic resolution. Thus, both woody and herbaceous species are classified together in the Leguminosae and Boraginaceae, reflecting the limited amount of distinction involved in the woody plant–herb transition in these groups. In other cases, enough change has occurred in the transition between the two groups to warrant the recognition of the woody ancestors and the herbaceous descendents as separate families, as in the case of the Umbelliferae and Araliaceae or the Cruciferae and the Capparidaceae. The most extreme case of differentiation is provided by the Compositae, which have undergone sufficient evolution that it is difficult to suggest a particular woody group as its source.

With the reappearance of small angiosperms, the selective forces of Cope's Rule were again called forth, and several classic cases of "arborescent herbs" are in the evolutionary and biogeographic literature, for example, the giant senecios of the African mountains (Hedberg 1969) and the lobelioids (Campanulaceae) of Hawaii (Carlquist 1974). Most of these examples are tropical or subtropical and frequently reflect the enhanced dispersal ability of herbs which permits them to attain isolated mountain tops or islands where they can diversify and occupy unfilled arborescent niches. In the temperate zone, the selective pressure for annual or biennial life cycles is apparently sufficient to prevent herbs from reverting to arborescent status. We are thus presented with a recurring pattern, a repetition in two cycles of the evolution of arborescent forms from herbaceous or nearly herbaceous progenitors. Paralleling this cycle of size may be one involving wind and insect pollination. The earliest angiosperms appear to have been insect pollinated in large part, and to have evolved wind pollination at a later date

(Doyle 1978a, 1978b), although some may have maintained a primitive condition of wind pollination (Dilcher 1979). The majority of the herbs which appeared in the Tertiary were also insect pollinated, judging from their extant relatives. While wind pollination in herbs (and thus the derivation of herbs from wind-pollinated arborescent ancestors) is limited by the small stature of the group which mitigates against effective pollen dispersal, certain herbs have reverted to wind pollination (Stebbins 1974), paralleling the trend in size change mentioned previously.

In both the case of the initial appearance of the angiosperms and the Tertiary appearance of the herbs, the group possessed an association of characters and life history traits that would be considered an "r" strategy relative to the dominant community of the time. The earliest angiosperms appeared at a time of low environmental stress and sought to escape from competition by living in unstable habitats. The group slowly evolved an association of interdependent adaptive characters through the Cretaceous. This permitted them to expand gradually from these limited environments until they came to dominate the forest community. The angiospermous herbs, by contrast, arose in response to strong climatic stimuli which simultaneously created widespread unstable environments appropriate for herbaceous plants of a short life cycle. Further, the angiospermous herbs were derived from established angiosperm lineages and thereby possessed an integrated set of features from the start, including established chemical defenses, pollination syndromes, and dispersal vectors. This contrast in evolutionary status and environmental conditions serves to explain why the first angiosperms took nearly 80 million years to achieve one-half of their greatest diversity in the fossil record, and only about another 35 million years to attain their ultimate diversity.

The present day hosts what appears to be the greatest diversity of land plants in the earth's history. This stems almost entirely from the evolution of the angiosperms which, with their developmental plasticity, have initiated a complex multilayer community. Further, the coevolutionary potential of the group has permitted the members of each layer to diverge in the establishment of a wide range of pollination and dispersal relationships, all of which serve to isolate individual species. In that the continued evolution of the group constantly creates new biotic environments and subdivisions, it can only be expected that land plant diversity will continue to increase, much as predicted by Whittaker (1977), although perhaps at rates which vary according to stimuli from the physical environment.

SUMMARY

An analysis of diversity patterns in land plants has provided the vehicle for an examination of some of the major aspects of land plant evolution.

While this embraces a host of evolutionary events, the three most important transitions are those involving the invasion of land, the origin of the seed, and the development of the angiosperms. In each case, a logical analysis suggests the presence of a series of evolutionary trends initiated by selective pressures and involving reduction, competitive displacement, transferal of function, and energy conservation. The success of both the early land plants and the angiosperms was dependent on an interactive multitude of characters, rather than on any single character. Often, the adaptations and evolutionary responses of the angiosperms may be seen as logical continuations of trends initiated in earlier land plants, rather than as *de novo* features. This serves to underline the continuity of selective pressures and available adaptive solutions which have characterized the evolution of land plants.

ACKNOWLEDGMENTS

Research partially supported by NSF grant DEB 7908052. Appreciation is due Andrew H. Knoll (Oberlin College) and Karl J. Niklas (Cornell University) for assistance and provocative discussion, without which this chapter would not have been written. Susan J. Mazer was instrumental in collecting data.

REFERENCES

Alvin, K. L., R. A. Spicer, and J. Watson. 1978. A *Classopollis*-containing male cone associated with *Pseudofrenelopsis*. *Palaeontology* 21:847–856.

Andrews, H. N. 1963. Early seed plants. *Science* 142:925–931.

Andrews, H. N., P. G. Gensel, and W. H. Forbes. 1974. An apparently heterosporous plant from the Middle Devonian of New Brunswick. *Palaeontology* 17:387–408.

Boucot, A. J. 1978. Community evolution and rates of cladogenesis. In *Evolutionary Biology* (M. K. Hecht, W. C. Steere, and B. Wallace, eds.). New York: Plenum. Vol. 11, pp. 545–655.

Bower, F. O. 1935. *Primitive Land Plants*. London: Macmillan, pp. 1–658.

Brauer, D. F. 1978. Two additional heterosporous, barinophytacean plants from the Fammenian of Pennsylvania. *Bot. Soc. Am. Misc.* Ser. Publ. 156:3 (Abstr.).

Buchardt, B. 1978. Oxygen isotope paleotemperatures from the Tertiary period in the North Sea area. *Nature* (Lond.) 275:121–123.

Carlquist, S. 1974. *Island Biology*. New York: Columbia, pp. 1–660.

Celakovsky, L. 1874. *Bedeutung des Generationswechsels der Pflanzen*. Prague.

Chaloner, W. G. 1967. Spores and land plant evolution. *Rev. Palaeobot. Palynol.* 1:83–94.

———. 1970. The rise of the first land plants. *Biol. Rev. Cambridge Philos. Soc.* 45:353–377.

Cornet, B. 1977. Angiosperm-like pollen with tectate-columellate wall structure from the Upper Triassic (and Jurassic) of the Newark Supergroup, U.S.A. *Am. Assoc. Strat. Palynol.* 10th Ann. Meet., Tulsa, pp. 8–9. (Abstr.).

Cox, C. B. 1974. Vertebrate paleodistributional patterns and continental drift. *J. Biogeog.* 1:75–94.

Crepet, W. L. 1979a. Insect pollination: a paleontological perspective. *BioScience* 29:102–108.

———. 1979b. Some aspects of the pollination biology of Middle Eocene angiosperms. *Rev. Palaeobot. Palynol.* 27:213–238.

Cronquist, A. 1968. *The Evolution and Classification of Flowering Plants.* Boston: Houghton Mifflin, pp. 1–396.

Dilcher, D. L. 1979. Early angiosperm reproduction: an introductory report. *Rev. Palaeobot. Palynol.* 27:291–328.

Doyle, J. A. 1973. The monocotyledons: Their evolution and comparative biology V. Fossil evidence on the early evolution of the monocotyledons. *Quart. Rev. Biol.* 48:399–413.

———. 1977. Patterns of evolution in early angiosperms. In *Patterns of Evolution* (A. Hallam, ed.). Amsterdam: Elsevier, pp. 501–546.

———. 1978a. Origin of angiosperms. *Ann. Rev. Ecol. Syst.* 9:365–392.

———. 1978b. Fossil evidence on the evolutionary origin of tropical trees and forests. In *Tropical Trees as Living Systems* P. B. Tomlinson and M. H. Zimmermann, eds.). Cambridge: Cambridge University Press, pp. 3–30.

Doyle, J. A., and L. J. Hickey. 1976. Pollen and leaves from the mid-Cretaceous Potomac Group and their bearing on early angiosperm evolution. In *Origin and Early Evolution of Angiosperms* (C. B. Beck, ed.). New York: Columbia University Press, pp. 139–206.

Florin, R. 1951. Evolution in cordaites and conifers. *Acta Horti Berg.* 15:285–388.

Hallé, F., R. A. A. Oldeman, and P. B. Tomlinson. 1978. Tropical Trees and Forests. Berlin: Springer-Verlag, pp. 1–441.

Hedberg, O. 1969. Evolution and speciation in a tropical high mountain flora. *Biol. J. Linn. Soc.* 1:135–148.

Hickey, L. J. 1977a. Changes in angiosperm flora across the Cretaceous-Paleocene boundary. *J. Paleontol.* 51 (suppl. part 3 to #2):14. (Abstr.).

———. 1977b. Stratigraphy and paleobotany of the Golden Valley Formation (Early Tertiary) of western North Dakota. *Geol. Soc. Am. Mem.* 150:1–183.

Hickey, L. J., and J. A. Doyle. 1977. Early Cretaceous fossil evidence for angiosperm evolution. *Bot. Rev.* 43:3–104.

Kennett, J. P. 1977. Cenozoic evolution of Antarctic glaciation, the circum Antarctic Ocean, and their impact on global paleoceanography. *J. Geophys. Res.* 82:3842–3860.

Kevan, P. G., W. G. Chaloner, and D. B. O. Savile. 1975. Interrelationships of early terrestrial arthropods and plants. *Palaeontology* 18:391–417.

Knoll, A. K., K. J. Niklas, and B. H. Tiffney. 1979. Phanerozoic land plant diversity in North America. *Science* 206:1400–1402.

Krassilov, V. A. 1975. Dirhopalostachyaceae—a new family of proangiosperms and its bearing on the problem of angiosperm ancestry. *Palaeontographica* (Abt. B.) 153:100–110.

——. 1977. The origin of angiosperms. *Bot. Rev.* 43:143–176.

——. 1978. Late Cretaceous gymnosperms from Sakhalin and the terminal Cretaceous event. *Palaeontology* 21:893–905.

Meeuse. A. D. J. 1976. Fundamental aspects of the evolution of the Magnoliophyta. In *Glimpses in Plant Research* (P. K. K. Nair, ed.) New Delhi: Vikas Publishing House. Vol. 3, pp. 82–100.

——. 1979. Why were the early angiosperms so successful? A morphological, ecological and phylogenetic approach. *Kon. Ned. Akad. Wetensk.* Ser C. 82: 343–369.

Miller, C. N. 1976. Early evolution in the Pinaceae. *Rev. Palaeobot. Palynol.* 21:101–117.

Mulcahy, D. L. 1979. The rise of the angiosperms: A genecological factor. *Science* 206:20–23.

Muller, J. 1970. Palynological evidence on early differentiation of angiosperms. *Biol. Rev. Cambridge Philos. Soc.* 45:417–450.

Niklas, K. J. 1978. Coupled evolutionary rates and the fossil record. *Brittonia* 30:373–394.

Niklas, K. J., B. H. Tiffney, and A. H. Knoll. 1979. Apparent changes in the diversity of fossil plants. In *Evolutionary Biology* (M. K. Hecht, W. C. Steere, and B. Wallace, eds.). New York: Plenum. Vol. 12, pp. 1–89.

Raup, D. M. 1972. Taxonomic diversity during the Phanerozoic. *Science* 177:1065–1071.

——. 1976a. Species diversity in the Phanerozoic: A tabulation. *Paleobiology* 2:279–288.

——. 1976b. Species diversity in the Phanerozoic: An interpretation. *Paleobiology* 2:289–297.

——. 1979. Biases in the fossil record of species and genera. *Bull. Carnegie Mus. Nat. Hist.* 13:85–91.

Russell, D. A. 1979. The enigma of the extinction of the dinosaurs. *Ann. Rev. Earth Planet. Sci.* 7:163–182.

Schuster, R. M. 1969. Problems of antipodal distribution in lower land plants. *Taxon* 18:46–91.

——. 1976. Plate tectonics and its bearing on the geographical origin and dispersal of angiosperms. In *Origin and Early Evolution of Angiosperms* (C. B. Beck, ed.). New York: Columbia University Press, pp. 48–138.

Schweitzer, H. J. 1977. Die Rhaeto-jurassischen Floren des Iran und Afghanistans. 4. Die rätische Zwitterblute *Irania hermaphroditica* nov. spec. und ihre Bedeutung für die Phylogenie der Angiospermen. *Palaeontographica* (Abt. B.) 161:98–145.

Scott, R. A., E. S. Barghoorn, and E. B. Leopold. 1960. How old are the angiosperms? *Am. J. Sci.* 258-A:284–299.

Simpson, G. G. 1944. *Tempo and Mode in Evolution.* New York: Columbia, pp. 1–237.

——. 1953. *The Major Features of Evolution.* New York, Columbia, pp. 1–434.

Stanley, S. M. 1973. An explanation for Cope's Rule. *Evolution* 27:1–26.

Stebbins, G. L. 1967. Adaptive radiation and trends of evolution in higher plants. In *Evolutionary Biology* (T. Dobzhansky, M. K. Hecht, and W. C. Steere, eds.).

New York: Appleton Century Crofts. Vol. 1, pp. 101–142.

———. 1970. Biosystematics: an avenue toward understanding evolution. *Taxon* 19:205–214.

———. 1974. Flowering Plants: Evolution above the Species Level. Cambridge: Belknap Press of Harvard University Press, pp. 1–399.

Swain, T. and G. Cooper-Driver 1981. Biochemical Evolution in Early Land Plants. In *Paleobotany, Paleoecology, and Evolution* (K. J. Niklas, ed.) New York: Praeger. Vol. 1.

Takhtajan, A. 1972. Patterns of ontogenetic alterations in the evolution of higher plants. *Phytomorphology* 22:164–171.

———. 1976. Neoteny and the origin of flowering plants. In *Origin and Early Evolution of Angiosperms* (C. B. Beck, ed.). New York: Columbia University Press, pp. 207–219.

Taylor, T. N., and M. A. Millay. 1979. Pollination biology and reproduction in early seed plants. *Rev. Palaeobot. Palynol.* 27:329–355.

Tiffney, B. H. 1977. Fossil angiosperm fruits and seeds. *J. Seed Technol.* 2:54–71.

Tiffney, B. H., and E. S. Barghoorn. 1974. The fossil record of the fungi. *Occ. Pap. Farlow Herb. Cryptog. Bot.* 7:1–42.

Tschudy, R. H. 1977. Palynological evidence for change in continental floras at the Cretaceous-Tertiary boundary. *J. Paleontol.* 51 (suppl. part 3 to #2):29. (Abstr.).

Valentine, J. W. 1973. Phanerozoic taxonomic diversity: A test of alternative models. *Science* 180:1078–1079.

Van Valen, L. 1973a. A new evolutionary law. *Evol. Theory* 1:1–30.

———. 1973b. Body size and numbers of plants and animals. *Evolution* 27:27–35.

Whittaker, R. H. 1977. Evolution of species diversity in land communities. *Evolutionary Biology* (M. K. Hecht, W. C. Steere, and B. Wallace, eds.). New York: Plenum. Vol. 10, pp. 1–67.

Wolfe, J. A. 1975. Some aspects of plant geography of the northern hemisphere during the Late Cretaceous and Tertiary. *Ann. Mo. Bot. Gard.* 62:264–279.

———. 1978. A paleobotanical interpretation of Tertiary climates in the northern hemisphere. *Am. Sci.* 66:694–703.

Zimmermann, W. 1952. Main results of the "Telome theory." *Palaeobotanist* (Lucknow) 1:456–470.

7

PALEOZOIC BIOGEOGRAPHY AND CLIMATOLOGY

Alfred M. Ziegler, Richard K. Bambach,
Judith Totman Parrish, Stephen F. Barrett,
Elizabeth H. Gierlowski, William C. Parker,
Anne Raymond, and J. John Sepkoski, Jr.

INTRODUCTION

Early Palezoic paleogeography is characterized by a low-latitude array of continents that were at times extensively flooded by shallow seas. Evidently, deep oceans between some of these continents were wide enough to prevent faunal exchange, as considerable longitudinal biogeographical differentiation is observable within single climatic zones. Toward the middle of the Paleozoic, continents began to converge and collide. Relative cosmopolitanism is observed when shallow seas were widespread, such as the Late Ordovician, Middle Silurian, and Late Devonian to Early Carboniferous. Conversely, provincial differentiation in the marine realm is observed during periods of continental collision, such as the Early Devonian and Late Carboniferous, and can be attributed to the formation of geographical barriers. In the late Paleozoic, most of the continents had collided to form the pole-to-pole land mass of Pangaea. As might be expected, terrestrial floras show progressive climatic zonation, while marine faunas on opposite sides of Pangaea show geographical isolation. With such a pervasive north-south land barrier, one would predict extreme east-west climatic asymmetry. Gulf Streams would have carried warm waters to high latitudes on the Tethyan margin, while return currents on the western Americas side would have produced the opposite effect. This temperature asymmetry is reflected in the wide latitudinal occurrence of Tethyan faunas in contrast to the restriction of warm-water faunas in the western United States.

In this chapter, we plot marine and terrestrial biogeographical units on the most recent set of paleogeographical maps for the Paleozoic periods (Scotese et al. 1979) and offer an interpretation of the distributional patterns in terms of climatic zones, ocean currents, and geographical barriers. Despite the appearance of several collections of papers on various aspects of paleobiogeography (Hallam 1973; Hughes 1973; C. A. Ross 1974a; Gray and Boucot 1979; Robison and Teichert 1979), there have been few attempts to integrate this diverse and complicated subject. Ziegler, Scotese, McKerrow, Johnson, and Bambach (1977), however, did present a series of biogeographical maps, showing the distribution patterns of the best known taxonomic group for each period. Also, Boucot and Gray (1979) provided a synthesis based on the assumption that the continents were grouped together for the whole of the Paleozoic. They prefer their "pangaeic" reconstruction to the "fixist-modern geography" and our "kaleidoscopic" maps. Gray and Boucot claim that "the Paleozoic data are permissive of a variety of possibilities," but we leave that judgment to the reader.

We accept as a primary constraint the current paleomagnetic data concerning the positions of the continents through the Paleozoic (Scotese et al. 1979). These data do not permit a pangaean configuration. Admittedly, the information is incomplete or unconfirmed for some paleocontinents. We have, in most cases, been able to check the orientation data by plotting the distribution of climatically sensitive sediments (Ziegler et al. 1979). Neither paleomagnetic nor paleoclimatic data provide much useful information on the longitudinal distance between continents nor the correct east to west order of continents. Biogeography does provide clues, however, about continental positions (Ziegler, Scotese, McKerrow, Johnson, and Bambach 1977), as in many cases provinces are shared by more than one paleocontinent, and this places some limits on longitudinal separation. Tectonic considerations are also helpful, particularly the timing of continental collisions and splits, in determining relationships prior to and following the events, respectively. Undoubtedly, the reconstructions we have used will be modified and we are currently making new paleomagnetic determinations where the biogeographical patterns are in seeming conflict with the available orientation data (see discussion in text on the Devonian reconstruction). The previous discussion serves to emphasize that there is no single line of evidence that can provide all the information on the relationships of the continents in the Paleozoic. Nor are the present maps correct in detail. We hope that by publishing our concept of Paleozoic biogeography at this time, we can stimulate others to improve on these reconstructions.

The distribution of floras and faunas is of course limited by a number of types of barriers and gradients that can operate on all scales from the local ecological level to the climatic or oceanographical realm level. This makes the problem of terminology and classification extremely difficult. Most paleobiogeographers apply the term *province* to their biogeographical entities and place their provincial boundaries along lines where there was very little interchange of taxa. It seems that most of these boundaries are placed in zones across which it is very difficult to correlate rocks, and many of these zones prove to be regions where once-separate continents have been brought together by plate tectonic motions. This procedure has resulted in the recognition of relatively few biogeographical entities in the Paleozoic. Thus, from three to ten marine provinces have been defined in Paleozoic rocks, whereas, in present oceans, well over 30 provinces are recognized (cf. Valentine, Foin, and Peart 1978; Schopf 1979). Many modern provinces are defined that share the majority of faunal elements in common with adjacent provinces. We conclude that most Paleozoic "provinces" are really realms in the modern biogeographical sense, and because of this discrepancy, we keep our terminology deliberately vague.

C. A. Ross (1974b, Fig. 2) has published a very simple world map of "marine faunal regions and subregions of the littoral zone and continental shelves." This map is useful as a model for the Paleozoic because the subdivisions are few and are based on climatic zones: (1) arctic, (2) boreal, (3) northern warm temperate, (4) tropical, (5) southern warm temperate, (6) antiboreal, and (7) antarctic. We accept the subdivisions, but change the terminology somewhat because geographical, as well as climatic, terms are combined; we use (1) polar, (2) temperate, (3) subtropical, and (4) tropical in both hemispheres. In applying these terms we note that the boundaries between zones occur at different latitudes on opposite sides of the present oceans. Thus, the tropical zone is confined to as little as $10°$ from the equator on the east side of the ocean, but extends to $30°$ north and south on the opposite side, as a result of the poleward deflection of equatorial waters by the subtropical gyres. Because oceanic currents are extremely important in modifying latitudinal temperature gradients and because circulation maps have appeared in the geological literature which appear to ignore the basic physics of circulation, we include a general section on oceanic circulation, and plot on our paleogeographical maps the inferred circulation patterns for each period.

OCEANIC CIRCULATION

Surface currents in the oceans are primarily wind driven so that the major current systems reflect the major atmospheric circulation systems. Thus, models of oceanic circulation must be founded on an understanding of atmospheric circulation. Following the approach used by Ziegler, Hansen, Johnson, Kelly, Scotese, and Van der Voo (1977) to model ocean currents for the Early Silurian, we have based our oceanic circulation maps on models of atmospheric circulation constructed for each time period. The principal driving mechanisms behind atmospheric circulation are fairly well known, at least qualitatively.

In the most general terms, the present atmospheric circulation is characterized by large-scale cells, or systems, consisting of ascending (low pressure) and descending (high pressure) air. These systems tend to be arranged in latitudinal (zonal) belts, whose integrity is affected by the presence of continents. Using present-day patterns as a guide, it is possible to apply the principles of circulation to past continental configurations.

Two important assumptions were made in the construction of the circulation models. The first assumption is that the present zonal component of the atmospheric circulation pattern has remained basically the same, despite the greater rotational speed of the earth during the Paleozoic (Wells 1963; Scrutton 1965; see also Mohr 1975). The validity of this assumption is supported by modeling experiments that suggest that the present zonal circulation would remain stable to rotational speeds considerably higher than those postulated for the early Paleozoic (Hunt 1979). The second assumption is that differences in certain thermal parameters, namely, the presence or absence of polar ice, the temperature at the sea surface, and the equator-to-pole gradient, will not change the overall circulation pattern. These factors, along with geography, affect the transfer of heat from one place to another over the surface of the earth. However, only geography can change the general cellular pattern of circulation; the remaining three factors affect only the intensity (Diester-Haass and Schrader 1979; Berger et al. 1978). This assumption is supported by models of circulation during the Pleistocene glaciations (CLIMAP 1976), which reveal that the intensity of the polar fronts was increased, but that their latitudinal positions were shifted only a few degrees. Furthermore, the major features of the oceanic circulation modeled for the Pleistocene were essentially the same as today's. With these assumptions, atmospheric circulation can be modeled with the data provided by paleogeographical studies, that is, positions of continents relative to latitude and to each other, size of continents, size of shallow seas and land areas, and height and orientation of mountain chains.

In modeling Paleozoic circulation, we started with the simplest aspect of the system. On an earth with an homogeneous surface, the wind systems would be wholly zonal, that is, parallel to latitude (Eady 1964; Lorenz 1967). The zonal circulation pattern is caused by the equator-to-pole temperature gradient and by the rotation of the earth, and consists of: (1) easterly winds at the equator (the tradewinds), (2) westerly winds between 40° and 60° latitude north and south, and (3) polar easterlies above 60° north and south. The winds are related to low-pressure belts at the equator and at 60° latitude and high-pressure belts at 30° latitude and the poles. Zonal oceans, that is, oceans that girdle the earth parallel to latitude, would be expected to have zonal circulation. The present-day example, the circum-Antarctic Ocean, does have zonal winds and currents. Zonal circulation is evident throughout the present oceans, although the clustering of continents in the Northern Hemisphere makes this imprint less apparent there. The primary zonal currents are associated with the wind belts, with an equatorial westward flowing current centered about the equator between 15° latitude north and south, eastward flowing currents between 40° and 60° latitude, and westward flowing currents above 60° latitude.

In the presence of continents, the zonal belts of high and low pressure and their associated winds are disrupted, owing to the differing thermal regimes of land and water. Particularly important are the locations of certain low- and high-pressure systems. Low-pressure systems tend to form when a continental margin lies near 60° latitude, as around the southern tip of Greenland and the margin of Antarctica. In such configurations, the land is colder than the adjacent ocean, and consequently the air over the land tends to sink relative to the air over the ocean. This effect locally intensifies the zonal belt of low pressure predicted for that latitude by the zonal model and creates cold-water oceanic gyres. In contrast, high-pressure systems are centered at about 30° latitude. At that latitude, the water is cooler than the adjacent land, and thus the high-pressure belt predicted for that latitude by the zonal model is intensified over the oceans. The oceanic gyres associated with these high-pressure systems are composed of the warm equatorial currents driven by the tradewinds and the cool westerly currents at midlatitudes. The gyres are completed adjacent to the continents by the so-called boundary currents. The western boundary currents, of which the Gulf Stream is an example, carry warm equatorial waters poleward. The eastern boundary currents, exemplified by the Peru Current, carry cool water from midlatitudes toward the equator. Thus, the boundaries between climatic zones occur at different latitudes on opposite sides of the oceans, and this, of course, has important biogeographical consequences.

Finally, it is important to note that the oceanic circulation patterns shown on our paleogeographical maps have been constructed independently of the biogeographical patterns. This procedure was adopted to avoid circularity in interpreting the biogeography. It is not uncommon to find in the literature oceanic current arrows based on supposed migration patterns of marine organisms. We feel that the successive appearance of taxa in different areas at different times is most likely the result of environmental shifts rather than an indication of prevailing currents. Many marine organisms have remarkable dispersal abilities (Scheltema 1977), so that with favorable currents, one would expect their appearance in different areas to be instantaneous on the geological scale. We therefore feel that it is hazardous to make predictions of ocean current directions based on biotic distribution patterns. Our inferred circulation systems, based on physical considerations, do help in many instances to explain latitudinal temperature anomalies and biotic ligations between otherwise distinct paleocontinents.

CAMBRIAN BIOGEOGRAPHY

Palmer's (1972) survey of Cambrian trilobite distributions remains the most comprehensive on the biogeography of this period. Although Palmer related the trilobite "provinces" to paleolatitude, R. J. Ross (1975) was the first to plot this information on a paleogeographical reconstruction and to relate the patterns to ocean surface currents. Ross's reconstruction differs somewhat from ours (Fig. 7.1), which has the advantage of being based on paleomagnetic data specific for the Upper Cambrian (Scotese et al. 1979). This is important because the apparent polar paths for the better-studied paleocontinents indicate that a significant shift in the pole occurred during the Cambrian, due possibly to true polar wander or to all of the continents having moved in concert. Therefore, averaging all Cambrian poles for a continent is an unsatisfactory procedure, resulting in an orientation that may be inaccurate for some, if not all, intervals of the period.

Cambrian workers have defined trilobite provinces within both shallow- and deep-water environments. "Differences in temperature and temperature stability between warmer shelf and colder adjacent slope habitats are among the more important factors that differentiated these trilobite provinces" (Taylor 1976, p. 679). Some of the within-habitat subdivisions can be related to geographical separation and some to differing latitudinal ranges (Palmer 1972).

The shallow-water subdivisions, herein termed realms, are four in number: American, Siberian, European, and Hwangho (Palmer 1979). Both the American, which is shared by Laurentia and the South American portion of Gondwana, and the Siberian realms are centered on the tropical zone. The differences between these two realms may have arisen through geographical separation or from the dominantly clastic nature of American environments as opposed to the carbonate-evaporite environments that prevailed in Siberia. The northern subtropical zone is occupied by the Hwangho realm of the shallow carbonate regions of China and the Australian and Antarctic portions of Gondwana. The European realm occupies temperate latitudes in the Southern Hemisphere and includes the portions of Gondwana from Morocco to Arabia.

The deeper-water trilobite faunas seem to be associated with the subtropical gyres of the northern and the southern oceans. The Chiangnan realm, as plotted by Taylor (1976), is associated with the basinal dark-colored, thinly-bedded limestones and shales of western North America and China. In the Cambrian these regions would have been encompassed by the northern subtropical gyres. Faunas of the deeper-water regions of western Europe and eastern North America from Newfoundland to Mexico constitute another biogeographical entity (Palmer 1972, p. 312) and would have been in the region of the southern subtropical gyres.

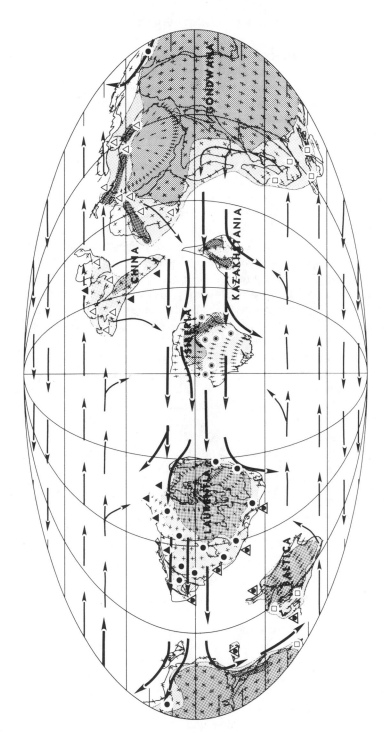

FIGURE 7.1. Middle Late Cambrian (Franconian) biogeography. Symbols for biogeographical units are as follows: △ = Hwangho, ▲ = Chiangnan, • = American, ⊙ = Siberian, ◬ = temperate. Symbol shape code for climatic assignments is as follows: ○ = tropical, △ = subtropical, □ = Acado-Baltic, ◫ = European. Arrows indicate inferred surface ocean currents: thick = warm, thin = cold. Topography is as follows: dark shading = mountains, intermediate shading = lowlands, light shading = shallow shelves, white = ocean basins.

237

ORDOVICIAN BIOGEOGRAPHY

Early Ordovician provinciality is similar to that of the Cambrian. Later during the Ordovician, provinciality in shallow marine faunas decreased from four to two provinces (Whittington and Hughes 1972), perhaps as a result of convergence of continents in the low-latitude regions and rising sea levels, which permitted free interchange of faunas.

The large Gondwana continent extended from the South Pole to the equator; throughout the Ordovician it moved southward across the pole. The remaining continents, except Baltica, lay in equatorial or low-latitude regions. Baltica apparently lay in southern midlatitudes, although this is not entirely certain since paleomagnetic data are few and conflicting. Our reconstruction employs a recent, but unconfirmed, pole which indicates that Baltica was centered on 70°S (Scotese et al. 1979).

Two sets of biogeographical names have been proposed for Ordovician shallow water faunas, one based on brachiopods (Williams 1973) and one based on trilobites (Whittington 1973). In our maps, we have used the regionally styled names of Williams—American, Baltic, and Anglo-French—rather than the taxonomic names of Whittington—Bathyurid, Asaphid, *Selenopeltis*, and Hungaiid-Calymenid—because the regional names are more in accordance with biogeographical practice for the Paleozoic. The American, or Bathyurid, fauna occupies the low-latitude carbonate platforms of Laurentia, Siberia, and Kazakhstania, and suggests closer ties between these paleocontinents than existed during the Cambrian. As in the Cambrian, however, there is a close tie during the Ordovician between China and Australia, as indicated by the low-latitude Hungaiid-Calymenid fauna, which extends along the north coast of Gondwana from Australia to South America. The high southerly latitudes of Gondwana, from Florida through Morocco to Arabia, are dominated by the *Selenopeltis*, or Anglo-French, faunas. Elements of the *Selenopeltis* fauna also occur in deeper-water sediments associated with the Hungaiid-Calymenid and Asaphid faunas.

Worldwide biogeographical summaries are available for four of the major Ordovician groups: trilobites (Whittington 1973; Whittington and Hughes 1972), brachiopods (Williams 1973), corals (Kaljo and Klaamann 1973), and graptolites (Berry 1979). Among shelf-dwelling organisms, the trilobites seem to be the best studied and most definitive, but the trilobite subdivisions are paralleled by the brachiopod and coral distributions. Provinciality existed in the pelagic realm, as seen in the graptolite distributions. This was best developed in the Early Ordovician and essentially disappeared by Llandeilo–early Caradoc time, the interval represented by our map (Fig. 7.2). Graptolite workers define two Early Ordovician pelagic realms, Pacific and Atlantic, and relate these to warm equatorial and cool southern temperate water masses, respectively (Berry 1979).

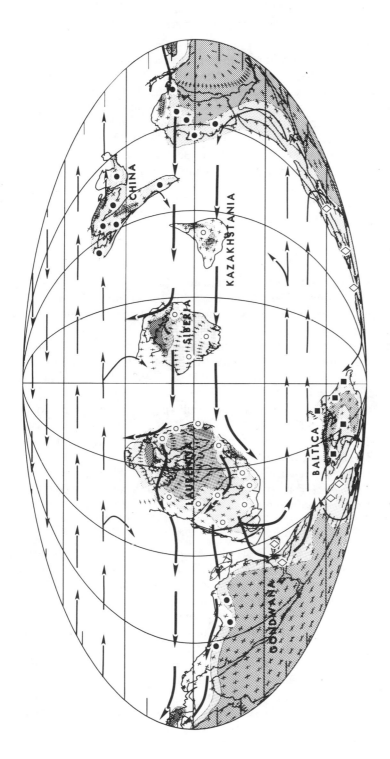

FIGURE 7.2. Middle Ordovician (Llandeilo-Caradoc) biogeography. Symbols for biogeographical units are as follows: ○ = American, ● = Chinese, □ = Baltic, ◇ = Anglo-French. Symbol shape code for climatic assignments is as follows: ○ = tropical, □ = temperate, ◇ = polar. Arrows indicate inferred surface ocean currents: thick = warm, thin = cold. Topography is as follows: dark shading = mountains, intermediate shading = lowlands, light shading = shallow shelves, white = ocean basins.

Ordovician biogeography may help to more accurately position Baltica and its associated Asaphid faunas. Traditionally, it has been thought that Baltica lay in low latitudes, situated between Laurentia and Siberia. This seems unlikely on biogeographical grounds, since faunal evidence suggests separation of Baltica from the equatorial continents and Gondwana. In addition, climatic indicators argue against tropical or subtropical environments. Carbonate sediments do occur on Baltica, but these are not of Bahamian type (until later in the period); Jaanusson (1971, p. 14) concluded that the evidence from the Baltica carbonate sediments actually supports a temperate-latitude origin.

SILURIAN BIOGEOGRAPHY

The trend toward decreasing provinciality, seen through the Ordovician, continued to about the middle of the Silurian. At this time, approximately 45 percent of the continental crust was inundated by shallow seas, which allowed free communication of marine faunas (Fig. 7.3). This may have been an important factor influencing the extreme cosmopolitanism of the mid-Silurian, together with the fact that most of the paleocontinents were relatively close together. In addition, all continents except Siberia and Gondwana were at low latitudes.

Although cosmopolitanism was generally high during the Silurian, some biogeographical units can be recognized. These subdivisions are based on brachiopods (Boucot and Johnson 1973; Boucot 1975), the most abundant and widespread components of Silurian faunas; however, distributional data are also available for corals (Kaljo and Klaamann 1973) and bivalves (Pojeta et al. 1976). Our map used the biogeographical terminology of Boucot (1979) and Boucot and Johnson (1973), and the locality data of Cocks and McKerrow (1973), Vinogradov (1968), and Amantov et al. (1970).

The Cosmopolitan Realm covers all low-latitude shallow-water regions. North of 40° is the low-diversity *Tuvaella* fauna of the Mongolo-Okhotsk region of the Siberian Platform. South of 40° there is a parallel situation with the *Clarkeia*, or Malvinokaffric, Realm.

A rise in coral diversity and reef development parallels the rise in sea level during the Early Silurian, with the acme occurring during the Wenlockian; coral diversity and reefs declined with the falling sea levels toward the end of the period (Kaljo and Klaamann 1973). European coral faunas show the greatest degree of endemism and the highest proportion of compound rugosids during the Silurian. This is not surprising in view of the low-latitude position of Baltica and its situation on the east side of the continent.

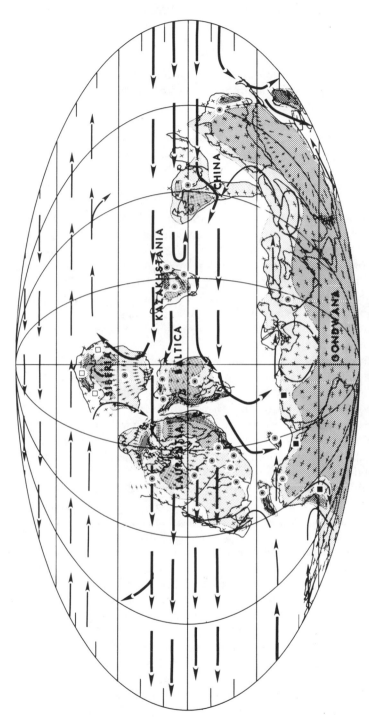

FIGURE 7.3. Middle Silurian (Wenlock) biogeography. Symbols for biogeographical units are as follows: □ = Mongolo-Okhotsk, ○ = Cosmopolitan, ■ = Malvinokaffric. Symbol shape code for climatic assignments is as follows: ○ = tropical, ⊡ = temperate. Arrows indicate inferred surface ocean currents: thick = warm, thin = cold. Topography is as follows: dark shading = mountains, intermediate shading = lowlands, light shading = shallow shelves, white = ocean basins.

DEVONIAN MARINE BIOGEOGRAPHY

Devonian marine biogeography is based on the distribution of benthic organisms, mostly brachiopods (Boucot, Johnson, and Talent 1969; Boucot 1975; Savage et al. 1979), and to a lesser degree trilobites (Eldredge and Ormiston 1979) and corals (Oliver 1977). Recognized biogeographical units at the time of greatest provinciality, the late Early Devonian, are numerous, as shown on Figure 7.4. Faunas of some of these units are sufficiently similar that the units are grouped generally into three realms: Malvinokaffric, Eastern Americas (Appalachian and Amazon-Colombian Provinces), and Old World (remaining units). Within the Old World Realm, Cordilleran-Uralian and Rhenish-Bohemian are also recognized. The Dzhungaro-Balkash Subprovince is usually considered a subset of the Uralian.

Certain "subprovinces" may have been zones of mixing between two adjoining biogeographical units. This is probably the case with the Amazonian part of the Amazon-Colombian unit (Appalachian with Malvinokaffric), and the New Zealand unit (Tasman with Malvinokaffric) (Boucot, Johnson, and Talent 1969). Other biogeographical entities (Mongolo-Okhotsk and Dzhungaro-Balkash) may simply be biofacies within a larger province (Boucot 1975).

Disjunct occurrences of certain taxa provide some insight into oceanic circulation and proximity of paleocontinents. Thus, occurrences of Appalachian brachiopod taxa in Kazakhstan (Dzhungaro-Balkash) and in Siberia (Mongolo-Okhotsk), but not in other Old World areas, suggests that dispersal was via equatorial currents. Appalachian corals (Oliver 1977) and brachiopods (Boucot 1974) in North Africa as early as the Emsian may indicate much greater proximity between North America and Gondwana than is shown in Figure 7.4.

The Middle and Late Devonian are characterized by decreasing provinciality, with Late Devonian faunas being generally cosmopolitan (House 1979; Oliver 1977; Boucot 1974). This is probably the result of continued coalescing of continents into Pangaea, as well as northward movement of Gondwana into warmer latitudinal belts. Marine transgression may also have played a role in the increasing cosmopolitanism (House 1979). The actual transition occurs during the Middle Devonian through breakdown of subprovinces within the Old World Realm and appearance of Old World Realm Rhenish–Bohemian taxa in eastern North America (Boucot 1974; Boucot, Johnson, and Talent 1969).

Figure 7.4 contains some apparent biogeographical anomalies: (1) The faunal similarity between eastern North America (Appalachian Province) and northwestern South America (Amazon-Colombian Province) suggests a closer geographical relationship between Laurussia and Gondwana. (2) Warm-water faunas (as well as indicators of warm climates, such as reefs,

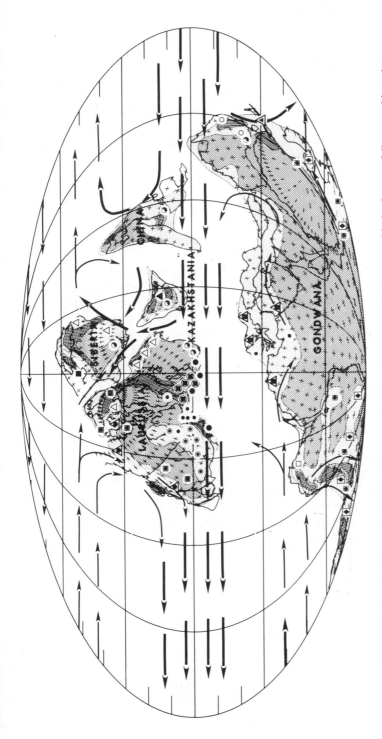

FIGURE 7.4. Late Early Devonian (Emsian) biogeography. Symbols for the marine biogeographical units are as follows: ■ = Mongolo-Okhotsk, ▲ = Cordilleran, ▵ = Uralian, ▴ = Dzungaro-Balkash, ⊙ = Appalachian, ● = Rhenish, ○ = Tasman, △ = Bohemian, ◮ = New Zealand, □ = Amazon-Columbian, ⊠ = Malvinokaffric. Symbols for the terrestrial biogeographical units are as follows: ▣ = Western North American, ● = Appalachian, ⊕ = Western European, ◕ = Eastern Equatorial, ⊗ = Southern Gondwanan. Symbol shape code for climatic assignments is as follows: ○ = tropical, ▵ = subtropical, ▵ = temperate. Arrows indicate inferred ocean surface currents. Topography is as follows: dark shading = mountains, intermediate shading = lowlands, light shading = shallow shelves, white = ocean basins.

243

thick carbonates, and evaporites) are widely distributed in the Devonian Northern Hemisphere. Warm western boundary currents can explain only part of this distribution. It is possible that the strongly asymmetric distribution of land and the completely open ocean at the North Pole may have caused asymmetry in climatic zonation, such that the Northern Hemisphere tropical and subtropical zones were displaced northward. (3) The distinctiveness of the equatorial Appalachian Province, separated from the equatorial Old World Realm Rhenish unit to the east and the Old World Realm Cordilleran Province to the west, was probably maintained by physiographical, thermal, and salinity barriers (Oliver 1977; Boucot 1975). However, this distinctiveness also may reflect environmental differences between epeiric (Appalachian) and marginal (Old World Realm) seas.

DEVONIAN TERRESTRIAL BIOGEOGRAPHY

Devonian floras reveal biogeographical patterns similar to those of marine invertebrates. Although previous workers have identified at most two Lower Devonian floras (Andrews 1961; Edwards 1973), a multivariate statistical analysis of a compilation of published data on Early Devonian plant assemblages (Gierlowski 1978) suggests that at least three intergrading phytogeographical units of probable provincial status existed during the Early Devonian (Fig. 7.5). The floras, like the marine faunas described in the previous section, reveal the following distinct biogeographical regions: Gondwana, Western North American, and Equatorial.

The cool south temperate Gondwana flora constitutes the most distinct of the three floral units. It includes presumably cool to cold regions near the Early Devonian South Pole, and comprises known plant localities in Brazil, Bolivia, Argentina, Falkland Islands, South Africa, and Antarctica. These localities have comparatively low described generic diversities (approximately three and one-half genera per assemblage, range: one to six. The absence or restriction of widespread genera such as *Psilophyton*, *Taeniocrada*, and *Zosterophyllum*, as well as the occurrence of the endemic *Haplostigma*, characterize these localities. Edwards (1973) also recognized these characteristics and suggested the existence of a separate Gondwana flora in the Devonian.

The temperate or subtropical Western North American flora comprises plant localities in Arizona, Wyoming, Alaska, Bathurst Island (Canadian Arctic), and Spitsbergen. These localities have low described generic diversities (approximately four genera per assemblage, range: two to six). All of these localities lay on the northwest side of Laurentia during the Early Devonian, within the influence of inferred cool eastern boundary currents. An absence or restriction of typical Equatorial genera, such as *Taeniocrada*

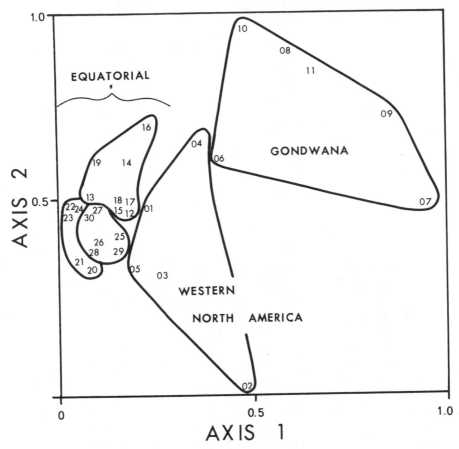

FIGURE 7.5. A polar ordination plot showing the generic relationships among Lower Devonian assemblages from different geographical areas. Distances between points were determined using the complement of the Dice (Sørenson) Coefficient and reflect the floral dissimilarities between localities. The first ordination axis was constructed by selecting the locality with the greatest variation in floral distances to all other points as one end point and the locality most dissimilar to this first one as the other end point. The end points of the second axis are the two localities which fit most poorly on the first axis. The two-dimensional ordination shown here encompasses 58 percent of the variation in the original floral distance matrix. Data were compiled by Gierlowski (1978). Due to sample bias arising from differences in preservation, only compression floral assemblages were considered in this ordination analysis. The list of references used by Gierlowski (1978) are available from Ziegler.

The locality number and paleolatitudes by biogeographical unit are as follows: *Western North American Province*—1, Alaska (42°N); 2, Arizona (2°N); 3, Wyoming (12°N); 4, Bathurst Island (29°N); 5, Spitsbergen (37°N); *Southern Gondwana Province*—6, Argentina (70°S); 7, Falkland Islands (70°S); 8, Brazil (65°S); 9, Bolivia (70°S); 10, South Africa (70°S); 11, Antarctica (48°S); *Equatorial Province*, Appalachian Subunit—12, Ontario (9°N); 13, Maine (0°N); 14, New Brunswick (2°N); 15, Scotland (15°N); 16, Newfoundland (2°N); 17, Portugal (28°S); 18, Gaspé (4°N); 19, Libya (40°S); Western European Subunit—20, South Wales (4°N); 21, Norway (6°N); 22, France (1°S); 23, Belgium (3°N); 24, Germany (1°N); Eastern Equatorial Subunit—25, Ukraine (1°N); 26, Kazakhstan (13°N); 27, Yunnan (15°N); 28, Victoria, Australia (22°S); 29, Siberia (32°N); 30, Poland (1°S).

and *Zosterophyllum*, characterize this flora. However, the Western North American flora is united with the Equatorial flora in the widespread occurrence of genera such as *Psilophyton*, which is absent from Gondwana.

The Equatorial phytogeographical unit includes all remaining localities, specifically those in eastern North America, Europe, Asia, and northern Gondwana (i.e., Portugal, Libya, and Victoria, Australia). These localities have comparatively diverse described floras (10 genera per locality on average, range: 2 to 23) and are unified by the widespread (although not universal) occurrences of *Drepanophycus*, *Zosterophyllum*, *Psilophyton*, and *Taeniocrada*. Still, the Equatorial flora is not entirely homogeneous, and three longitudinally restricted subunits can be discerned on the ordination plot (Fig. 7.5). The westernmost subunit, the Appalachian region (including northern Scotland), appears intermediate in character between the Western North American flora and the more typical Equatorial flora of Europe and Asia. Widespread Equatorial genera, such as *Prototaxites*, *Sporogonites*, and *Zosterophyllum*, seldom occur in these localities. Western Europe contains all of these genera, as well as *Sciadophyton*, which is much more widespread there than it is in other parts of the equatorial region. Finally, the eastern equatorial subunit, which includes China, Victoria, Kazakhstan, Siberia, the Ukraine, and, perhaps, Poland, is typified by numerous endemics (possibly, however, reflecting a taxonomic bias), and the common occurrence of *Cooksonia*, which is much less widespread in other parts of the equatorial region. The locality in northern Siberia, which was near a latitude of 30°N in the Early Devonian, clusters with the eastern equatorial subunit possibly because Siberia enjoyed a warm climate produced by equatorial currents deflected northward by Europe (see Fig. 7.4). Portugal and Libya, which were in the southern Temperate Zone on northern Gondwana at this time, also fall within the Equatorial phytogeographical unit, but cluster with the comparatively low-diversity Appalachian localities in the ordination (Fig. 7.5). Marine invertebrate faunas of Libya and Portugal also reveal a close relationship with Laurussian faunas. Both floral and faunal assemblages of Libya and Portugal suggest a closer geographical relation between Gondwana and Laurussia than is indicated in Figure 7.4.

This type of analysis is subject to three sources of error: (1) The reports used range from old to very recently published and come from paleobotanists of many nationalities, possibly creating historical and national taxonomic biases. To avoid this, we used recent studies that noted synonymies whenever possible. We hope much of the taxonomic bias is random with respect to our analysis. (2) In order to include as many localities as possible, we included reports from areas that have not been thoroughly studied. This may create false diversity and distribution gradients within the data. (3) The age of many "Lower" Devonian plant localities, particularly those from

Gondwana and the Appalachians, is debated (e.g., Andrews et al. 1978). Thus, the possibility remains that observed gradients are time-transgressive and do not represent contemporaneous spatial distributions.

EARLY CARBONIFEROUS MARINE BIOGEOGRAPHY

C. A. Ross (1979a, p. A283) has summarized Early Carboniferous marine biogeography by stating, "Faunas of Tournaisian and Visean times were relatively widespread; many were nearly cosmopolitan and the remainder had low levels of provincialism." Ross related this to high sea level stands and broad epicontinental carbonate seas, which were disposed to allowing free migration of marine forms. The continental platforms must have been close together, since the collisions that resulted in the supercontinent Pangaea began in the Late Carboniferous.

Corals show some biogeographical differentiation and provide the basis for the faunal subdivisions on our map (Fig. 7.6), which shows the situation in the Visean. Hill (1948) defined four coral provinces: (1) the European (*Dibunophyllum*) Province, which occurs from the Urals to southern Europe, Morocco, and Nova Scotia; (2) the Chinese (*Kueichouphyllum*) Province, which overlaps the European Province in Kazakhstan; (3) the Australian (*Amygdalophyllum*) Province, which is restricted to eastern Australia; and (4) the North American (*Lithostrotionella*) Province. Recently, Sando, Bamber, and Armstrong (1975) subdivided North American coral faunas into a number of provinces and subprovinces, based on analyses of endemism and similarity. In the Visean, the greatest endemism occurs in the Southeastern Subprovince, which was partially separated from the Western Interior Subprovince by the Transcontinental Arch. Finally, Vinogradov (1969) proposed a twofold biogeographical subdivision for northern Eurasia, based on corals and other faunal elements, between the Mediterranean Realm (= the European Province of Hill) and the Boreal Realm of the northerly paleocontinent of Siberia.

Geographical isolation, latitudinal temperature gradients, and oceanic current patterns all seem to be important in accounting for the biogeographical differentiation observed in the Early Carboniferous. The European and Southeastern North American faunas both occupy the tropical zone, and the land mass along the eastern seaboard of North America apparently acted as a barrier. Two faunas also occupy the northern subtropical zone, the Chinese and the Western Interior North American faunas. Here, a wide ocean basin was probably the isolating factor. A single southern subtropical fauna, constituting the Australian Province, has been defined and predictably can be related to the warm limb of the subtropical gyre. Much of Gondwana was in high southern latitudes, and one would predict cold-water faunas in these regions. Siberia occupied the high northern latitudes and was evidently biogeographically distinct for this reason.

EARLY CARBONIFEROUS TERRESTRIAL BIOGEOGRAPHY

In contrast to the decreased endemism of Visean marine invertebrates, Visean plant distributions evidence strong biogeographical separation. Chaloner and Meyen (1973) presented only two Lower Carboniferous floras, the Angaran and the Euramerian, and suggested that the broad generic concepts used by workers to classify Lower Carboniferous plants have resulted in many apparently cosmopolitan genera. The three Visean spore suites presented by Sullivan (1967) for the Northern Hemisphere suggest that redefinition of fossil genera or different methods of analysis might reveal more differentiation among Lower Carboniferous compression plant assemblages. New statistical analyses of published Visean and questionably Visean floral assemblages reveal three phytogeographic units distinguishable at the generic level: an Angaran unit (northern polar, high-latitude temperate); a Gondwanan unit (southern temperate); and a Circum-Tethyan unit (equatorial).

The Angaran unit includes Siberian localities (Kuznetsk Basin, Gorlovsk Basin, Minusinsk Basin, and Tom River area), as well as the Bayanhongor Province of Mongolia. These localities possess an average of eight genera per described assemblage (range: 6 to 16). The fern genus *Chacassopteris* is the only form that occurs in all five localities and nowhere else. The lycopods *Tomiodendron* and *Angarodendron* and the foliage genus *Angaropteridium* occur only in Angara, although not in all Angaran localities. The latitudinal distribution of these localities corresponds to the northern Temperate Zone.

The Gondwanan unit consists of the following localities: Teresina, Brazil; the Sinai Peninsula; Paracas and Garhuamayo, Peru; the San Juan and Rioja Provinces of Argentina; Hunter Valley, New South Wales, Australia; and Nova Scotia. Rigby (1969) reviewed all "pre-Gondwana" Carboniferous floras of the Southern Hemisphere and India and accepted a Visean or younger age for these localities. Since some Australian localities have been redated as Upper Carboniferous, Rigby's generic list for Australia has been modified according to Morris (1975) in our analysis. Rigby suggested that a number of shared endemic species differentiates these localities from Northern Hemisphere Lower Carboniferous floras. They contain an average of five and one-half genera per described assemblage (range: three to eight). The lycopod *Cyclostigma* occurs in all but the two most southern Gondwana unit localities, San Juan and Rioja, Argentina, and in Nova Scotia. The foliage genera *Triphyllopteris and Rhacopteris* occur in seven of the eight Gondwanan unit localities. However, *Triphyllopteris* also occurs in the eastern United States and *Rhacopteris* occurs in Silesia, Arkansas, and Spiti on the Indian subcontinent. These points, except for Nova Scotia, correspond to a southern temperate distribution. Nova Scotia may very well have been farther south than shown on Figure 7.6, as it is now known that there was considerable strikeslip motion in maritime Canada in the Late Carboniferous (Ziegler et al. 1979, p. 193).

The remaining localities ordinate near one another and form the Circum-Tethyan unit. These localities possess an average of 11 genera per described assemblage (range: 4 to 20). The Spiti, India, sample is the only locality from mainland Gondwana that ordinated within the equatorial-subequatorial unit. Although the paleolatitude shown for Spiti is 35°S, an exact paleolatitude for this point is difficult to reconstruct as a result of deformation caused by the Cenozoic collision of India and Asia. Spiti must have been farther north in the Visean than is shown on the map (Fig. 7.6). The lycopod genus *Lepidodendron* occurs in all of the Circum-Tethyan localities except Spiti; this genus does not occur in Angara (Meyen 1976) or elsewhere in Gondwana (Rigby 1969). The foliage genera *Sphenopteris* and *Rhodea* occur in 14 of the 20 Circum-Tethyan localities, but also occur widely in Gondwana. (See Fig. 7.7.)

Within the Circum-Tethyan unit, samples from Scotland, Spitsbergen, Silesia, the Donetz Basin, Kazakhstan, the Malayan Peninsula, and South China consistently ordinate above the other Circum-Tethyan localities. With the exception of Silesia and the eastern Asian localities, these localities correspond to the European portions of Sullivan's (1967) *Monilospora* and Kazakhstan spore suites. They also correspond to Radchenko's "Scottish-Kazakhstanian" flora (Meyen 1976, and Sveshnikova and Budantsev 1969). The foliage genera *Sphenopteridium* and *Neuropteris* occur in most of these "northern" and Asian localities. *Sphenopteridium* also occurs in Australia (Gondwana) and France (southern Laurussia), and *Neuropteris* also occurs in the Karaganda and Minusinsk Basins (Angara). All of these northern and Asian assemblages, except the Donetz Basin and one Spitsbergen sample, share genera with Angaran assemblages (e.g., *Angaropteridium, Cariopteridium, Neuropteris, Cordaites,* and *Asterocalamites*). The distribution of these points corresponds to the tropical and northern subtropical zone. The eastern position of Silesia relative to other southern Laurussian localities may explain the presence of the Silesian sample within this subunit.

The remaining Circum-Tethyan sample—Utah, Arkansas, eastern United States, the Loire Basin, Mont du Mâconnais and Vosges in France, and Spiti, India—form a loosely associated group, which corresponds to Radchenko's "northern European" flora and Sullivan's *Grandispora* spore suite. The only endemic genus that occurs in more than one of these

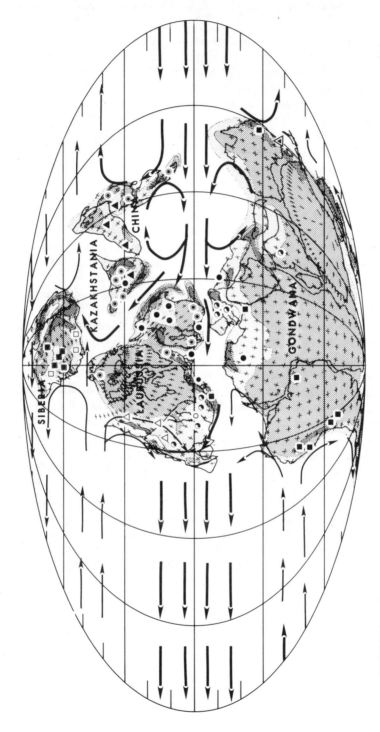

FIGURE 7.6. Late Early Carboniferous (Visean) biogeography. Symbols for marine biogeographical units are as follows: □ = Siberian, △ = Western Interior, ▲ = Chinese, ○ = Southeastern, ● = European, ◬ = Australian. Symbols for terrestrial biogeographical units are as follows: ▣ = Angaran, ◉ = Northern Circum-Tethys, ◎ = Southern Circum-Tethys, ■ = Gondwanan. Symbol shape code for climatic assignments is as follows: ○ = tropical, ▵ = subtropical, ▢ = temperate. Arrows indicate inferred surface ocean currents: thick = warm, thin = cold. Topography is as follows: dark shading = mountains, intermediate shading = shallow shelves, white = ocean basins.

localities is *Lepidophyllum*, which occurs in three of the eight localities. All of these assemblages, except Utah, share genera with Gondwanan assemblages (e.g., *Sphenopteridium*, *Adiantites*, *Triphyllopteris*, and *Rhacopteris*). The distribution of these localities corresponds to the tropical zone.

LATE CARBONIFEROUS MARINE BIOGEOGRAPHY

Marine provinciality increased during the Carboniferous owing to the collision between Gondwana and Laurussia that resulted in a north-south land barrier in tropical latitudes (C. A. Ross 1979a, p. A286). This barrier is reflected in the differentiation of shallow-water fusilinid faunas into an Eurasian-Arctic Realm and a Midcontinent-Andean Realm (Ross 1967). The Eurasian-Arctic Realm extends along the east side of the Uralian seaway to Arctic Canada, while the Midcontinent-Andean Realm is limited to a narrow latitudinal belt on the west side of the Americas (Fig. 7.8).

There was a concomitant decrease in coral faunas (Hill 1948, p. 125) at a time when many carbonate seas were overwhelmed by the influx of clastics and the development of coal swamps in the tropical rainy zone. The Chinese coral faunas show considerable endemism that serves to differentiate them from the European faunas. The latitudinal position of China is uncertain at this time, as is its relationship to Siberia. The endemism shown by Chinese coral faunas adds weight to the argument that this paleocontinent was not associated closely with the rest of Eurasia during the Late Paleozoic.

Siberia and most of Gondwana were in relatively high latitudes in the Late Carboniferous. Southern high-latitude faunas have yet to be described, while northern Siberian faunas have been described as Boreal by Soviet geologists (Vinogradov 1969). These seem to be differentiated clearly from the faunas of European Russia, to which the term Mediterranean Realm has been applied.

LATE CARBONIFEROUS TERRESTRIAL BIOGEOGRAPHY

Westphalian floral provinces (Chaloner and Meyen 1973; Read 1947) mirror the distribution of Westphalian marine provinces with a few important exceptions. The Siberian and Chinese marine faunal provinces have exact floral counterparts in the Angaran and Cathaysian floras, respectively. The marine faunal Mid-Continent–Andean and Uralian Provinces, which are separated by Pangaea, have only one floral counterpart, the Euramerian flora, which stretches from the east to west coast of equatorial and subequatorial Pangaea. There is no described marine faunal equivalent of the south temperate Gondwana floral province during the Westphalian.

The Angaran flora occurs throughout Siberia and in northern Kazakh-

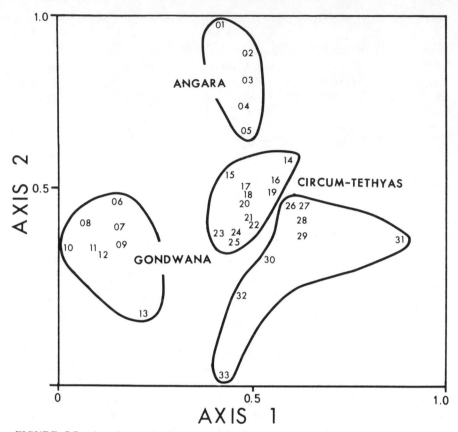

FIGURE 7.7. A polar ordination plot showing the generic relationships among Early Carboniferous plant compression assemblages from different geographical areas. Distances between points were determined using an arccosine transformation of the Otsuka Coefficient and reflect the floral dissimilarity between localities. Axial end points were chosen as in Figure 7.5. This two-dimensional ordination encompasses 41 percent of the variation in the original floral distance matrix. Data for this ordination came from the following sources: Angara—Gorlova (1978), Draber (1972); Kazakhstan—Radchenko (1961), Tchirkova (1937), Zalessky (1937a,b); China and Southeast Asia—Asama (1973), Asama et al. (1975), Gothan and Sze (1933); North America—Bell (1960), Arnold and Sadlick (1962), Read (1955), White (1937); Europe—Bureau (1914), Sveshnikova and Budantsev (1969); Gondwana—Morris (1975), Read (1941), Rigby (1969). Due to biases in taxonomy arising from differences in preservation, only compression floral assemblages exclusive of fructification genera were used.

The locality number and paleolatitudes by biogeographical unit are as follows: *Angaran Province*—1, Kuznetsk (62°N); 2, Minusinsk (65°N); 3, Gorlovsk (60°N); 4, Tom River (60°N); 5, Bayanhongor, Mongolia (70°N); *Gondwanan Province*—6, Sinai Peninsula (22°S); 7, Rioja, Argentina (63°S); 8, Teresina, Brazil (45°S); 9, New South Wales, Australia (30°S); 10, Paracas, Peru (45°S); 11, San Juan, Argentina (65°S); 12, Garhuamayo, Peru (47°S); 13, Nova Scotia, Canada (5°S); *Circum-Tethys Province*, Northern Circum-Tethys Subunit—14, South China (40°N); 15, Kiangsi, China (28°N); 16,19, Karaganda Basin (30–33°N); 17,20, East Urals (30–33°N); 18, Gambang, Malaysia (7°N); 22, Donetz Basin (0°); 21,23, Spitsbergen (45°N); 24, Silesia (1°S); 25, Scotland (15°N); Southern Circum-Tethys Subunit—26, Virginia, USA (2°S); 27, Vosges Mts., France (12°S); 28, Loire Bassin, France (10°S); 29, Mâconnais Mts., France (10°S); 30, Arkansas, USA (1°S); 31, Utah, USA (10°N); 32, Pennsylvania, USA (2°S); 33, Spiti, India (37°S).

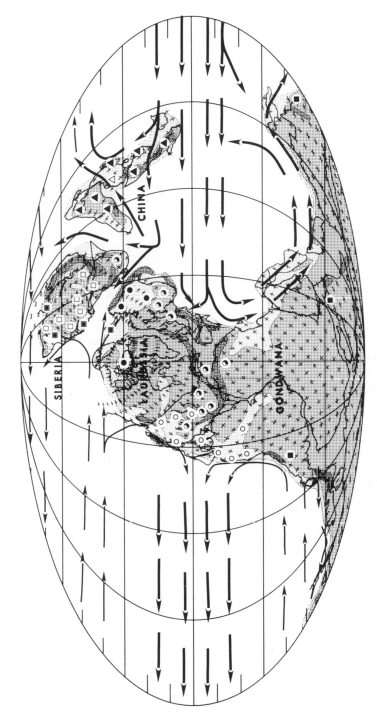

FIGURE 7.8. Middle Late Carboniferous (Westphalian CD) biogeography. Symbols for marine biogeographical units are as follows: ▣ = Siberian, △ = Chinese, ○ = Mid-Continent-Andean, ● = Eurasian-Arctic. Symbols for terrestrial biogeographical units are as follows: ▫ = Angaran, ▲ = Cathaysian, ◕ = Euramerian, ■ = Gondwanan. Symbol shape code for climatic assignments is as follows: ○ = tropical, △ = subtropical, ▫ = temperate. Arrows indicate inferred ocean currents: thick = warm, thin = cold. Topography is as follows: dark shading = mountains, intermediate shading = lowlands, light shading = shallow shelves, white = ocean basins.

253

stan. The lycopod *Angarodendron*, the sphenopsid *Paracalamites*, the foliage genera *Angaridium*, *Paragondwanidium*, and *Angaropteridium*, and the cordaite *Rufloria* characterize this flora (Chaloner and Meyen 1973). The distribution of this flora corresponds to the North Temperate Zone. Petrified wood from the Upper Carboniferous of Angara shows growth rings indicating a seasonally dry or cold environment.

The Cathaysian flora, which occurs in North China, South China, and the Malay Peninsula, contains three endemic genera, *Tingia*, *Konchophyllum*, and *Kaipingia*. This flora also shares the genera *Neuropteris*, *Linopteris*, and *Cordaites* with the Euramerian flora. The latitudinal distribution of this flora (Fig. 7.8) corresponds to the northern subtropical zone.

Two types of floral assemblages characterize the Euramerian flora of equatorial and subequatorial Pangaea: (1) a lowland and swamp flora composed of the lycopods *Lepidodendron*, *Lepidophlois*, *Sigillaria*, and *Bothrodendron*, the sphenopsid *Calamites*, the foliage genera *Alethopteris*, *Neuropteris*, *Pecopteris*, and *Mariopteris* and the arborescent gymnosperm *Cordaites*; and (2) an upland assemblage with the conifer *Walchia*, but no lycopods. The upland assemblage occurs only in the North American Cordillera, which led Read (1947) to suggest that this assemblage could represent a subprovince of the Euramerian Province. The occurrence of a lowland Euramerian flora in nearby Mexico (Silva-Pineda 1970) suggests that the Cordillera assemblage represents an environmentally correlated variant of the Euramerian flora. The Euramerian flora has an equatorial and subequatorial distribution. Petrified wood associated with Westphalian-age coal swamps in this region has no growth rings, indicating a climate without marked seasonality (Chaloner ànd Meyen 1973).

Gondwana floras of Westphalian age occur in Argentina, Brazil, and Australia. In Argentina and Brazil, the genera *Glossopteris*, *Gangamopteris*, *Phyllotheca*, and *Brachyphyllum* characterize these floral assemblages. These genera also occur in shales associated with coals in India, dated as Upper Carboniferous to Permian (Surange 1966). Most occurrences of the *Glossopteris-Gangamopteris* flora are associated with tillites.

In the Hunter Valley of New South Wales, Morris (1975) reported a Westphalian floral assemblage containing *Rhacopteris*, *Cardiopteris*, *Cyclostigma*, *Lepidodendron*, and *Sigillaria* (subsigillaria) and suggested that this Westphalian assemblage occupied nonglaciated areas of Australia. The latitudinal distribution of both the *Glossopteris-Gangamopteris* and the *Rhacopteris* assemblages corresponds to the south temperate climatic zone.

PERMIAN MARINE BIOGEOGRAPHY

The biogeography of the Permian can be best understood in terms of the unique continental configuration and oceanic circulation of that time.

Our reconstruction of Late Permian geography in Figure 7.9 shows the continents grouped tightly together to form Pangaea, which extended nearly from pole to pole. This left a single world ocean, Panthalassa, with a partly enclosed body, the Tethys Sea, on the east side of Pangaea. China served to block free circulation between Tethys and the cooler, northern part of Panthalassa; Australia, at the eastern edge of Gondwana, did the same in the Southern Hemisphere. The result was that only warm equatorial currents freely entered and circulated through Tethys, making this a warm-water sea from 50°S to 40°N latitude. The rest of Panthalassa had circulation rather like the modern Pacific (see discussion in text on oceanic circulation). Equatorial currents flowed westward across Panthalassa, while eastward-flowing currents flowed in the midlatitudes of both hemispheres; cool counter currents circulated westward in the polar regions. Along the west coast of Pangaea, cool eastern boundary currents flowed from midlatitude regions toward the equator, presumably creating a strong pole-to-equator climatic gradient in each hemisphere; this was in marked contrast to the much more uniformly warm Tethyan region on the other side of Pangaea. The only connections between these two oceanic regions were across the extremely wide Panthalassan Ocean or through the Uralian epicontinental seaway.

Permian benthic biogeography closely reflects this climatic and oceanographical regime. The biogeographical units are more numerous than realized generally. There are at least seven major marine units: the Tethyan, Uralian (Mordvinian), Boreal, Cordilleran, Grandian, Tasman, and one or possibly more South American units. In addition, the Tethyan unit is possibly divisible into three subunits, and Siberia, Central Europe (the Zechstein), and Andean South America probably also should be recognized as distinct units. This would total at least a dozen marine benthic biogeographical units for the Late Permian.

The Tethyan Sea is a large, but distinctive, faunal region (Adams and Ager 1967), as might be predicted from the paleogeographical reconstruction. The entire Tethyan area is characterized by high taxonomic diversity (Stehli 1971; Humphreville and Bambach 1979), as well as endemic lineages of fusulinids (C. A. Ross, 1979b), ectoprocts (J. R. P. Ross 1979), brachiopods, smaller Foraminifera, and corals (Yancey 1979). There are some indications that the Tethyan region may be divisible into several biogeographical subunits. Ross (1979) distinguished three subdivisions, the southern, central, and northern Tethys, on the basis of ectoproct distributions. This may be matched by a gradient in brachiopod generic diversity, with highest diversities in the equatorial (central) Tethys and somewhat lower diversities to the north and south (Humphreville and Bambach, in preparation). Waterhouse and Bonham-Carter (1975) have already demonstrated a distinct diversity history for the southern Tethys (western Australia, northern India), and Runnegar and Campbell (1976) discussed the difficulties of

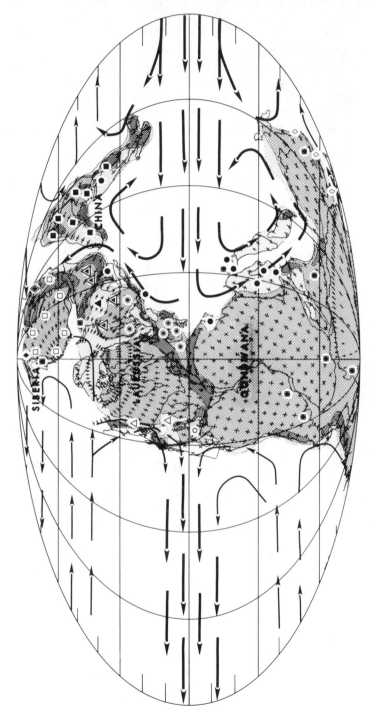

FIGURE 7.9. Early Late Permian (Kazanian) biogeography. Symbols for marine biogeographical units are as follows: ◆ = Transbaikalian. ▣ = Taymyr-Kolyma, △ = Cordilleran, ▲ = Mordvinian, ○ = Grandean, ● = Tethyan, ◇ = Tasman. Symbols for terrestrial biogeographical units are as follows: ◈ = Far Eastern, ▫ = Eastern, ⬛ = Angaran, ◨ = Petchora, ■ = Cathaysian, ◮ = Eastern European, ◉ = Eduramerian, ⊠ = Gondwanan. Symbol shape code for climatic assignments is as follows: ○ = tropical, ◿ = subtropical, ◻ = temperate, ◇ = polar. Arrows indicate surface ocean currents: thick = warm, thin = cold. Topography is as follows: dark shading = mountains, intermediate shading = lowlands, light shading = shallow shelves, white = ocean basins.

256

correlating Australian faunas with other regions, which implies a considerable degree of biogeographical distinctness. This differentiation may have been related to glacial environments in southern Pangaea during the Early Permian, followed by a shift to warm, more normal Tethyan conditions in the Late Permian.

The Uralian Seaway and Russian Platform to the north of Tethys have a peculiar fauna that Yancey (1979) recently named the Mordvinian fauna. It contains a mixture of Tethyan and non-Tethyan elements, as well as a number of endemics. J. R. P. Ross (1979) has documented a diverse and distinctive ectoproct fauna from this region.

The non-Tethyan western coast of Pangaea appears to have a sequence of biogeographical units that parallels a marked, latitudinally related generic diversity gradient for brachiopods (Humphreville and Bambach, 1979). Distinctive low-diversity faunas are well defined in both the northern and southern high latitudes, presumably cold regions. Boreal faunas are well known from Siberia, Arctic Canada, and Alaska at relatively high northern latitudes (Yancey, 1979). Waterhouse and Bonham-Carter (1975) and J. R. P. Ross (1979) have documented endemic Southern Hemisphere faunas of brachiopods and ectoprocts in the Tasman area of southeastern Australia, which, they note, were isolated from Tethyan Faunas to the north.

Yancey (1975, 1979) recognized in somewhat lower latitudes a Cordilleran fauna of moderate taxonomic diversity on the western craton of North America. This fauna, which probably represents temperate climatic conditions, differs from both the Boreal fauna to the north and the faunas of Texas and Mexico to the south. In a comparable position in the Southern Hemisphere there are several South American faunas of Permian age whose affinities are not yet well understood. Although these are not shown on our map (Fig. 7.9), it seems likely that they represent one or more separate temperate biogeographical units. The brachiopod diversities in South America are intermediate between those of inferred cold and warm-water areas.

The high-diversity faunas of west Texas and Mexico were located in tropical environments near the Permian equator. Yancey (1975, 1979) cited a variety of biotic and sedimentological reasons for regarding the fauna as representing tropical conditions. He also justified recognizing it as a separate biogeographical unit, the Grandian, which is distinct from the tropical Tethyan fauna. C. A. Ross (1979b) also distinguished this fauna from the Tethyan and Boreal faunas on the basis of fusulinids.

A number of anomalous occurrences of Tethyan faunas at places in the Pacific coastal zone of the Americas, northern New Zealand, and the eastern Soviet Union were noted by Yancey (1979). Structurally, these occurrences all seem to be incorporated in accretionary prisms rather than being part of the cratonic blocks of these areas. Yancey's suggestion that they are exotic blocks carried from their original location in the Tethyan Sea area by rifting and sea-floor spreading (plate motion) seems reasonable, and is supported by detailed structural and paleomagnetic studies (Jones, Silberling, and Hillhouse 1977, 1978; Jones and Silberling 1979; Brown 1979; Stone, Packer, and Panuska 1979).

PERMIAN TERRESTRIAL BIOGEOGRAPHY

The Late Permian was a time of increased provinciality for both marine invertebrates and terrestrial floras. Chaloner and Meyen (1973) and Plumstead (1973) divided Late Permian plant assemblages into seven separate floras: the Euramerian flora (tropical), the Eastern European flora (northern subtropical), the Petchoran flora (northern temperate), the Cathaysian flora (northern subtropical or low-latitude temperate), the Angaran flora (north temperate), the Far Eastern flora (polar), and the Gondwana flora (south temperate).

The Euramerian flora occurs in Britain, central Europe, Hungary, northern Italy, and Poland. The sphenopsid *Neocalamites*, the peltaspermacean *Lepidopteris*, the cycadophyte *Pseudoctenis*, the ginkgoalian *Sphenobaeira*, and the conifers *Ullmannia* and *Pseudovoltzia* characterize this flora (Chaloner and Meyen 1973). The paleolatitudes of these localities range from 2°N to 27°N and correspond to a tropical distribution.

The Eastern European flora occurs on either side of the shallow seaway separating eastern Europe from the Urals. According to Chaloner and Meyen (1973), the Eastern European flora contains a mixture of Angaran and European genera. European forms in this flora include the conifer *Ullmannia* and the Peltaspermaceae, a seed fern family. Angaran-Petchoran forms include *Phylladoderma*, the foliage genera *Callopteris* (Petchoran) and *Tartarina* (Petchoran), a problematical conifer (Petchoran), and the sphenosid foliage genus *Annularia* (Angaran). The distribution of this flora corresponds to the northern subtropical zone.

The Petchoran flora occurs only in the Petchora Basin located in the northern Urals. Beside the genera *Callipteris*, *Tartarina*, and *Phylladoderma*, which it shares with the Eastern European flora, this flora contains *Rhipidopsis*, a ginkgophyte, and the lycopods *Paichoria* and *Tundrodendron*, which occur only in Petchora. It shares the gymnosperm *Cordaites* with Angara (Chaloner and Meyen 1973). This flora lay slightly north of the Eastern European flora in the northern low-latitude temperate zone.

The Cathaysian flora occurs in China, Korea, Laos, and Thailand. The foliage genus *Gigantopteris*, a possible noeggerathippsid *Tingia*, the sphenopsid *Lobatannularia*, and the cycadophyte *Nilssonia* characterize this flora

(Chaloner and Meyen 1973). It shares the genus *Callipteris* with the Eastern European and Petchoran northern subtropical and low-latitude temperate floras. On our Late Permian reconstruction, the Cathaysian flora lies in subtropical and low north temperate latitudes.

The Angaran flora, which occurs on the Siberian continent, is characterized by the gymnosperms *Rufloria* and *Cordaites*, the sphenopsid genera *Tschernovia* and *Annularia*, the fern *Prynadaeopteris*, and the mosses *Polyssaieva* and *Protosphagnum* (Chaloner and Meyen 1973). The distribution of this flora corresponds to the North Temperate Zone.

The Far Eastern flora occurs on the eastern edge of Siberia. It is allied with the Angaran flora, but contains fewer sphenopsids and cordaitalians and more ferns (Chaloner and Meyen 1973). It shares the foliage genus *Taeniopteris* with the Cathaysian and European floras. This flora lies in the northern polar zone.

The Late Permian Gondwana flora, described by Plumstead (1973), occurs in South America, Africa, India, Australia, and New Zealand. *Glossopteridophyta*, the lycopod genus *Lycopodiopsis*, the sphenopsid genus *Schizoneura*, and the conifers *Walkomia*, *Walkomiella*, and *Burriadia* characterize this flora. The distribution of Late Permian Gondwana floral localities corresponds to a southern temperate distribution.

Wagner (1971) described a mixed Cathaysian-Gondwanan assemblage of Late or Middle Permian age from Hazro, Turkey. On a generic level, this flora shares *Gigantopteris* and *Lobatannularia* with the Late Permian Cathaysian flora and *Glossopteris* with the Late Permian Gondwana flora. Hazro has a paleolatitude of 15°S in the Late Permian and marks the northernmost extension of the Gondwanan flora, which may account for the presence of Cathaysian subtropical genera within the assemblage.

CONCLUSIONS

Faunas and floras must be placed in their paleogeographical and paleoclimatic context before evolutionary questions concerning speciation, diversity, and extinctions can be resolved. Paleogeography, employing plate tectonic concepts, is in itself such a new and uncertain field that a review of paleobiogeography at this time is bound to provoke as many questions as it provides answers. Clearly, paleontologists have barely scratched the surface in defining biogeographical provinces; the major faunal and floral discontinuities are as obvious as their causes, but the more subtle gradations await detailed taxonomic and statistical analysis. We hope that our maps will be used as a framework to predict regions where biogeographical gradients can be expected.

A primary need in biogeography is the development of a set of biological criteria for recognizing the major climatic regimes. Our climatic "assignments" are based more on latitude and orientation of coastlines and land masses than on direct evidence based on the fossils. Of course, we have paid attention to such obvious features as the distribution of coral reefs, the occurrence of diversity gradients, and the presence of seasonal growth rings in trees in confirming our climatic assignments. Terms such as tropical and boreal have been applied to faunas as old as the Late Paleozoic, but a rigorous set of criteria for differentiating tropical, subtropical, temperate, and polar faunas and floras awaits elaboration.

Another major need is for community level studies to sort faunas and floras by ecological category. Within every province an environmental spectrum exists, and until this is accounted for, statistical approaches will be flawed. For instance, we are certain that some of the variations detected in our polar ordination plots are due to local changes in habitat, rather than to biogeographical barriers.

Certain patterns do emerge from an analysis on the level presented here. Sea level changes strongly influence the degree of provinciality. High sea level stands, such as in the Late Ordovician, the Middle Silurian, and the Late Devonian to Early Carboniferous, seem to have resulted in migration routes for marine organisms and relative cosmopolitanism. Conversely, major continental collisions and accompanying regressions, such as the Caledonian and Acadian Orogenies in the Late Silurian to Early Devonian and the Hercynian Orogenies in the Late Carboniferous resulted in new barriers and provinciality.

Also, the predictable east-to-west asymmetry in ocean current and temperature regimes is observable, particularly in the Late Paleozoic, with the wide latitudinal extent of the Tethyan faunas, as opposed to the greater differentiation of faunas along the western Americas. Factors such as these must be taken into account in latitudinal diversity studies. Continents and oceans are very asymmetric in terms of temperature and precipitation patterns, and this must have an important effect on biogeography.

The maps we have presented can be improved in all respects—paleogeographical, climatic, and biogeographical. We feel that the basic relationships are correct, but the details need refinement and elaboration. All aspects of the faunas and floras of each period must now be considered and integrated from one period to the next before a dynamic biogeography of the Paleozoic can emerge.

AUTHORSHIP RESPONSIBILITIES

The marine biogeographical boundaries were drawn by Bambach, Barrett, and Ziegler, and the oceanic currents were reconstructed by Parrish.

The terrestrial biogeographical boundaries were determined by Raymond, Parker, and Ziegler. The new data summaries for the Devonian and Early Carboniferous plant distributions were prepared by Gierlowski and Raymond, respectively, with invaluable comments and suggestions from Stephen E. Scheckler. Sepkoski, Parker, and Raymond performed the computer analyses.

ACKNOWLEDGMENTS

We would like to thank Leah Haworth, Carol Kazmer, and Jean Pasdeloup for their technical and artistic endeavors. General support for our paleogeographical research comes from the National Science Foundation (NSF Grant EAR 7915133 to Ziegler), Shell Development Company, Amoco International Oil Company, Mobil Exploration and Producing Services Inc., Chevron Oil Field Research Company, Exxon Production Research Company, and the Exxon Education Foundation.

REFERENCES

Adams, C. G., and D. V. Ager. 1967. *Aspects of Tethyan Biogeography.* Systematic Assoc. Publ. No. 7. London.

Amantov, V. A., V. A. Blagonravov, Yu. A. Borzakovsky, M. V. Durante, L. P. Zonenshain, B. Luwsandarsan, P. S. Matrosov, O. D. Suetenko, I. B. Fulunnova, and R. A. Khasin. 1970. *Stratigraphy and Tectonics of the Mongolian Peoples Republic.* Academia Nauk, Trans. Vol. 1, The Joint Soviet Mongolian Scientific Research Geological Expedition, 1970.

Andrews, H. N., Jr. 1961. Some Paleozoic and Mesozoic floras. In *Studies in Paleobotany* (H. N. Andrews, ed.). New York: John Wiley and Sons, pp. 408–412.

Arnold, C. A., and W. Sadlick. 1962. A Mississippian flora from northeastern Utah and its faunal and stratigraphic relations. Contr. Mus. of Paleont. XVII (11): 241–263. University of Michigan.

Asama, K. 1973. Lower Carboniferous Kuantan flora, Pahang, West Malaysia. In *Geology and Paleontology of Southeast Asia* (T. Kobayashi and R. Toriyama, eds.) 11:109–128.

Asama, K., A-ng. Hongnusonthi, J. Iwa, E. Kon'no, S. S. Rajah, and M. Veeraburas. 1975. Summary of the Carboniferous and Permian plants from Thailand, Malaysia and adjacent areas. In *Geology and Paleontology of Southeast Asia* (T. Kobayashi and R. Toriyama, eds.).15:77–101.

Bell, W. A. 1960. Mississippian Horton Group of type Windsor-Horton District, Nova Scotia. *Geol. Surv. Can. Mem.* 314:1–112.

Berger, W. H., R. Diester-Haass, and J. S. Killingley. 1978. Upwelling off north-west

Africa: The Holocene decrease as seen in carbon isotopes and sedimentological indicators. *Oceanol. Acta* 1:3–7.

Berry, W. B. N. 1979. Graptolite biogeography: a biogeography of some lower Paleozoic plankton. In *Historical Biogeography, Plate Tectonics, and the Changing Environment* (J. Gray and A. J. Boucot, eds.). Corvallis: Oregon State University Press, pp. 105–115.

Boucot, A. J. 1974. Silurian and Devonian biogeography. In *Paleogeographic Provinces and Provinciality* (C. A. Ross, ed.). *Soc. Econ. Petrol. Mineral. Spec. Pub.* 21:165–176.

———. 1975. *Evolution and Extinction Rate Controls.* New York: Elsevier.

———. 1979. Silurian. In *Treatise on Invertebrate Paleontology.* Part A. (R. A. Robison and C. Teichert, eds.). Geological Society of America and University of Kansas, Lawrence, pp. A167–A182.

Boucot, A. J., and J. Gray. 1979. Epilogue: a Palezoic pangaea? In *Historical Biogeography, Plate Tectonics, and the Changing Environment* (J. Gray and A. J. Boucot, eds.). Corvallis: Oregon State University Press, pp. 465–482.

Boucot, A. J., and J. G. Johnson. 1973. Silurian brachiopods. In *Atlas of Palaeobiogeography* (A. Hallam, ed.). New York: Elsevier, pp. 59–65.

Boucot, A. J., J. G. Johnson, and J. A. Talent. 1969. Early Devonian brachiopod zoogeography. *Geol. Soc. Amer. Spec. Pap.* 119.

Brown, R. L. 1979. Convergence of Shuswap arc and North American cratonic margin, southern British Columbia. *Geol. Soc. Amer. (Abstr. Prog.)* 11(7):395.

Bureau, M. E. 1914. Bassin de la Basse Loire. Etudes des Gites Mineraux de la France. Vol. II(2), Fascicule 2, pp. 1–417. Ministere des traveux publics.

Chaloner, W. G., and S. V. Meyen. 1973. Carboniferous and Permian floras of the northern continents. In *Atlas of Palaeobiogeography* (A. Hallam, ed.). New York: Elsevier, pp. 169–186.

CLIMAP. 1976. The surface of the ice-age earth. *Science* 191:1131–1144.

Cocks, L. R. M., and W. S. McKerrow. 1973. Brachiopod distributions and faunal provinces in the Silurian and Lower Devonian. In *Organisms and Continents Through Time* (N. F. Hughes, ed.). *Palaeont. Assoc. Spec. Pap.* 12:291–304.

Diester-Haass, L., and H.-J. Schrader. 1979. Neogene coastal upwelling history off northwest and southwest Africa. *Mar. Geol.* 29:39–53.

Eady, E. T. 1964. The general circulation of the atmosphere and oceans. In *The Planet Earth* (R. Bates, ed.). Oxford: Pergamon, pp. 141–163.

Edwards, D. 1973. Devonian floras. In *Atlas of Palaeobiogeography* (A. Hallam, ed.). New York: Elsevier, pp. 105–116.

Eldredge, N., and A. R. Ormiston. 1979. Biogeography of Silurian and Devonian trilobites of the Malvinokaffric Realm. In *Historical Biogeography, Plate Tectonics, and the Changing Environment.* (J. Gray and A. J. Boucot, eds.). Corvallis: Oregon State University Press, pp. 147–167.

Gierlowski, E. H. 1978. On the geographic distribution of vascular plants in the Devonian. University of Chicago: Unpublished Senior Thesis.

Gorlova, S. G. 1978. The flora and stratigraphy of the coal-bearing Carboniferous of middle Siberia. *Palaeontographica* (Abt. B) Band 165, Lfg. 1–3, pp. 53–77.

Gothan, W., and H. C. Sze. 1933. Uber die Palaeozoische Flora der Provinz Kiangsu. Chung Yang Yen Chiu Yuan, Institute of Geology Memoirs XIII (13):1–40.

Gray, J., and A. J. Boucot (eds.). 1979. *Historical Biogeography, Plate Tectonics, and the Changing Environment.* Corvallis: Oregon State University Press.

Hallam, A. 1973. Distributional patterns in contemporary terrestrial and marine animals. *Organisms and Continents Through Time* (N. F. Hughes, ed.). *Palaeont. Assoc. Spec. Pap.* 12:93–105.

Hill, D. 1948. The distribution and sequence of Carboniferous coral faunas. *Geol. Mag.* LXXXV:11–38.

House, M. R. 1979. Devonian in the eastern hemisphere. In *Treatise on Invertebrate Paleontology.* Part A (R. A. Robison and C. Teichert, eds.). Geological Society of America and University of Kansas, Lawrence, pp. A183–A217.

Hughes, N. F. (ed.) 1973. *Organisms and Continents Through Time. Palaeont. Assoc. Spec. Pap.* 12.

Humphreville, R., and R. K. Bambach. 1979. Influence of geography, climate, and ocean circulation on the pattern of generic diversity of brachiopods in the Permian. *Geol. Soc. Amer. (Abstr. Prog.)* 11:447.

Hunt, B. G. 1979. The effects of past variations of the Earth's rotation rate on climate. *Nature* 28:188–191.

Jaanusson, V. 1971. Aspects of carbonate sedimentation in the Ordovician of Baltoscandia. *Lethaia* 6:11–34.

Jones, D. L., and N. J. Silberling. 1979. Mesozoic stratigraphy–The key to tectonic analysis of southern and central Alaska. *U.S. Geol. Sur. Open-File Rept.* 79-1200.

Jones, D. L., N. J. Silberling, and J. Hillhouse. 1977. Wrangellia—a displaced terrane in northwestern North America. *Can. J. Earth Sci.* 14:2565–2577.

———. 1978. Microplate tectonics of Alaska—Significance for the Mesozoic history of the Pacific Coast of North America. In *Mesozoic Paleogeography of the Western United States* (D. G. Howell and K. A. McDougall, eds.). Pac. Coast Paleogeogr. Symp. 2, Soc. Econ. Paleont. Min. Pacific Section, pp. 71–74.

Kaljo, D., and E. Klaamann. 1973. Ordovician and Silurian Corals. In *Atlas of Palaeobiogeography* (A. Hallam, ed.). New York: Elsevier, pp. 37–45.

Lorenz, E. N. 1967. *The Nature and Theory of the General Circulation of the Atmosphere.* World Meteorological Organization.

Meyen, S. V. 1976. Carboniferous and Permian Lepidophytes of Angaraland. *Paleontographica* (Abt. B) Band 157, Lfg. 5–6, pp. 112–157.

Mohr, R. E. 1975. Measured periodicities of the Biwabik (Precambrian) stromatolites and their geophysical significance. In *Growth Rhythms and the History of the Earth's Rotation* (G. D. Rosenberg and S. K. Runcorn, eds.). London: John Wiley and Sons, pp. 43–56.

Morris, N. 1975. The Rhacopteris Flora in New South Wales. In *Papers from the Third Gondwana Symposium, Canberra, Australia, 1973,* (K. S. W. Campbell, ed.). Canberra: Australian National University Press, pp. 99–108.

Oliver, W. A., Jr. 1977. Biogeography of Late Silurian and Devonian rugose corals. *Palaeogeog., Palaeoclim., Palaeoecol.* 22:85–135.

Palmer, A. R. 1972. Problems in Cambrian biogeography. 24th Int. Geol. Congr., Sec. 7, Paleontology, pp. 310–315.

———. 1979. Cambrian. In *Treatise on Invertebrate Paleontology.* Part A (R. A. Robison and C. Teichert, eds.). Geologic Society of America and University of Kansas, Lawrence, pp. A119–A135.

Plumstead, E. P. 1973. The Late Paleozoic Glossopteris flora. In *Atlas of Palaeobiogeography* (A. Hallam, ed.). New York: Elsevier, pp. 18–206.

Pojeta, J., J. Kriz, and J. M. Berdan. 1976. Silurian-Devonian pelecypods and Paleozoic stratigraphy of subsurface rocks in Florida and Georgia and related Silurian pelecypods from Bolivia and Turkey. *Prof. Pap. U.S. Geol. Sur.* 879.

Radchenko, M. I. 1961. Paleophytological basis for the stratigraphy of the Carboniferous of Kazakhstan. Quatrième Congrès pour l'avancement des Études de Stratigraphie et de Géologie du Carbonifère, Compte Rendu Tome II, pp. 562–599. Heerlen, 1958.

Read, C. B. 1941. Plantas Fosseis do Neo-Paleozoico do Paraná E Santa Catarina. Ministerio da Agricultura, Departmento Nacional da Producão Mineral, Divisão de Geologia E Mineralogia Monografia XII:1–96.

———. 1947. Pennsylvanian floral zones and floral provinces. *J. Geol.* 55:271–279.

———. 1955. Floras of the Pocono Formation and Price Sandstone in parts of Pennsylvania, Maryland, West Virginia, and Virginia. *Prof. Pap. U.S. Geol. Surv.* 262:1–32.

Rigby, J. F. 1969. A reevaluation of the Pre-Gondwana Carboniferous flora. *Anais da Acad. Brazil. de Cien.* 41:391–413.

Robison, R. A., and C. Teichert (eds.). 1979. *Treatise on Invertebrate Paleontology.* Part A. Geol. Surv. Amer. and Univ. of Kansas, Lawrence.

Ross, C. A. 1967. Development of fusulinid (Foraminiferida) faunal realms. *J. Paleont.* 41:1341–1354.

———. (ed.). 1974a. Paleogeographic Provinces and Provinciality. *Soc. Econ. Paleont. Mineral. Spec. Pub. 21.*

———. 1974b. Paleogeography and provinciality. In *Paleogeographic Provinces and Provinciality* (C. A. Ross, ed.). *Soc. of Econ. Paleont. and Mineral., Spec. Pub. 21,* pp. 1–17.

———. 1979a. Carboniferous. In *Treatise on Invertebrate Paleontology.* Part A (R. A. Robison and C. Teichert, eds.). Geological Society of America and University of Kansas, Lawrence, pp. A254–290.

———. 1979b. Evolution of Fusilinacea (Protozoa) in late Paleozoic space and time. In *Historical Biogeography, Plate Tectonics, and the Changing Environment* (J. Gray and A. J. Boucot, eds.). Corvallis: Oregon State University Press, pp. 215–226.

Ross, J. R. P. 1979. Permian ectoprocts in space and time. In *Historical Biogeography, Plate Tectonics and the Changing Environment.* (J. Gray and A. J. Boucot, eds.). Corvallis: Oregon State University Press, pp. 259–276.

Ross, R. J., Jr. 1975. Early Paleozoic trilobites, sedimentary facies, lithospheric plates, and ocean currents. *Fossils and Strata* 4:307–329.

Runnegar, B., and K. S. W. Campbell. 1976. Late Paleozoic faunas of Australia. *Earth-Scien. Rev.* 12:235–257.

Sando, W. J., E. W. Bamber, and A. K. Armstrong. 1975. Endemism and similarity indices: Clues to the zoogeography of North American Mississippian corals. *Geology* 2:661–664.

Savage, N. M., D. G. Perry, and A. J. Boucot. 1979. A quantitative analysis of Lower Devonian brachiopod distribution. In *Historical Biogeography, Plate Tectonics, and the Changing Environment* (J. Gray and A. J. Boucot, eds.). Corvallis: Oregon State University Press, pp. 169–200.

Scheltema, R. S. 1977. Dispersal of marine invertebrate organisms: paleobiogeographic and biostratigraphic implications. In *Concepts and Methods of Biostratigraphy*, (E. G. Kauffman and J. E. Hazel, eds.). Stroudsburg: Dowden, Hutchinson, and Ross, pp. 73–107.

Schopf, T. J. M. 1979. The role of biogeographic provinces in regulating marine faunal diversity through geologic time. In *Historical Biogeography, Plate Tectonics and the Changing Environment* (J. Gray and A. J. Boucot, eds.). Corvallis: Oregon State University Press, pp. 449–457.

Scotese, C. R., R. K. Bambach, C. Barton, R. Van der Voo, and A. M. Ziegler. 1979. Paleozoic base maps. *J. Geol.* 87:217–277.

Scrutton, C. T. 1965. Periodicity in Devonian coral growth. *Palaeontology* 7:552–558.

Silva-Pineda, A. 1970. Plantas del Pensilvanico de la Región de Tehuacán, Puebla. Universidad Nacional Autonoma de Mexico, Instituto de Geologia. *Paleontologia Mexicana* 29:1–78.

Stehli, F. G. 1971. Tethyan and Boreal Permian faunas and their significance. *Smithsonian Contrib. to Paleobiol.* 3:337–345.

Stone, D. B., D. R. Packer, and B. Panuska. 1979. Mesozoic and Cenozoic paleomagnetic paleolatitudes for southern Alaska. *Geol. Soc. Amer. (Abstr. Prog.)* 11:524.

Sullivan, H. J. 1967. Regional differences in Mississippian spore assemblages. *Rev. Paleobot. and Palynol.* 1:185–192.

Surange, K. R. 1966. Distribution of Glossopteris Flora in the Lower Gondwana Formations of India. Symposium on Floristics and Stratigraphy of Gondwanaland. Special Session of the Paleobotanical Society, Birbal Sahni Institute of Paleobotany, Lucknow, pp. 55–68.

Sveshnikova, I. N., and L. Y. Budantsev. 1969. *Florulae Fossiles Arcticae.* Leningrad: Nauka Filia Leningradensis.

Taylor, M. E. 1976. Indigenous and redeposited trilobites from Late Cambrian basinal environments of central Nevada. *J. Paleont.* 50:668–700.

Tchirkova, H. Th. 1937. Contribution novelle à la flore Carbonifere inférieure du versant Oriental de l'Oural. In *Problems of Paleontology*. (A. Hartmann-Weinberg, L. M. Kretschetovitsch, and Th. M. Kusmin, eds.). Publications of the Lab. of Paleontol., Moscow University. Vol. II, pp. 207–231.

Valentine, J. W., T. C. Foin, and D. Peart. 1978. A provincial model of Phanerozoic marine diversity. *Paleobiology* 4:55–66.

Vinogradov, A. P. (ed.). 1968. Atlas of the lithological-paleogeographical maps of the U.S.S.R. Vol. 1. Min. Geol. U.S.S.R., Acad. Sci. U.S.S.R., Moscow.

———. (ed.). 1969. Atlas of the lithological-paleogeographical maps of the U.S.S.R. Vol. 1. Min. Geol. U.S.S.R., Acad. Sci. U.S.S.R., Moscow.

von Draber, R. 1972. Abbildungen und Beschreibungen unterkarbonischer Pflanzenreste aus der Mongolischen Volksrepublik. Palaeontologische Abhandlungen, Abteilung B, Palaeobotanik, Band III, Heft 5, pp. 867–885.

Wagner, R. H. 1961. On a mixed Cathaysia and Gondwana flora from SE. Anatolia (Turkey). Quatrième Congrès pour l'avancement des Études de Stratigraphie et de géologie du Carbonifère. Compte Rendue, Tome III, pp. 745–751. Heerlen, 1958.

Waterhouse, J. B., and G. F. Bonham-Carter. 1975. Global distribution and character of Permian biomes based on brachiopod assemblages. *Can J. Earth Sci.* 12:1085-1146.

Wells, J. W. 1963. Coral growth and geochronometry. *Nature* 197:948-950.

Whittington, H. B. 1973. Ordovician trilobites. In *Atlas of Palaeobiogeography* (A. Hallam, ed.). New York: Elsevier, pp. 13-18.

Whittington, H. B., and C. P. Hughes. 1972. Ordovician geography and faunal provinces deduced from trilobite distribution. *Phil. Trans. Royal Soc.* (Lond.) 263(B):235-278.

Williams, A. 1973. Distribution of brachiopod assemblages in relation to Ordovician paleogeography. In *Organisms and Continents Through Time* (N. F. Hughes, ed.). *Palaeont. Assoc. Spec. Pap.* 12:241-269.

White, D. 1937. Fossil flora of the Wedington ss member of the Fayetteville shale. *Prof. Pap. U.S. Geol. Sur.* 186:1-56.

Yancey, T. E. 1975. Permian marine biotic provinces in North America. *J. Paleo.* 49:758-766.

——. 1979. Permian positions of the northern hemisphere continents as determined from marine biotic provinces. In *Historical Biogeography, Plate Tectonics, and the Changing Environment* (J. Gray and A. J. Boucot, eds.). Corvallis: Oregon State University, pp. 239-247.

Zalessky, M. D. 1937a. Sur Quelques végétaux fossiles nouveaux des terrains carbonife res et permiens due Bassin de Donetz. In *Problems of Paleontology* (A. Hartmann-Weinberg, L. M. Kretschetovitsch, and Th. M. Kusmin, eds.). Publ. of the Lab. of Paleontol., Moscow University. Vol. 2, pp. 155-193.

——. 1937b. Sur une division de l'anthracolithique continental du bassin de Kousnetzk d'apres sa flore fossile. In *Problems of Paleontology* (A. Hartmann-Weinberg, L. M. Kretschetovitsch, and Th. M. Kusmin, eds.). Publications of the Lab. of Paleontol., Moscow University. Vol. 2, pp. 103-122.

Ziegler, A. M., K. S. Hansen, M. E. Johnson, M. A. Kelly, C. R. Scotese, and R. Van der Voo. 1977. Silurian continental distributions, paleogeography, climatology, and biogeography. *Tectonophysics* 40:13-51.

Ziegler, A. M., C. R. Scotese, W. S. McKerrow, M. E. Johnson, and R. K. Bambach. 1977. Paleozoic biogeography of continents bordering the Iapetus (Pre-Caledonian) and Rheic (Pre-Hercynian) Oceans. In *Paleontology and Plate Tectonics* (R. M. West, ed.). Milwaukee Public Museum, *Spec. Publ. Biol. Geol.* No. 2, pp. 1-22.

——. 1979. Paleozoic paleogeography. *Ann. Rev. Earth Planet. Sci.* 7:473-502.

INDEX

Acaciaephyllum spatulatun, 50
Acer, 49
Acer carpinifolium, 80
Acer macrophyllum, 80
Acer oregonianum, 80
Acerites multiformis, 37
Actinostachys, 149
Adaptive radiation, angiosperm-like
    pollen, 45
Adelanthus, 178
Adiantites, 251
Agathis, 142, 168
Alcicornopteris, 16
Alethopteris, 254
Alfaroa-Oreomunnea-Engelhardia
    complex, 122
  megafossil record of, 106–107
Allisonia, 145
Alnus, 83
Alternation of generations, 202
Amentiferae
  distribution and insect pollinators,
    121–122
  palynological problems of, 105
  pollination methods, 104–105
  status during Middle Eocene, 106–115
  temperature distribution of, 122–123
Amygdalophyllum, 247
Anastrophyllum, 176–178
Anemophily, 104
Aneura pinguis, temperature tolerance
    in, 140
Angaran flora, 258–59
Angaridium, 259
Angarodendron, 248, 254
Angaropteridium, 248, 249, 254
Angiosperm(s), 214–226
  adaptation, 52, 220
  adaptations for dominance over
    conifers, 97

adaptive value of individual
    characteristics of, 214
additive evolution of, 218
center of origin for, 53
characteristics of, 214
chemical defenses, 217
chenopodiaceous, 61
of coastal streams, 40–41
coevolution with insects, 220–221
competitive replacement of
    preexisting vegetation, 60–63
Cretaceous-Tertiary, 222
and dispersal agents, 27, 222
diversity, 220–221
dominance, 64
Early Cretaceous, 219
early dispersal of, 51–60
  in central Australia, 59
  in English coastal lagoons, 53–56
  in Gondwanaland, 53
  in the interior North America,
    56–59
  in southeastern U.S.S.R., 59
environmental amelioration by,
    167–169
in Eocene-Oligocene, 223
fossils, 28
and Hepaticae evolution, 141–142
herbaceous aquatic, 49–50
and insect diversity, 220
in lagoon debris, 30–31
of lake margins, 41
during Late Paleocene-Early Eocene,
    222
Mesozoic explosion of, 166–169
microgametophytic efficiency, 214
in Oligocene-Miocene, 223
origin of, 129
origin of wind pollination in, 118
of paleosols, 35–39

Bryophyte gametophytes,
  characteristics, 149
Burriadia, 259

Calamites, 254
Callandrium, 14
Callipteris, 258
Callistophytales
  pollen evolution, 18
  pollen organ evolution, 14–15
Callopteris, 258
Calypogeia, 145
Cambrian biogeography, 236
  trilobite distributions, 236
Carboniferous-Cretaceous period,
  212–213
  environmental fluctuations during,
  213
  speciation during, 213
Cardiopteris, 254
Cariopteridium, 249
Carpolithus conjugatus, 47
Carpolithus fascicularis, 47
Carpolithus geminatus, 47
Carpolithus karatcheensis, 47
Carpolithus ternatus, 47
Carpolithus virginiensis, 47
Carya, 107
Caspiocarpus paniculiger, 47, 49, 50
Castanopsis, 91
Casuarina, 50
Cathaysian flora, 258–259
Cavicularia, 145
Cedrus, 95, 96
Cephalozia, 145, 152
Cercidiphyllum, 48
Chacassopteris, 248
Chaleuria, 209
Chamaecyparis, 142, 168
Chandler Formation, 58
Cheirorhiza, 145, 166
Chiangnan realm, 236
Chondrites, 35
Claiborne Formation, 106, 107, 109
Clarno Formation, 107
Classoidites, 61
Classopollis, 53, 61

Clavatipollenites, 45, 46
Clavatipollenites hughesii, 45, 46
Clavatipollenites rotundus, 46
Clay pits (Hoisington), 31
Clematis, 93
Climatic gradient, pole-to-equator, 255
Climatic zones, longitudinal
    biogeographical differentiation,
    231
  Paleozoic, 231–261
Coal swamps, 251
Coastal streams, 40–41
  angiosperms of, 40–41
Coevolution, plant-insect, 218
Collawash floras, 89
Cololejeunea, 142, 168
Compression-impression specimens, 5
Conifer mangroves, 61
Coniferous forests, effects of climate on,
    95–96
Conifers
  halophytic, 53
  physiological characters of and
    drought, 95
Continental collision, 231
Continental drift, and angiosperm
    speciation, 52
Continental environments, adaptation
    to, 213
Continental fragments, 170
Continent(s)
  collisions, 232
  convergence, 238
  Gondwana, 236, 238
  longitudinal separation, 232
  pangaean configuration, 232
Convergent evolution, and
    environmental pressures, 218
Cooksonia, 204, 246
Cope's Rule, 224, 225
Coprolites, of pollen-rich, 22
Coral
  decrease in, 251
  provinces, 247
Coral diversity, and rising sea levels, 240
Corbicula, 35
Cordaitales, pollen evolution, 18–20

Walkomia, 259
Walkomiella, 259
Webbing, 157
Weichselia, 28, 58, 60, 61, 64
Wettsteinia, 178
Wind pollination, 21, 45–48
  adaptation for, 115
  adaptations to, 121
  and climate conditions, 115–123
  evolution in dicotyledonous
       angiosperms, 115–123
  in seasonally dry tropics, 121
  selection for, 117
  selective advantage in, 120

Winged fruits, 107
Woodlands, angiosperm, 39–40
Woody to herb transition, 225

Xanthorrhea, 63
Xenothallus, 145

Zelkova, 96
Zimmermannitheca, 16
Zimmermannitheca cupulaeformis, 3
Zonal circulation patterns, 234
Zosterophyllum, 244, 246